Agricultural Extension

Global Status and Performance in Selected Countries

Library of Congress Cataloging-in-Publication Data

Names: Davis, Kristin, editor. | Babu, Suresh Chandra, editor. | Ragasa, Catherine, editor.
Title: Agricultural extension : global status and performance in selected countries / Kristin Davis, Suresh C. Babu, Catherine Ragasa.
Description: Washington, DC : International Food Policy Research Institute, [2020] | Includes bibliographical references and index.
Identifiers: LCCN 2020021712 (print) | LCCN 2020021713 (ebook) | ISBN 9780896293755 (paperback) | ISBN 9780896293779 (epub)
Subjects: LCSH: Agricultural extension work—Developing countries. | Agricultural extension work—Evaluation. | Agricultural extension work—Brazil. | Agricultural extension work—Uganda. | Agricultural extension work—Ethiopia. | Agricultural extension work—Malawi. | Agricultural extension work—Congo (Democratic Republic)
Classification: LCC S544.6 .A26 2020 (print) | LCC S544.6 (ebook) | DDC 630.71/5—dc23
LC record available at https://lccn.loc.gov/2020021712
LC ebook record available at https://lccn.loc.gov/2020021713

Cover Design: Jason Chow/IFPRI
Project Manager: John Whitehead
Book Layout: BookMatters

About IFPRI

The International Food Policy Research Institute (IFPRI), a CGIAR Research Center established in 1975, provides research-based policy solutions to sustainably reduce poverty and end hunger and malnutrition. IFPRI's strategic research aims to foster a climate-resilient and sustainable food supply; promote healthy diets and nutrition for all; build inclusive and efficient markets, trade systems, and food industries; transform agricultural and rural economies; and strengthen institutions and governance. Gender is integrated in all the Institute's work. Partnerships, communications, capacity strengthening, and data and knowledge management are essential components to translate IFPRI's research from action to impact. The Institute's regional and country programs play a critical role in responding to demand for food policy research and in delivering holistic support for country-led development. IFPRI collaborates with partners around the world.

About IFPRI's Peer Review Process

IFPRI books are policy-relevant publications based on original and innovative research conducted at IFPRI. All manuscripts submitted for publication as IFPRI books undergo an extensive review procedure that is managed by IFPRI's Publications Review Committee (PRC). Upon submission to the PRC, the manuscript is reviewed by a PRC member. Once the manuscript is considered ready for external review, the PRC submits it to at least two external reviewers who are chosen for their familiarity with the subject matter and the country setting. Upon receipt of these blind external peer reviews, the PRC provides the author with an editorial decision and, when necessary, instructions for revision based on the external reviews. The PRC reassesses the revised manuscript and makes a recommendation regarding publication to the director general of IFPRI. With the director general's approval, the manuscript enters the editorial and production phase to become an IFPRI book.

Agricultural Extension

Global Status and Performance in Selected Countries

Edited by
Kristin Davis, Suresh C. Babu,
and Catherine Ragasa

A Peer-Reviewed Publication

International Food Policy Research Institute
Washington, DC

Recommended citation: Davis, K., S. C. Babu, and C. Ragasa. 2020. *Agricultural Extension: Global Status and Performance in Selected Countries.* Washington, DC: International Food Policy Research Institute. https://doi.org/10.2499/9780896293755.

International Food Policy Research Institute, 1201 Eye Street, NW, 12th floor, Washington, DC 20005 USA, Telephone: +1-202-862-5600, www.ifpri.org

ISBN: 978-0-89629-375-5
DOI: https://doi.org/10.2499/9780896293755

Library of Congress Cataloging-in-Publication Data may be found on page ii.

CONTENTS

Part 2: Performance of Extension Systems

LIST OF TABLES, FIGURES, AND BOXES

Tables

Figures

Boxes

EXECUTIVE SUMMARY

To address gaps in knowledge about agricultural extension, this book provides a global overview of the status of agricultural extension and advisory services, compares extension systems at national and regional levels, examines performance of extension approaches in a selected set of country cases, and shares lessons and policy insights. Drawing on both primary and secondary data, the book contributes to the literature by applying a common and comprehensive framework—the "best-fit" approach—to the data analysis and by updating previous global extension services assessments. The conceptual framework used assists in standardizing the principles on which assessments are conducted and allows some comparability of characteristics and results across geographies.

"The Best Fit Framework for Analyzing Agricultural Advisory Services Worldwide" by Regina Birner and colleagues, published by the *Journal of Agricultural Education and Extension* in 2009, looks at characteristics of extension and advisory services in the context of the frame conditions that affect the services. The key characteristics of extension include governance structures, capacity, management, and advisory methods. These characteristics are variables that can be changed by implementers in response to the prevailing contextual factors (the frame conditions) such as the policy environment, institutional linkages, and production systems. For instance, in a country with an extension policy advocating pluralism that assigns government the main role in coordinating the various actors, governance structures would assign government an oversight role, with a large mix of nongovernmental and private actors participating.

Part 1 of the book provides a landscape view of extension globally. It compares the recent status of extension and advisory services to the situation 20 to 30 years ago, using primary data from several global assessments that have not been analyzed before. In particular, the global landscape assessment examines best-fit characteristics, including governance and legal status of extension providers, capacity in terms of staffing, and advisory methods and clientele. It also looks at trends and recent developments by comparing the earlier global assessments with a more recent one and by examining relevant literature.

This global picture is complemented with descriptive secondary and primary data on country and regional extension systems using the best-fit characteristics as the basis for analysis. The cases come from Latin America and Africa as well as Central Asia, South Asia, and Southeast Asia. Country and regional cases were chosen purposively based on availability of secondary and primary data relevant to the best-fit characteristics and how recently the data were collected (most data are no older than 2011).

To go a step beyond the global picture and the descriptive cross-country analyses, Part 2 presents a set of in-depth country case studies that examine the performance of extension services provision and the impact of extension approaches using new primary and secondary data from Brazil, Uganda, Ethiopia, Malawi, and the Democratic Republic of the Congo. These country cases follow a similar structure, with the best-fit framework characteristics guiding the analysis. Due to different research teams and varied research designs, the specific indicators describing the conditions and elements of the extension system as well as those measuring performance and impacts have both similarities and differences. The book concludes with a set of findings organized by the main characteristics of the best-fit framework (that is, governance structures, capacity, management, and advisory methods), which provide a means to diagnose the status of extension services and compare countries and regions.

The analysis of global extension systems in Part 1 shows that extension and advisory services have become much more pluralistic; systems in most countries include both public and private agents (including civil society). There are greater numbers of extension agents today (more than one million), and they belong to a variety of different types of organizations from the private, nongovernmental, and public sectors, and even include lead farmers (essentially as volunteer extension agents). Reaching women and youth continues as a goal over the decades covered by the analysis. Newer trends include the use of information and communications technologies, a focus on nonrural clients, and market-oriented extension.

Regarding governance structures, in most countries the public sector—usually the Ministry of Agriculture, Food, or Rural Development—is responsible for the overall coordination and regulation of extension services. As extension systems become increasingly pluralistic, governments must address coordination and regulation of the many providers; Brazil (Chapter 4) and Malawi (Chapter 7) have established mechanisms to deal with such coordination. The linkages between extension services, research, and farmer feedback mechanisms remain weak in many developing countries. However, newer forms of innovation platforms that focus on market access and center on specific value chains are showing some promise toward integration in several countries.

Although more organizations are providing extension services, the country cases show that pluralism of service delivery is far from complete. Service providers are dominated by international nongovernmental organizations (NGOs), often with limited human resources and relying heavily on government extension workers. It is common to find international NGOs addressing government resource limitations by providing mobility and operating funds for government extension agents, while the government provides the human resources. As a result of this limited pluralism, there is little competition among service providers and hardly any expansion of options or choices for farmers.

For pluralistic extension to work, policies must provide an overall framework in which multiple actors can operate. However, many countries do not have a specific policy for extension and advisory services; rather, they are usually rooted in broader agriculture sector development policies. Brazil and Ethiopia, on the one hand, stand out as having good extension policies in place accompanied by appropriate implementation mechanisms. Both countries showed commitment to using public funding for extension provision and have invested significantly in extension services. Malawi, on the other hand, has a national extension policy but struggled to fund and implement the policy. Malawi focused most of its agricultural investment on the Farm Input Subsidy Programme, neglecting extension services and other rural services.

Decentralization was tried in most of the country cases examined in this book as a way to increase accountability by bringing the services closer to the clientele. However, the legal transfer of responsibilities for agricultural services is often not matched by sufficient fiscal decentralization, and the autonomy of local governments is undermined by continuing dependence on the transfer of funds from the central government. Where such transfers have been reduced, and taxes and revenues collected by the local government remain meager, the promises of decentralization often are not realized. For example, attempted reforms toward district-level agricultural extension services systems in Malawi lacked the resources and capacity to coordinate among different service providers and engage with farmers. This is consistent with public expenditure studies in other countries; for example, in Ghana analysis of district budgets over time indicated that total average agricultural expenditures at the subnational level decreased from 11 percent to 6 percent of spending between the introduction of decentralization in 2012 and 2015.[1]

In most cases, financing of the services was insufficient and often donor-driven with limited thought to financial sustainability. The shortage of funds hampered the performance of extension services, especially in terms of meeting operational costs and sufficient staffing numbers. However, financing of extension services was shown to be quite diverse in Latin America. In Africa and Central Asia and the Caucasus, financing was donor-dependent and public sector–led, with some slight variations. For example, while Rwanda and Ethiopia did use donor funds, their extension programs were driven by national extension and agriculture strategies rather than by donor priorities. Country studies indicated a lack of cost-effective ways to increase coverage and provide services to many unserved or underserved areas, but found extension services were duplicated in some areas. Nevertheless, there were promising moves toward cofinancing and farmer contributions in some Latin American countries and in Senegal. Certain advisory methods, such as the use of information and communications technologies and farmer advisers, were used to enhance sustainability in many countries.

Capacity of extension services includes the staff numbers, training level, skills, infrastructure, and financial resources. Despite the presence of more than one million extension agents worldwide, the capacity of relevant individuals and organizations—especially in Africa (with the exception of Ethiopia)—was, in general, low. Numbers of extension agents from the public sector were seldom sufficient for the job at hand, and there were high vacancy rates and turnover in some countries. Most countries had a poor extension agent-to-farmer ratio, and many of the extension agents in Africa struggled with mobility and poor transport infrastructure. Salaries were low, especially in the public sector, and there were few rewards or recognition. The foundational training for extension agents was usually focused on technical topics

1 D. Resnick. 2018. *The Devolution Revolution: Implications for Agricultural Service Delivery in Ghana*. IFPRI Discussion Paper 1714. Washington, DC: IFPRI.

and thus may have missed the functional skills needed for extension. Some exceptions were Central and South American countries that appeared to have well-established foundational and continuing education systems for extension providers. Several countries, such as India and Malawi, became interested in professionalizing their extension services by requiring certification and registration, continuing education, and other elements of quality control, but implementation was challenging. More important than adding more extension staff is the need to invest in greater operating budgets for carrying out extension services. Without funds to support mobility and operations, additional staffing would be expensive and counterproductive.

With regard to management, a number of mechanisms were used to ensure better management, such as market orientation or demand-driven services, and provide monitoring and evaluation (M&E). Many countries were putting in place demand-driven or client-oriented mechanisms, as well as M&E systems, to ensure that extension services are more participatory and thus more relevant to farmers, in particular the underserved, women (for example, Latin America), and indigenous groups (for example, Brazil). The increasing use of multistakeholder networks, innovation platforms, and other demand strategies contributed to these efforts. Chapter 2 and Chapter 3 discuss the huge gaps in data, capacities, and funding necessary for these management tasks. Chapter 3 and Chapter 9 ("Conclusions and Policy Implications") present selected indicators to provide guidance to policymakers to operationalize the collection of best-fit indicators at country or provider level.

An increasingly wide range of methods was used to provide advisory services (see Table 3.1), and the use of digital approaches has increased. Farmer field schools—a participatory, group-based adult education approach—was widely used (Chapter 5 and Chapter 8). Group methods and demonstrations continued to be widely used, and private-sector extension and the use of lead farmers appeared to be growing and used worldwide. Market-oriented extension has frequently and appropriately become an integral part of extension activities (Chapter 2). This was a big focus in Latin America and increasingly in Africa south of the Sahara. However, capacities and information to provide market-oriented extension were often lacking; Central Asian and Caucasus countries in particular struggled with the ability to provide such services.

The conclusions provide a number of policy implications across the best-fit characteristics—implementing policies and programs to improve governance, reforming curricula to strengthen capacity, providing incentives to extension staff at various levels to improve management, and using appropriate advisory methods to enhance financial sustainability and achieve greater scale. Specific policy insights include the following:

1. Explicit policy or strategy for extension services is a key first step for better governance, funding, coordination, decentralization, and for overall effective design and implementation of extension services. Capacity for developing such country-specific extension strategies is weak or does not exist and thus needs strengthening in the national systems.

2. Public funding for extension services has been declining, current funding remains unstable, and most of the poorer countries continue to depend on donor funding to run their extension systems. Although private-sector, NGOs, and farmer-based organizations (FBOs) play an increasingly important role in extension services, the public role cannot be overestimated for reaching remote areas and marginal groups. A combination of funding sources and mixing innovative modern approaches with traditional mechanisms can increase the sustainability of extension services.

3. Given that extension and advisory services have become more pluralistic in developing countries, the need for coordination, quality control, and avoiding duplication of efforts becomes a key policy and programmatic challenge for the public sector. In addition, identifying the gaps left by the multiplicity of actors as they focus on specific target groups or value chains, and defining and coordinating the roles and responsibilities of these actors, is a major challenge for policymakers. Investment in such coordination capacity needs immediate attention in the public extension system.

4. The organizational and institutional capacity to effectively reach the farmers in a country context is important to improve the efficiency and effectiveness of the extension services provided by the pluralistic actors. The first set of policy and programmatic concerns relate to the supply-driven nature of the extension services in several countries studied. To make them more demand-driven, there is need for identifying information needs of the producers in an organizational context, setting priorities for the extension services, and sharing the goals and approaches among the extension services providers. This implies building capacities of producers and especially producer organizations to identify and prioritize their needs. These activities will further help in the management

of the limited resources for extension that is currently split among the pluralistic providers.

5. Assessing the performance of extension in meeting their clientele's needs requires sound monitoring and evaluation systems in place, which remain weak in many countries. The learning and improvement and the trade-off between the quantity and quality of extension professionals can be addressed only through feedback provided by a well-capacitated monitoring and evaluation system.

6. Organizational and institutional innovations are needed on a regular basis for improving the capacity of the extension professionals, improving their focus on the farmers' problems, reducing their overburden with multiple objectives, and increasing the reach of extension in different agroecological zones. Continuous improvement in the technical, managerial, and leadership capacities of extension professionals that goes beyond foundational training should be a key part of the national extension strategy.

7. Innovative policy and programmatic approaches are needed to reduce the top-down approaches to information sharing. Increasing farmer participation in decision-making on what they would like the extension services to provide will help move from a technology-transfer orientation to market-driven approaches to the extension services. Use of private sector, traders, volunteer farmers, lead farmers, and youth as entrepreneurs to increase the reach of advisory services will require adequate attention at the policy level.

8. Moving toward digital technology and its use for reaching producers on topics such as weather, technology, markets, prices, and other real-time information for solving farmers' problems also require adequate policy and institutional arrangements at various levels. Digital technologies and use of the Internet of Things can save resources at the extension system and at the farmers' level. Their use can be intensified both as extension delivery tools and as mechanisms for demand articulation, monitoring, and greater transparency and accountability.

9. Finally, the future of extension systems crucially depends on how they are built to meet such emerging challenges as climate change, precision agriculture, nutrition and health goals, youth and gender, and other challenges related to the transformation of food systems and

to resilience building. In this context, the extension worker is seen more as a problem solver and a facilitator of services in the rural areas. Developing a policy environment to strengthen the capacity of the extension system to meet these emerging needs remains the most important development concern.

The concluding chapter highlights several other policy, institutional, and programmatic recommendations for improving the global extension services.

ABBREVIATIONS AND ACRONYMS

ADD	agricultural development division (Malawi)
ADLI	Agriculture Development–Led Industrialization (Ethiopia)
ADPLAC	Agriculture Development Partners Linkage Advisory Council (Ethiopia)
AEA	agricultural extension agent (Uganda)
AEDC	agricultural extension development coordinator (Malawi)
AEDO	agricultural extension development officer (Malawi)
AEMO	agricultural extension methodology officer (Malawi)
AEO	agricultural extension organization (DRC)
AES	agricultural extension service
AFO	association field officers (Malawi)
AGP	Agricultural Growth Program (Ethiopia)
APES	Agricultural Production Estimates Survey (Malawi)
ARDPLAC	Agriculture and Rural Development Partners Linkage Advirsory Council (Ethiopia)
ARTP	Agricultural Research and Training Project (Ethiopia)
ASP	area stakeholder panels (Malawi)
ASTI	Agricultural Science and Technology Indicators
ASWAp	Agriculture Sector Wide Approach (Malawi)
ATA	Agricultural Transformation Agency (Ethiopia)
ATMA	Agricultural Technology Management Agency (India)
ATVET	Agricultural Technical and Vocational Education and training (Ethiopia)
AU	African Union

CAADP	Comprehensive Africa Agriculture Development Programme
CAEO	chief agricultural extension officer (Malawi)
CAETS	Controller of Agricultural Extension and Training Services (Malawi)
CARG	Agricultural and Rural Management Council (DRC)
CIALCA	Consortium for Improving Agriculture-Based Livelihoods in Central Africa
CLAD	Censored Least Absolute Deviations (Uganda)
CNONGD	National Council of Development NGOs (DRC)
COCO	Connect Online—Connect Offline (Ethiopia)
COPEMECO	Confederation of Small and Medium Enterprises of the Congo (DRC)
CRE	correlated random effects (Malawi)
CSA	Central Statistical Agency (Ethiopia)
CSO	Civil Society Organization (Uganda)
DA	development agent (Ethiopia)
DAC	district agriculture committee (Malawi)
DADO	district agricultural development officer (Malawi)
DAECC	District Agricultural Extension Coordinating Committee (Malawi)
DAES	Department of Agricultural Extension Services (Malawi)
DAES	Directorate of Agricultural Extension Services (Uganda)
DAESS	District Agricultural Extension Services System (Malawi)
DAO	district agriculture office (Malawi)
DC	district commissioner (Malawi)
DHS	Demographic Health Survey (Malawi)
DLEC	Developing Local Extension Capacity
DRC	Democratic Republic of the Congo
DSP	district stakeholder panels (Malawi)
EARO	Ethiopian Agricultural Research Organization
EMATER	Minas Gerais Technical Assistance and Rural Extension Company (Brazil)
EPA	extension planning area (Malawi)
FAO	Food and Agriculture Organization of the United Nations
FARA	Forum for Agricultural Research in Africa
FBO	farmer-based organization
FBS	farmer business schools (Malawi)
FEC	Federation of Private Enterprises in Congo (DRC)

FFS	farmer field school
FISP	Farm Input Subsidy Programme (Malawi)
FOLECO	Federation of Laic and Economic NGOs (DRC)
FPPM	Food Production, Processing, and Marketing (DRC)
FSR	Farming System Research (Ethiopia)
FTC	farmer training center (Ethiopia)
GFRAS	Global Forum for Rural Advisory Services
GoE	Government of Ethiopia
GoM	Government of Malawi
GTP	Growth and Transformation Plan (Ethiopia)
GVAC	group village agricultural committee (Malawi)
IAR	Institute of Agricultural Research (Ethiopia)
IAR4D	Integrated Agricultural Research for Development (DRC)
ICT	information and communications technologies
IECAMA	Imperial Ethiopian College of Agriculture and Mechanical Arts
IFAD	International Fund for Agricultural Development
IFPRI	International Food Policy Research Institute
IHS3	Integrated Household Survey (Malawi)
IITA	International Institute of Tropical Agriculture (DRC)
IP	innovation platform (DRC)
IPMS	Improving Productivity and Market Success (Ethiopia)
INERA	National Agronomic Research Institute (DRC)
ISFM	integrated soil fertility management (DRC)
IVR	Interactive Voice Response (Uganda)
LSMS	Living Standard Measurement Survey (Uganda)
M&E	monitoring and evaluation
MAAIF	Ministry of Agriculture, Animal Industry and Fisheries (Uganda)
MaFAAS	Malawi Forum for Agricultural Advisory Services
MDA	Ministry of Agrarian Development (Brazil)
MEAS	Modernizing Extension and Advisory Services Project
MGDS	Malawi Growth and Development Strategy
MINAGRI	Ministry of Agriculture, Livestock and Fisheries (DRC)
MINRD	Ministry of Rural Development (DRC)
MINREST	Ministry of Research, Science, and Technology (DRC)
MKW	Malawi kwacha
MoAIWD	Ministry of Agriculture, Irrigation and Water Development (Malawi)

MoANR	Ministry of Agriculture and Natural Resources (Ethiopia)
MPP	Minimum Package Project (Ethiopia)
MoFA	Ministry of Food and Agriculture (Ghana)
MoLGRD	Ministries of Local Government and Rural Development (Malawi)
NAADS	National Agricultural Advisory Services (Uganda)
NABCO	Nation Builders Corps (Ghana)
NACDC	National Agriculture Content Development Committee (Malawi)
NAEIP	National Agricultural Extension Intervention Program (Ethiopia)
NAEP	National Agricultural Extension Policy (Malawi)
NAIP	National Agricultural Investment Plan (DRC)
NAP	National Agricultural Policy (Malawi)
NARO	National Agricultural Research (Uganda)
NASFAM	National Association of Smallholder Farmers of Malawi
NEPAD	New Partnership for Africa's Development (DRC)
NGO	nongovernmental organization
NRC	National Resources College (Malawi)
OECD	Organization for Economic Co-Operation and Development
PADEP	Peasant Agricultural Development Program (Ethiopia)
PADETES	Participatory Demonstration and Training Extension System (Ethiopia)
PNATER	National Policy for Technical Assistance and Rural Extension (Brazil)
PRONAF	National Program for Family Agriculture (Brazil)
PSNP	Productive Safety Net Program (Ethiopia)
RCBP	Rural Capacity Building Project (Ethiopia)
RED	Research–Extension Division (Ethiopia)
REFAC	Research–Extension–Farmers Linkage Advisory Council (Ethiopia)
RELC	Research–Extension Liaison Committee (Ethiopia)
RPO	rural producer organization
SANE	Strengthening Agricultural and Nutrition Extension project (Malawi)
SAPP	Sustainable Agriculture Promotion Programme (Malawi)
SENAFIC	National Fertilizer Agency (DRC)
SENAMA	National Mechanization Agency (DRC)

SENASEM	National Seed Agency (DRC)
SMS	subject matter specialist
SNNP	Southern Nations, Nationalities and Peoples Region (Ethiopia)
SNV	national extension service (DRC)
SSA-CP	Sub-Saharan Africa Challenge Program (DRC)
SSES	Single Spine Agricultural Extension System (Uganda)
TSBF-CIAT	Tropical Soil Biology and Fertility Research Area of the International Center for Tropical Agriculture
UNDP	United Nations Development Programme
USAID	United States Agency for International Development
VAC	village agriculture committee (Malawi)
VDC	village development committee (Malawi)
YEA	Youth Employment Agency (Ghana)
ZARDIs	Zonal Agricultural Research and Development Institutes (Uganda)

FOREWORD

By Gary Alex

At the global level, it is well recognized that social resilience, poverty reduction, natural resource conservation, and adaptation to climate change should be safeguarded. Achieving this, however, depends on a substantial transformation of agricultural extension and advisory services which, in turn, will rely on high-quality information and knowledge sharing for farmers in rural communities, particularly with reference to farm productivity. The agenda, therefore, must be such that it incorporates an evaluation of extension services among the world's more than 500 million family farms.

Growing up in a farming community in the United States in the mid-twentieth century was an exciting experience. It inspired in me a strong responsibility to cultivate the land. In the United States, change came about with speed when surpluses constrained farm incomes and food prices. While famine remained a threat to many parts of the globe, we on the farm in the United States were proud to be helping "feed the world." Information on new technologies and innovations in management methods were brought home by way of various sources—from magazines on farming, to radio, to input and equipment dealers, to county fairs, and to vocational agriculture courses. A particularly effective source of information comes to mind: a series of evening seminars, presented by agriculture specialists (that is, mostly input dealers) and coordinated into extension services. This, as well as the many other sources of information relating to farming, left in me a lasting impression, especially in terms of the potential that extension services can provide.

Robert Bertram, chief scientist of the United States Agency for International Development, noted in a 2017 workshop that "extension is all about providing farmers with options." The novel technological development of the Green Revolution brought about investment in extension services, providing farmers information on the ways to increase productivity (for example, new seed varieties and fertilizers). Donor funding further helped to establish or expand many such services in the public domain. Although the overall effort was successful, the metrics to assess the impact of extension services were often challenging.

Assessing impacts and effectiveness of extension systems is a daunting challenge. They differ in the categories of farmers and other clients served; use different approaches and strategies; deal with different crops, livestock, rural enterprises, and farming systems; and face scale issues in serving the whole rural sector. The history of extension services programs has been fairly turbulent. They initially had been considered as public, top-down, technology-transfer activities. While sometimes neglected, there are diverse approaches that also include participatory methods such as facilitation, adult education, and local capacity development. "Silver bullet" solutions have proved detrimental, whereby specific approaches are widely replicated without regard to local situations.

Many public extension services programs have relied on poorly trained field staff, dysfunctional public-sector bureaucracies, and an overdependency on funding from donors. A few sustainable extensions have nevertheless emerged, especially in Asia; while others have declined as a result of shrinking donor funding. As a result, the overall state of extension services—difficult to assess and document—is at a critical crossroads. In the 1990s the informal cooperation of the Neuchâtel Group provided a means to rethink the nature of extension services programs, an effort that led to the general consensus that they should be participatory, market focused, pluralistic, decentralized, client oriented, well monitored, and sustainable. This, in turn, has led to the concept of best practices as well as the need to recognize the essentiality of tailoring extension services to local needs and conditions—that is, "best-fit" practices. This book forms the basis for these.

The contributions in this volume build on the experiences of the Food and Agriculture Organization, the Neuchâtel initiative, and the Global Forum for Rural Advisory Services in seeking the way forward to foster revitalized rural and extension services. The ability to systematically and objectively assess the function of such services and their impact is, however, fundamental to this effort. While clearly a challenge, rigorous assessments and evaluations will strengthen the case for further investments in rural and extension services as well as identify systems in need of reform and their effective management.

Evaluations of extension and advisory services, while understood today as being pluralistic knowledge and information services, tend to be empirical. They are undertaken in rural areas, as little more than a farmer-to-farmer information exchange or as more complex multi-institutional networks in commercialized agricultural systems. Few doubt the need for the development of effective extension services programs, although serious effort to strategically foster this has been scarce.

Many countries have adopted the concept of a best-practice design for extension services programs, while at the same time embracing the principles generally considered for effective and sustainable rural extensions. Nevertheless, the implementation of these has fallen behind. As reflected throughout this book, most countries have neither moved to dynamically develop extension services to serve the various client groups, agricultural systems, and off-farm stakeholders and participants, nor have they encouraged the effectiveness and sustainability of agricultural knowledge and information services. The information and cases described herein provide a wealth of experience in terms of understanding and evaluating extension services. There is, as yet, no simple set of dashboard indicators for such assessments. Nevertheless, the tools provided in this book will complement those that are applied to assess the level of satisfaction of and the effectiveness of extension services for the small farmer, the sustainability of financing by constituency, and the extent of social benefits.

PREFACE

Public extension remains as relevant to increasing agricultural productivity in developing countries today as it was decades ago, despite the emergence of other types of rural advisory services providers and recent developments in digital information and communications. As the authors of this informative global review of extension services point out, even where the private sector or nongovernmental organizations are involved in extension efforts, they all work with and rely on public extension. Yet public extension systems have been woefully neglected by most national governments and the global development community for the last three decades. This book offers an excellent mix of case studies that remind us of the consequences of this neglect while also showing us a more hopeful vision of how effective extension systems could be, if better financed and coordinated.

The authors understand the budgetary constraints facing governments and international financiers and thus emphasize the need for "best-fit" solutions. With cost considerations paramount, significant attention is paid to private sector provision of extension, voluntary or lead farmer approaches, and digital extension methods. Another important consideration for improving cost-effective extension systems is to make them more demand-driven—that is, providing advice on the issues most salient to farmers and thus more likely to be acted upon.

The enumeration by the authors of policy and program options to improve public extension systems is comprehensive—reflecting the variability in challenges faced and in the feasibility of improvements across countries. A strong message of the book is the need for improved coordination of rural advisory services, which can help to address increasing pluralism in providers, decentralization of services, and the current disconnect between research and extension.

This book serves as an excellent source of information and inspiration for those interested in how the world's smallholder farming sector can meet pressing challenges, continuing to provide itself with food and income while contributing to meeting global challenges such as climate change adaptation and mitigation, land degradation, water scarcity, and malnutrition in all forms. Public extension will be called upon to play this connecting role.

Frank Place
Director, CGIAR Research Program
on Policies, Institutions, and Markets

ACKNOWLEDGMENTS

This book was made possible, in part, through support from the United States Agency for International Development (USAID) and the CGIAR Research Program on Policies, Institutions, and Markets. The work relating to extension systems in Latin America was supported by the Inter-American Development Bank and the Family Farming and Rural Territorial Development for Latin America and the Caribbean regional program of the Food and Agriculture Organization of the United Nations. The information from Central Asia was the result of collaboration between the Global Forum for Rural Advisory Services (GFRAS), team members of the Modernizing Extension and Advisory Services project, and the Central Asia and the Caucasus Association of Agricultural Research Institutions.

The case study on Brazil in Chapter 4 was funded by GFRAS with the support of USAID and was carried out in collaboration with the Latin American Network for Rural Extension Services. The case study on Ethiopia (Chapter 6) was funded by USAID through the country strategy support program of the International Food Policy Research Institute (IFPRI). The case study on Malawi (Chapter 7) was funded through the Government of Flanders, the German Society for International Cooperation, Ltd., USAID Strengthening Agricultural and Nutrition Extension, and the CGIAR Research Program on Policies, Institutions, and Markets (PIM), done in partnership with Wadonda Consult and the Lilongwe University of Agriculture and Natural Resources. For the Democratic Republic of the Congo (Chapter 8), funding came from USAID through IFPRI. The opinions expressed herein are those of the authors and do not necessarily reflect the views of USAID.

For their significant contributions to the information and studies within various chapters, we would like to acknowledge the following people:

- Chapter 2: Andrea Bohn and Burton Swanson, who compiled some of the data and contributed to the data collection and analysis, and an anonymous reviewer who suggested the chapter.
- Chapter 3: Botir Dosov (Central Asia and the Caucasus); John Preissing, Sergio Ardila, Francisco Aguirre, and Julián Buitrón (Latin America); Suresh C. Babu and Mahika Shishodia (India); Sam Oeurn Ke and Suresh C. Babu (Cambodia); and Pham Hoang Ngan and Suresh C. Babu (Viet Nam). PIM also supported preparation of the India, Viet Nam, and Cambodia papers.
- Chapter 3: Three major projects contributed many of the country cases, including Modernizing Extension and Advisory Services (MEAS, https://meas.illinois.edu/), Integrating Gender and Nutrition in Agricultural Extension Services (INGENAES, https://ingenaes.illinois.edu/), and Developing Local Extension Capacity (DLEC, www.digitalgreen.org/usaid-dlec/).

We also thank two anonymous reviewers for helpful comments and suggestions. Most important, we are grateful to all the people who have shared their precious time with us during the surveys and interviews: the women and men farmers, community and group leaders, extension agents, extension organizations' officers and heads, development agencies' representatives, and government officials.

We dedicate this book to our partners in the different countries we work with. This volume is particularly dedicated to the late Dr. Ephraim Chirwa, for his devotion and legacy for quality teaching, research, and data collection for research-based policy solutions to development, and to the late Dr. Burton Swanson for his passionate lifelong work in extension globally.

Chapter 1

INTRODUCTION AND MOTIVATION

Kristin Davis, Suresh C. Babu, and Catherine Ragasa

1.1 Need for Better Understanding of the Status and Performance of Extension and Advisory Services

Agricultural development is critical to the livelihoods of more than a billion small-scale farmers and other rural populations in developing countries. Challenges such as low productivity, persistent food insecurity and malnutrition, food price crises, natural resource depletion, changing and uncertain markets, environmental degradation, and climate change directly impact (and are impacted by) the agriculture sector and rural development. Agricultural extension and advisory services can help address these challenges by assisting farmers with advice and information, brokering and facilitating innovations and relationships, dealing with risks and disasters, and in many other ways (Babu and Joshi 2019; Kilelu et al. 2011; Hoffmann et al. 2009; Boteler 2007; van den Ban and Hawkins 1996).

In the past, agricultural extension and advisory services (defined in Box 1.1) have shown significant and positive effects on knowledge, skills, income levels, technology adoption, crop and livestock yield, and productivity (Davis 2008), especially during the Green Revolution in Asia (Swanson and Rajalahti 2010). Studies of rates of return to extension generally show high numbers (85 percent on average), despite wide variation (Alston et al. 2000). More recently, several country-specific results have emerged as well. Empirical literature documents the positive effects of public expenditures on extension services compared to those on input subsidies (Benfica, Cunguara, and Thurlow 2015; Armas, Gomez Osorio, and Moreno-Dodson 2012; Allcott, Lederman, and Lopez 2006; Rosegrant, Kasryno, and Perez 1998), and other investments, such as rural feeder roads or irrigation infrastructure in Uganda (Pauw and Thurlow 2015) and irrigation in Mozambique (Benfica, Cunguara, and Thurlow 2015). Rigorous time-series analyses show high returns and poverty reduction impacts of extension services in Ethiopia (Dercon et al. 2009). Nationally representative panel data in Malawi show that quality extension and advisory services have contributed to greater farm productivity and increased food security (Ragasa and Mazunda 2018). Yet there has not been adequate documentation of policy and program constraints that extension services are facing, and there is limited evidence on what policy and program reforms are needed to increase the performance and impact of extension services on development outcomes.

Extension and advisory services are complex to study because they are often part of other agricultural intervention programs. In part, due to the nature of their design and implementation, which includes technical, social, and educational elements, their assessment on development outcomes remains a challenge. Extension systems can be quite complicated even within national boundaries. This is because countries have different agroecological zones with varying production systems, natural resource endowments, and extension service needs. Moreover, most developing countries are moving from sole dependency on public extension systems to pluralistic ones, in which extension services are also provided by a plethora of actors, including nongovernmental organizations (NGOs), farmer organizations, and the for-profit private sector. In a pluralistic extension system, service providers may have very different objectives, goals, and motivations. For example, in some cases, extension services aim to primarily increase yields and income, while in others they may strive for greater women's empowerment through education and skill development. In still others, they may focus on increasing food security and nutrition outcomes.

The increasing pluralism of extension in terms of types of organizations providing these services (and of their objectives, methods, and approaches) means that extension programs are becoming even more difficult to evaluate methodologically and to attribute causal impact on development outcomes (Ragasa and Mazunda 2018; Faure et al. 2016; Birner et al. 2009; Anderson 2007). Basic data are difficult to obtain because of the increasingly pluralistic and decentralized nature of today's extension services. Public extension services are often subject to changes of policies of different governments because provision of extension services is not institutionalized through legislation.

While policy and development communities recognize the importance of understanding the issues, challenges, and constraints facing extension and advisory services (Pye-Smith 2012), very few studies are available to gain a regional and global perspective and a view of the global status of extension services. Major data-collection efforts were undertaken in 1981, 1989, and 2009 (Swanson and Rassi 1981; Swanson, Farner, and Bahal 1990; GFRAS 2012; see Chapter 2). These studies provide basic detailed information, but this type

BOX 1.1 Defining agricultural extension and advisory services

Over time, the term "agricultural extension," while still commonly used, is gradually being replaced by the term "agricultural advisory services," indicating a less top-down approach that views farmers and other producers as clientele. Some have extended the term even more broadly to "rural advisory services," to include sources of livelihood other than agriculture and greater focus on the facilitation and brokerage role beyond technology transfer (Davis and Heemskerk 2012; Faure, Desjeux, and Gasselin 2012; World Bank 2012; Swanson and Rajalahti 2010). Some refer to "nutrition and agricultural extension and advisory services" to include nutrition information provision and behavior change communication for better health and nutrition outcomes (Fanzo et al. 2015; Kuyper and Schneider 2016). There are several definitions of extension or advisory services (see Faure et al. 2016). Other authors prefer to use the terms "communication" (Leeuwis 2004) or "facilitative approach" (Ingram 2008) to emphasize the role extension plays in facilitating the negotiation between different actors to solve problems.

For the purposes of this book, following Birner et al. (2009: 342), we define agricultural extension and advisory services as "the entire set of organizations that support and facilitate people engaged in agricultural production to solve problems and to obtain information, skills, and technologies to improve their livelihoods and well-being." Throughout the book we use "extension services," "extension," "agricultural extension," and "advisory services" interchangeably to mean this broader definition of these services, and we use "extension organizations" or "service providers" and "extension agents" or "extension workers" to refer to the set of organizations providing extension and advisory services and frontline workers with direct contact to the rural communities, respectively.

of information goes out of date nearly as soon as it is collected. However, it is a valuable set of information that allows comparison over time and across regions. This is especially the case because the information has never been adequately analyzed. Thus this book is an attempt to fill this gap.

A second gap in the extension services assessment literature is the lack of common framework and comparison of assessments on national or regional extension systems. While many country-level assessments have been conducted, most of them are in unpublished and informal reports, and they have never had a common framework applied to enable cross-country comparisons. Chapter 3 addresses this second gap. Information regarding the performance of extension services, linking national- or system-level assessments and program- and approach-level evaluations (as in Figure 1.1) is scanty. Many national-level diagnostic assessments are available and program- and approach-level assessments are emerging (see summaries by Faure et al. 2016; Kondylis 2019; Ragasa 2019; Ragasa, Spielman, and Place 2019). Chapters 4–8 look at performance of extension services in five countries and link three different levels of assessments using primary and secondary data—in some cases, including evaluation of their impacts.

There are many ways to assess agricultural extension and advisory services, from simply describing what exists in a country to measuring and explaining variations of performance and impact. Figure 1.1 shows a continuum of different assessments and evaluations on extension services. The first set of assessments are at a system level (national or subnational) and provide diagnostic assessment of the contextual factors and policies, conditions, and characteristics of an extension system, usually conducted using descriptive, narrative, and qualitative assessment methods. The second set of assessments are at the program or project level, in which extension services are usually provided and bundled with other services and interventions. These are usually conducted using quantitative and/or qualitative impact evaluation methods. The third set are small-scale evaluations of pilots or field experimentation of a specific extension approach or design of an extension approach and are usually conducted using quantitative and/or qualitative assessment methods. Rigorous randomized controlled trials are emerging that provide useful insights on which design or approach of extension service provision works or does not work, and which has the greater impact or is more cost-effective (see summaries by Kondylis 2019; Ragasa 2019; and Ragasa, Spielman, and Place 2019).

The objectives of this book are to assess extension and advisory services in a cross-country comparative context in the following two ways:

1. Provide a description and comparison of the existing extension services in the regions and countries under consideration focused on a set of characteristics; and

2. Provide an assessment of the performance of extension services provision and impact evaluation of extension approaches in selected case study countries.

Objective 1 addresses specific extension characteristics from the best-fit framework (these include governance structures, capacity, management, and advisory methods) and offers a global overview based on primary data and regional and country cases based on both primary and secondary data (Part 1

FIGURE 1.1 Different levels of assessment of extension services

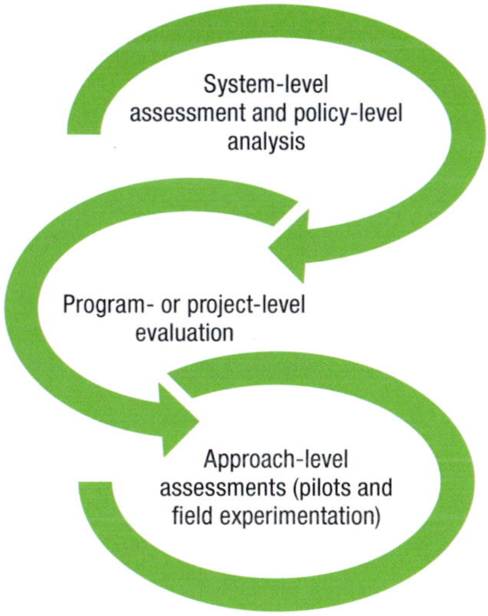

Source: Authors.

of the book). Objective 2 undertakes in-depth country case studies based on both primary and secondary data (Part 2 of the book). Part 2 attempts to illustrate the three levels of assessments in Figure 1.1 in a particular country. Some country cases have richer illustrations of second and third sets of assessments than others due to the availability of primary and secondary data and past evaluation studies.

1.2 The Best-Fit Framework for Assessing Extension Services

The importance of a comprehensive and reliable system of assessing extension systems and their reforms cannot be debated. Such studies help practitioners decide which factors influence the effectiveness of particular extension programs, and what modifications need to be made for future improvement. In fact, it is important to build the capacity of practitioners to design extension systems that are most suited to contextual realities and harness the potential of NGOs, the private sector, and other stakeholders (Birner et al. 2009; Babu, Sette, and Davis 2015; Babu and Joshi 2019).

To better analyze extension systems, a comprehensive framework that takes into account heterogeneity within and between countries is necessary to guide their assessment. Birner and colleagues produced such a framework in 2006 (Birner et al. 2009). The framework was designed for analyzing and designing pluralistic extension systems. Although it has often been referred to in the literature (for example, Herrera et al. 2019; Davis and Spielman 2016; Faure et al. 2016; Álvarez-Mingote and McNamara 2018), it has not been systematically applied to evaluate and compare country extension systems or programs. The best-fit framework provides an impact chain to comprehensively analyze extension and advisory services by examining the overall environment (frame conditions), the characteristics of the extension that are affected by the frame conditions, extension performance, primary outcomes in terms of farmer behavior, and the ultimate impact (Figure 1.2). The framework can be used to develop assessment tools for agricultural advisory services, to inform processes of reforming the services, and to guide interdisciplinary research. It is unique because it uses insights from different disciplines, which were previously treated separately in the literature. The framework can help policymakers and analysts to move from theoretical discussions to an evidence-based best-fit approach. It helps users to disentangle elements of advisory services by distinguishing between the following extension characteristics: (1) governance structures, (2) capacity, (3) management, and (4) advisory methods (Birner et al. 2009).

These four characteristics are all variables that can be manipulated by implementers in response to the prevailing contextual factors such as policy environment and production systems. Thus they are important factors that can be changed by governments and implementing agencies to affect the performance of the system—and thus the ultimate impact. The "governance structures" refer to the institutional setup of extension services. They include the role of the public sector in extension services, level of privatization, degree of decentralization, funding mechanisms, and coordination and linkages. "Capacity" refers to the human and organizational competencies in the system as well as the financial and physical assets. "Management" is the way in which extension services are managed within the respective governance structures. This includes training and retraining efforts, organizational management procedures carried out, incentives and methods of performance assessment of individual agents, and monitoring and evaluation of the services. Organizational capacity and management are combined in several of the country chapters. "Advisory methods" are used by extension services field staff in interactions with farmers and other clientele. Our analysis focuses on these four choice variables because they are the factors that can be directly changed by extension agencies. It is rather more difficult to effect changes at the wider

FIGURE 1.2 Best-fit framework for analyzing and designing pluralistic extension services

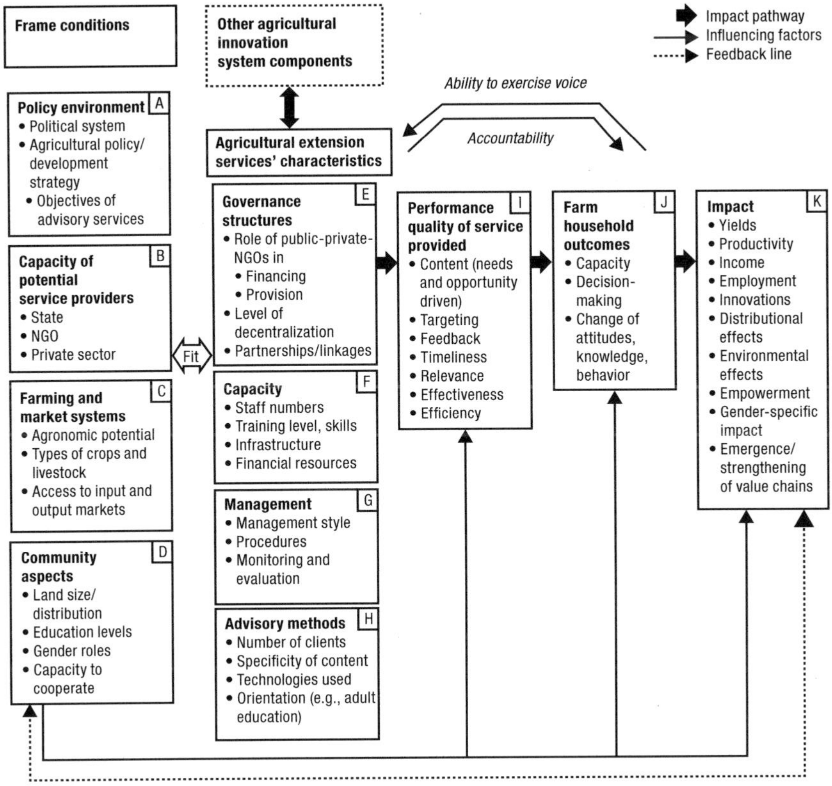

Source: Birner et al. (2009).

policy environment or farming systems (contextual factors that affect the choice of extension characteristics).

The logic of the best-fit framework is that users must first consider contextual factors or framework conditions (Boxes A–D in Figure 1.2) that influence how extension should be structured and organized (Boxes E–H) to reach optimum performance (Box I). The ultimate impact of the services (Box K), however, depends on changes (in attitudes, opinions, knowledge, etc.) at the farm household level (Box J). The contextual factors, or frame conditions, are outside the control of extension and include the policy environment, capacity of potential service providers from all sectors, farming and market systems, and community aspects such as education levels and land size. The frame conditions affect the characteristics of the extension. Characteristics include governance structures, capacity, management, and advisory methods used. The characteristics lead to the performance of the system. The performance can be judged by indicators such as relevance, timeliness, efficiency, and so forth. Performance leads to behavior of farm households, such as changing attitudes or adoption of techniques.

Based on the best-fit framework, this book analyzes and synthesizes new data that has been collected over the past five to ten years from a series of case studies examining the status and characteristics and, in some cases, the performance of agricultural extension and advisory services around the world. It provides practitioners and policymakers with descriptive information on extension services, makes cross-country comparisons using the best-fit framework, and overviews impact assessments of extension. The publication offers a common framework with which to examine the practices of countries and to make cross-regional and cross-country comparisons using the best-fit characteristics. Recommendations and policy implications are provided for improving extension globally. Using the best-fit framework for analysis, the book provides overall guidance on the process of evaluating advisory services systems while exemplifying global practices through various case studies. This publication is less an academic book than a guide on the process of assessments and a showcase of results. Not only is it intended as a general reference guide; the book sets the standard for policy analysts and managers worldwide who wish to apply similar systems-level evaluation methods in their country.

1.3 Organization of the Book

This book is divided into two parts. Part 1 (Chapters 2 and 3) focuses on the global assessment of extension systems and the features of extension services that are referred to as "characteristics" in the best-fit framework. An overview of the global landscape is presented in Chapter 2, followed by a comparison of these best-fit characteristics at the country and regional level in Chapter 3. Chapter 2 analyzes governance structures, including the legal status of providers, capacity (especially in terms of staffing), management, and advisory methods and clientele. Chapter 3 provides a synthesis of information from in-depth country and regional assessments that were conducted mainly since 2015. These assessments took place with various partners and projects; to pull out comparisons of the national systems, we applied the best-fit approach. This analysis provides a snapshot of extension services, zooming in at country or regional level. It does not address the performance or the impact of extension services. The country cases in Chapter 3 were chosen purposively based on availability of relevant data and how recently the data were collected (within the past 5 to 10 years). For the country descriptions of extension and advisory services, the studies had to contain recent and in-depth analysis on agricultural extension and advisory services at the country or regional level.[1] Furthermore, the information had to include relevant data about the best-fit characteristics.

Part 2 focuses on the performance of extension systems using in-depth country case studies from Brazil (Chapter 4), Uganda (Chapter 5), Ethiopia (Chapter 6), Malawi (Chapter 7), and the Democratic Republic of the Congo (Chapter 8). These countries were chosen purposively based on (1) major knowledge gaps and relatively few assessments available; (2) availability of primary and secondary data to analyze; (3) empirical application of the best-fit framework; and (4) diversity of systems to enable comparisons and contrasts. "Brazil" (Chapter 4) offers evidence related to implementation of a new extension policy and system. "Uganda" (Chapter 5) provides a rich body of evidence on extension reforms over the years. The Ethiopia case (Chapter 6) represents the largest extension cadre and largest investments in extension systems in the continent and among the largest in the world, while "Malawi" (Chapter 7) and the "Democratic Republic of the Congo" (Chapter 8) represent cases with limited public investment in extension systems and an increasing role of non-government actors. The Democratic Republic of the Congo case presents a postconflict country with enormous agricultural potential. The DRC and Malawi also represent cases with limited availability of past assessments but recent new large-scale data within the best-fit framework in the past five years, which enabled the detailed analyses presented in this book.

Overall, the countries selected present a wide spectrum of different systems operating in two different continents. The book attempts to compare and contrast them and to synthesize the lessons learned. Table 1.1 shows the empirical basis of the included studies. These in-depth country cases thus provide a good illustration and comparison of the types of methods, datasets, indicators, and analyses to conduct systems-level assessments of extension services. Furthermore, we include cases from various parts of Asia and Latin America in Chapter 3. Chapter 2 is global in scope and covers all regions. Admittedly, the regional balance, especially for the country case study chapters in Part 2, is not ideal, since African countries have a heavy focus and only

1 The oldest reports are from 2009; however, these country cases also have data from 2014 and later.

TABLE 1.1 Empirical basis of the studies included in the book

Book section	Focus area/region focus	Type of empirical data	Dates
Part 1			
Chapter 2	Global assessment of current extension programs. This part focuses on the features of extension services that are referred to as "characteristics" in the best-fit framework described in this chapter: governance structures and legal status of providers, capacity in terms of staffing, and extension methods and clientele.	Primary data from a 1980 survey, which was an update of an earlier 1975 survey, compiling information on extension staff numbers and qualifications, objectives, mechanisms for public participation in governance, and client groups served.	Swanson and Rassi (1981)
		Primary surveys from the FAO 1991 data were gathered through a mail-type questionnaire sent to 154 FAO member countries. Some 132 organizations in 113 countries responded.	FAO 1991 (data collected in 1989)
		Questionnaires developed in English, French, and Spanish were sent out to key informants (heads of departments) in government extension departments in more than 160 countries. Data were obtained from 347 organizations in 81 countries.	GFRAS 2012 (data collected 2009–2012)
Chapter 3	Comparison of these best-fit characteristics at the country and regional levels. Provides a synthesis of information from in-depth country and regional assessments that were conducted over the past five years or so. These assessments took place with various partners and projects; to pull out comparisons of the national systems, we applied the best-fit approach described in this chapter. It is a snapshot of extension services zooming in at the country or regional level.	Number of in-depth reports, mainly out of extension projects and manuscripts and five country cases described in Chapters 4–8 in this book. Mainly secondary sources of information (except for the five country cases in Part 2, which used primary data).	Various between 2012–2018; see Chapter 3 for more details
Part 2			
Chapter 4	Brazil	Survey of 1,000 farmers and 87 extensionists in 5 territories in 3 states (purposively selected for low Human Development Index, high concentration of family farmers and rural settlements, presence of black farmer communities and indigenous populations, and municipalities with low economic dynamism).	2014–2015

Book section	Focus area/region focus	Type of empirical data	Dates
Chapter 5	Uganda	Survey of 208 agricultural extension agents done by IFPRI and Makerere University.	2007
		Living Standard Measurement Survey, nationally representative data collected by Uganda Bureau of Statistics.	2016–2017
Chapter 6	Ethiopia	Household survey of 7,500 households in selected regions (where the Agricultural Growth Program was implemented).	2011, 2013
		Central Statistical Agency annual survey	2005–2017
		Survey of 237 extension agents	2009
		Survey of 896 extension agents	2017
Chapter 7	Malawi	Nationally representative survey of 3,001 households, 2 rounds	2016, 2018
		Survey of 30 service providers and 71 extension agents, focus group discussions in selected communities	2017, 2019
		Malawi integrated household panel surveys	2010, 2013, 2016
Chapter 8	Democratic Republic of the Congo	Interviews with 45 key informants	2010–2011
		Survey of 55 Agricultural Management Councils	2011
		Survey of 181 community organizations	2011
		Survey of 107 extension organization heads	2011
		Survey of 163 extension agents	2011
		Survey of 3,110 households in randomly selected treatment and control communities as part of the midline assessment of the Food Production, Processing, and Marketing project	2014

Source: Authors.

the Brazil case comes from outside the African continent. Ideally we would have included country cases from all major regions of the world. However, the authors and editors were constrained by lack of available primary and secondary data, available literature, and authors and writing teams who could write up their work. Recent lessons from large countries such as India and China are documented elsewhere (Glendenning and Babu 2011; Babu et al. 2013; Babu et al. 2015). A recent volume (Babu and Joshi 2019) covers the current trends in extension reforms in the South Asia region.

1.4 Overview of Findings

The global analysis of extension systems provided throughout this book points to a number of policy implications across the best-fit characteristics, implementing policies to improve governance and coordination, undertake curricula reform, provide incentives to staff, and use appropriate methods to enhance financial sustainability and to achieve greater scale. The analysis and case studies show that with regard to governance structures, extension and advisory services have become much more pluralistic. There are more than 1 million public and private extension officers today, but their roles are becoming less clearly defined with the outreach of information and communications technologies, the use of lead farmers (essentially volunteer extension agents), and the use of other individual service providers in various agricultural value chains. The public sector, usually ministries of agriculture, is responsible for the overall coordination and regulation of extension. This coordination is becoming crucial to take advantage of the different types of providers and deal with the emerging challenges of the day such as climate change and malnutrition of all forms. The coordination and regulation of the many providers is now an issue that governments must address; "Brazil" (Chapter 4) and "Malawi" (Chapter 7) show innovative mechanisms to deal with such coordination. However, the linkages between extension services, research, and farmer feedback mechanisms remain weak (Chapter 3).

For pluralism to work, policies are needed to provide the overall framework within which multiple actors can work. But many countries do not have a specific policy for extension and advisory services; these are usually rooted in broader agricultural sector development policies (Chapter 3). The findings indicate that Brazil and Ethiopia stand out as having in place the best extension services policies and accompanying implementation mechanisms. Most countries lack an explicit extension policy, causing them to rely solely on general agricultural strategies to guide extension services provision.

Financing of extension services is, in most cases, insufficient and often donor-driven with limited thought given to financial sustainability. The performance of extension services is hampered by a shortage of funding, especially in terms of operational costs and sufficient staffing numbers. Country studies indicate that there is a lack of cost-effective ways to increase coverage and provide services to many unserved or underserved areas. Furthermore, extension services may be duplicated in some areas by different sets of agents serving the same group of farmers. Nevertheless, there are promising moves toward cofinancing and farmer contributions in Latin America and Senegal. Certain advisory methods, such as use of information and communications technologies and farmer advisers, can also enhance efficiency of service provision and thus financial sustainability.

Capacity of extension services includes the staff numbers, training level, skills, infrastructure, and financial resources. Capacity levels in all the case studies were low in general. Numbers of extension agents from the public sector were seldom sufficient for the job at hand, and there were high vacancy rates and turnover in some countries. Most countries have a poor extension agent-to-farmer ratio, and many of the extension agents in Africa struggle with mobility and poor transport infrastructure. Salaries are normally low, especially in the public sector, and there are few rewards or recognition. The foundational training for extension agents usually focuses on technical topics and may miss out on functional skills that are also needed for extension.

With regard to management, there are a number of mechanisms that can be used to ensure better management such as market-orientation or demand-driven services as well as monitoring and evaluation (M&E). Many countries are putting in place demand-driven or client-oriented mechanisms and M&E systems to ensure that extension services are more participatory and thus more relevant to farmers, in particular to the underserved, women (for example, in Latin America), and indigenous groups (for example, in Brazil). The increasing use of multistakeholder networks, innovation platforms, and other demand strategies are contributing to these efforts. Chapters 2 and 3 address the huge gaps in data, capacities, and funding to do so. Chapter 3 and Chapter 9 provide guidance to policymakers to operationalize the collection of best-fit indicators at the country or provider level, with some recommended indicators.

An increasingly wide range of methods was used to provide advisory services (see Table 3.1), and the use of digital approaches has increased. Farmer field schools—a participatory, group-based adult education approach—are widely used today (Chapter 5 and Chapter 8). Group methods and demonstrations remain effective approaches; and private-sector extension and the use of lead farmers is growing and used worldwide. Market-oriented extension has frequently and appropriately become the focus for extension activities, particularly along value chains (Chapter 2). This was a big focus in Latin America. However, capacities and appropriate information to meet farmers' needs and provide market-oriented extension were often lacking; Central Asian and Caucasus countries in particular struggled with relevant ability to provide extension services.

Reaching women and youth continue to emerge as trends over the past few decades in the analysis. Evidence from Uganda (Chapter 5) furthermore suggested the need to increase the number of female agents to exploit their great potential to reach women and the poor. Other newer trends in extension include the use of information and communications technologies, the Internet of Things, market-orientation, and urbanization of extension services.

The concluding chapter draws several major policy insights. First, there is a need for an explicit policy or strategy for extension service provision that identifies design and implementation issues along with funding and coordination mechanisms for effective delivery of extension services to clientele. Second, funding mechanisms must be optimized to increase the sustained financing of the extension services and to reduce donor dependency and funding uncertainty from political shifts. Combining traditional and innovative modern approaches to extension service provision can reduce the cost and increase efficiency of services. Third, increasing the pluralistic nature of extension services calls for better coordination of players in the system to avoid duplication and maintain quality of services.

Fourth, sharing of organizational strategies and goals, jointly setting priorities for extension services, and managing resources are key programmatic aspects of running an extension system effectively. Fifth, for effective management it is important to have functioning monitoring and evaluation systems, both for assessing performance and for continuous learning and improvement of the extension services. Sixth, the technical, managerial, and leadership capacities of the extension personnel should be continuously updated for improving institutional innovations and to increase the reach to clientele in various agroecological zones.

Seventh, the advisory method needs to be context-specific and should take into account the information needs of producers and their ability to absorb and use the information. While traditional means such as on-farm demonstrations are still effective, sharing technologies through digital technologies has been shown to be successful. Adopting a mixed-method means to increase knowledge access by farmers requires constant updating of advisory methods.

Finally, the concluding chapter highlights several other policy, institutional, and programmatic recommendations for improving extension services globally. These include cross-cutting issues such as climate change, nutrition, youth, and gender, and other challenges related to the transformation of food systems that call for a new generation of extension worker who will be a problem solver and broker of knowledge rather than just a technology disseminator.

References

Allcott, H., D. Lederman, and R. Lopez. 2006. *Political Institutions, Inequality, and Agricultural Growth: The Public Expenditure Connection*. World Bank Policy Research Working Paper 3902. Washington, DC: World Bank.

Alston, J. M., C. Chan-Kang, M. C. Marra, P. G. Pardey, and T. J. Wyatt. 2000. *A Meta-Analysis of Rates of Return to Agricultural R&D: Ex Pede Herculum?* IFPRI Research Report 113. Washington, DC: International Food Policy Research Institute (IFPRI).

Álvarez-Mingote, C., and P. McNamara. 2018. "Evaluating Agricultural Extension and Advisory Services through a Governance Lens." *Journal of International Agricultural and Extension Education* 25: 71–86. DOI: 10.5191/jiaee.2018.25206.

Anderson, J. R. 2007. *Agricultural Advisory Services*. Background Paper for the 2008 World Development Report Agriculture for Development. Washington, DC: World Bank.

Armas, E. B., C. Gomez Osorio, and B. Moreno-Dodson. 2012. *Agriculture Public Spending and Growth: The Example of Indonesia*. World Bank Policy Research Working Papers. Washington, DC: World Bank.

Babu, S. C., J. Huang, P. Venkatesh, and Y. Zhang. 2015. "A Comparative Analysis of Agricultural Research and Extension Reforms in China and India." *China Agricultural Economic Review* 7 (4): 541–573.

Babu, S. C., and P. K. Joshi. 2019. *Agricultural Extension Reforms in South Asia: Status, Challenges, and Policy Options*. London: Academic Press.

Babu, S. C., P. K. Joshi, C. J. Glendenning, K. Asenso-Okyere, and R. V. Sulaiman. 2013. "The State of Agricultural Extension Reforms in India: Strategic Priorities and Policy Options." *Agricultural Economics Research Review* 26 (2): 159–172.

Babu, S., C. Sette, and K. Davis. 2015. "Private Technical Assistance Approaches in Brazil: The Case of Food Processing Company Rio de Una." In *Knowledge Driven Development: Private Extension and Global Lessons*, edited by S. Babu and Y. Zhou, 105–124. Public Policy and Global Development series. London: Elsevier.

Benfica, R., B. Cunguara, and J. Thurlow. 2015. "Distributional Effects of Public Investments in Mozambique." Draft paper. Accessed August 9, 2017. www.gtap.agecon.purdue.edu/resources/download/8636.pdf.

Birner, R., K. Davis, J. Pender, E. Nkonya, P. Anandajayasekeram, J. Ekboir, A. Mbabu, D. Spielman, D. Horna, and S. Benin. 2009. "From Best Practice to Best Fit: A Framework for Analyzing Agricultural Advisory Services Worldwide." *Journal of Agricultural Extension and Education* 15 (4): 341–355. DOI: 10.1080/13892240903309595.

Boteler, F. E. 2007. "Building Disaster-Resilient Families, Communities, and Businesses." *Journal of Extension* [Online] 45 (6). Article 6FEA1.

Davis, K. 2008. "Extension in Sub-Saharan Africa: Overview and Assessment of Past and Current Models and Future Prospects." *Journal of International Agricultural and Extension Education* 15 (3): 15–28.

Davis, K., and W. Heemskerk. 2012. "Investment in Extension and Advisory Services As Part of Agricultural Innovation Systems." Module 3 of *Agricultural Innovation Systems: An Investment Sourcebook*. Washington, DC: World Bank. http://siteresources.worldbank.org/INTARD/Resources/335807-1330620492317/9780821386842_ch3.pdf.

Davis, K., and D. J. Spielman. 2016. "A Global Update on the Status of Agricultural Extension." Unpublished manuscript. IFPRI, Washington, DC.

Dercon, S., D. Gilligan, J. Hoddinott, and T. Woldehanna. 2009. "The Impact of Agricultural Extension and Roads on Poverty and Consumption Growth in Fifteen Ethiopian Villages." *American Journal of Agricultural Economics* 91 (4): 1007–1021.

Fanzo, J., Q. Marshall, D. Dobermann, J. Wong, R. Merchan, M. Jaber, A. Souza, N. Verjee, and K. Davis. 2015. "Integration of Nutrition into Extension and Advisory Services: A Synthesis of Experiences, Lessons, and Recommendations." *Food and Nutrition Bulletin* 36 (2): 120–137.

FAO (Food and Agriculture Organization of the United Nations). 1991. *International Directory of Agriculture Extension Organizations*. Rome: FAO.

Faure, G., K. E. Davis, C. Ragasa, S. Franzel, and S. C. Babu. 2016. *Framework to Assess Performance and Impact of Pluralistic Agricultural Extension Systems: The Best-Fit Framework Revisited.* IFPRI Discussion Paper 1567. Washington, DC: IFPRI.

Faure, G., Y. Desjeux, and P. Gasselin. 2012. "New Challenges in Agricultural Advisory Services from a Research Perspective: A Literature Review, Synthesis and Research Agenda." *Journal of Agricultural Education and Extension* 18 (5): 461–492.

GFRAS (Global Forum for Rural Advisory Services). 2012. *Master Worldwide Extension Study Database.* Excel File. Lindau, Switzerland: GFRAS.

Glendenning, C. J., and S. C. Babu. 2011. *Decentralization of Public-Sector Agricultural Extension in India: The Case of the District-Level Agricultural Technology Management Agency (ATMA).* IFPRI Discussion Paper 1067. Washington, DC: IFPRI.

Herrera, B., M. Gerster-Bentaya, I. Tzouramani, and A. Knierim. 2019. "Advisory Services and Farm-Level Sustainability Profiles: An Exploration in Nine European Countries." *Journal of Agricultural Education and Extension* 25 (2): 1–21. DOI: 10.1080/1389224X.2019.1583817.

Hoffmann, V., M. Gerster-Bentaya, A. Christinck, and M. Lemma, eds. 2009. *Rural Extension, Volume 1: Basic Issues and Concepts.* Weikersheim, Germany: Margraf Publishers.

Ingram, J. 2008. "Agronomist-Farmer Knowledge Encounters: An Analysis of Knowledge Exchange in the Context of Best Management Practices in England." *Agriculture and Human Values* 25: 405–418.

Kilelu, C. W., L. Klerkx, C. Leeuwis, and A. Hall. 2011. "Beyond Knowledge Brokering: An Exploratory Study on Innovation Intermediaries in an Evolving Smallholder Agricultural System in Kenya." *Knowledge Management for Development Journal* 7 (1): 84–108. DOI: 10.1080/19474199.2011.593859.

Kondylis, F. 2019. "What Have We Learnt from Evaluations of Extension Interventions for NRM Technologies?" World Bank Research Group. Draft PowerPoint presentation. https://ispc.cgiar.org/sites/default/files/files/events/NRM%20Workshop/Kondylis_lessonslearnt_extension.pdf.

Kuyper, E., and L. Schneider. 2016. *Conceptualizing the Contribution of Agricultural Extension Services to Nutrition.* Integrating Gender and Nutrition within Agricultural Extension Services (INGENAES) Discussion Paper. Urbana-Champaign, IL, US: INGENAES.

Leeuwis, C. 2004. *Communication for Rural Innovation. Rethinking Agricultural Extension.* Wageningen, Netherlands: Blackwell Publishing.

OECD (Organisation for Economic Co-operation and Development) Development Assistance Committee. 1991. *Principles for Evaluation of Development Assistance.* Paris: OECD.

Pauw, K., and J. Thurlow. 2015. "Prioritizing Rural Investments in Africa: A Hybrid Evaluation Approach Applied to Uganda." *European Journal of Development Research* 27 (3): 407–424.

Pye-Smith, C. 2012. *Agricultural Extension: A Time for Change. Linking Knowledge to Policy and Action for Food and Livelihoods.* Wageningen, Netherlands: CTA.

Ragasa, C. 2019. *Modeling the Effectiveness of the Lead Farmer Approach in Agricultural Extension Service Provision Nationally Representative Panel Data Analysis in Malawi.* IFPRI Discussion Paper 1848. June. http://ebrary.ifpri.org/cdm/ref/collection/p15738coll2/id/133285.

Ragasa, C., and J. Mazunda. 2018. "The Impact of Agricultural Extension Services in the Context of a Heavily Subsidized Input System: The Case of Malawi." *World Development* 105: 25–47.

Ragasa, C., D. Spielman, and F. Place. 2019. "Innovation in Agricultural Extension and Advisory Services: Lessons Learned from Ongoing Research." Presentation during the Self-Help Africa (SHA) learning event, Blantyre, Malawi, July 8, 2019.

Rosegrant, M., F. Kasryno, and N. D. Perez. 1998. "Output Response to Prices and Public Investment in Agriculture: Indonesian Food Crops." *Journal of Development Economics* 55 (2): 333–352.

Swanson, B. E., B. J. Farner, and R. Bahal. 1990. "The Current Status of Agricultural Extension Worldwide." In *Report of the Global Consultation on Agricultural Extension*, edited by B. E. Swanson, 43–76. Rome: Food and Agriculture Organization of the United Nations (FAO).

Swanson, B. E., and R. Rajalahti. 2010. *Strengthening Agricultural Extension and Advisory Systems: Procedures for Assessing, Transforming, and Evaluating Extension Systems*. Washington, DC: World Bank.

Swanson, B. E., and J. Rassi. 1981. *International Directory of National Extension Systems*. Bureau of Educational Research, College of Education, University of Illinois at Urbana-Champaign.

Van den Ban, A. W., and H. S. Hawkins. 1996. *Agricultural Extension*, 2nd ed. Oxford, UK; Malden, MA, US: Blackwell Science.

World Bank. 2012. *Agricultural Innovation Systems: An Investment Sourcebook*. Washington, DC. http://siteresources.worldbank.org/INTARD/Resources/335807-1330620492317/9780821386842_ch3.pdf.

PART 1

Global Assessment of Extension Characteristics

Part 1 focuses on several features of extension services that are referred to as "characteristics" in the best-fit framework (Chapter 1): governance structures and legal status of providers, capacity in terms of staffing, management, and extension methods and clientele. Chapter 2 presents a global landscape of extension services, followed by a comparison of the best-fit characteristics at country and regional levels in Chapter 3.

Chapter 2

GLOBAL TRENDS IN EXTENSION PROVISION, STAFFING, AND METHODS

Kristin Davis and Gary Alex

2.1 Introduction

This chapter is a landscape view of extension globally. It compares the recent status to the situation several decades ago, when earlier global assessments were conducted. In particular, we look at best-fit characteristics, including legal status and governance of extension providers, staffing, and advisory methods and clientele. We also look at trends and recent developments using the global assessments and recent literature. There have been several major efforts to collect worldwide extension data. Surveys were conducted in 1975 and 1980 by the University of Illinois (Swanson and Rassi 1981). Follow-up work was conducted by the Food and Agriculture Organization (FAO) in 1988–1989, and by the International Food Policy Research Institute (IFPRI) and Global Forum for Rural Advisory Services (GFRAS) in 2009–2012. However, there were never any narrative reports produced or much analysis conducted of this data collection effort. Swanson and Rassi (1981) and FAO (1991) were simply directories. GFRAS (2012) is a database; the only report produced was Swanson and Davis (2014), which was a 14-page document giving some key details on the status of extension by region and with a table providing numbers of agents.

The University of Illinois completed an *International Directory of National Extension Systems* based on a 1980 survey, which was an update of an earlier 1975 survey (Swanson and Rassi 1981). The survey (in English, French, and Spanish) compiled information on extension staff numbers and qualifications, objectives, mechanisms for public participation in governance, and client groups served. The survey covered only public extension organizations whose primary objective was extension; organizations with extension as a secondary objective were not included. The 105 responding countries reported 290,246 extension staff, of which 241,962 were field-level extensionists, split between "extension agents" and "extension assistants" (see Table 2.1 and 2A.1 for a summary of the studies and total numbers of staff, respectively).

FAO conducted a worldwide study of extension organizations in preparation for the Global Consultation on Agricultural Extension held in December 1989. The global consultation was conducted to better understand extension systems of FAO member countries, with the view that extension was "a major knowledge and technology transfer system for farmers" that improves productivity and living standards of farm people. The data were collected in November 1988 through December 1989. This study is referred to as the FAO study or the 1991 study (for the publication date). The resulting *International Directory of Agriculture Extension Organizations* was published in 1991 by FAO as a 496-page book. The purpose of the directory was to share information and promote technical cooperation in developing countries. It was also meant as a benchmark to measure future progress and trends in extension and provide basic data and information to compare programs and resource allocation in national extension systems. However, there was no global-level follow-up of this study until 2009.

The FAO study survey followed the Swanson and Rassi (1981) survey methodology. The FAO 1991 data were gathered through mail-type questionnaires sent to 154 FAO member countries. Most of the surveys were answered by the ministries or departments of agriculture, which normally oversee national agricultural extension programs. The questionnaire had three parts: extension organizations and activities; personnel and human resource development, and financial resources for extension. Some 132 organizations in 115 countries responded (Table 2.1). Using the FAO country offices allowed for a greater country response (that is, more countries). The *International Directory of Agriculture Extension Organizations* is the most comprehensive database of agricultural extension organizations ever compiled. However, completeness was variable.

Twenty years later, the "Worldwide Extension Study" effort was undertaken by IFPRI, funded by the United States Agency for International Development and in collaboration with FAO, GFRAS, and regional organizations including the Inter-American Institute for Cooperation on Agriculture and the French Agricultural Research Centre for International Development. Because GFRAS undertook to manage and house the data that were collected from 2009 through 2012, we refer to it as the GFRAS study or the 2012 study. The GFRAS results were never published except as an online database; no analysis was undertaken.

The objectives of the GFRAS study were to assess and provide empirical data on the current status of pluralistic extension systems worldwide for use in planning future efforts to modernize and strengthen these rural extension and

TABLE 2.1 Summary of major global extension studies

Indicator	University of Illinois (1981)	FAO (1991)	GFRAS (2012)
Number of countries	105	115	81
Number of organizations	151 (all public but from various departments)	132 (mainly public)	347 (all sectors)
Number of extension personnel reported	290,246	438,695	1,050,861
Data available	Hard copy in libraries; online at https://pdf.usaid.gov/pdf_docs/PNAAQ985.pdf	Hard copy in libraries	Online at www.g-fras.org/en/world-wide-extension-study.html

Source: Authors' compilation based on Swanson and Rassi (1981); Swanson, Farner, and Bahal (1991); and GFRAS (2012).
Note: There may be disparities between numbers due to changed country names; some data came in late and were not incorporated for the 1981 survey; and not all data were usable in all the surveys even when reported.

advisory services. The study focused on collecting useful empirical data on the human and financial resources of agricultural extension and advisory systems worldwide, as well as other important data and information, including (1) the primary extension service providers in each country (for example, public, private, and/or nongovernmental organizations); (2) which types and groups of farmers were the primary target groups (for example, large, medium, and/or small-scale farmers, including rural women) for each extension organization; (3) how each organization's resources were allocated to key extension and advisory service functions; (4) each organization's information and communications technology resources and capacity; and (5) what role, if any, different categories of farmers played in setting extension's priorities and/or assessing performance.

Questionnaires were developed in English, French, and Spanish, and sent to key informants (heads of departments) in government extension departments in more than 160 countries. The questionnaires were based on the FAO study but included additional information (Table 2.2). The data were put online on the GFRAS website, and secondary data were used to write up country profiles. Information collected was shared in the form of international directory pages created for each institution (public, private, NGOs, and farmer-based organizations) in each country with 15 or more extension staff members. The names of identified institutions and—to the extent available—their organizational structure, institutional resources, and mandate for providing agricultural advisory services were listed on the respective country pages on the Worldwide Extension Study on the GFRAS website (www.g-fras.org/en/world-wide-extension-study.html).

Data were obtained from 347 organizations in 81 countries (see Table 2.1). Much like the earlier FAO study, the information was not always complete

TABLE 2.2 Data collected by University of Illinois, Food and Agriculture Organization, and Global Forum for Rural Advisory Services

University of Illinois (1981)	FAO (1991)	GFRAS (2012)
Name and [physical] address	Name and [physical] address	
	Legal status	Legal status
	Type of organization [and methods used]	
		Primary management authority [level]
		Primary source(s) of funding for 2009
Clientele served	Clientele served	Clientele served
Distribution of personnel by program area and sex Size of program staff Type of position Educational qualifications	Human resources (1988) (gender disaggregated) Administrators/supervisors Specialists Field officers/assistants	Human resources (2005–2010) (gender disaggregated) Senior management Subject matter specialists Field agents Level of education Other support staff Subcontracts Staff performance Areas covered by subject matter specialists
Source of financial support	Financial resources (1970–1988)	Annual expenditure (2005–2007)
		Time allocation
		Primary methods used
		Staff working in ICT (information and communications technology)
		Use of mass media and ICT for dissemination
		Internet access by staff
		Preservice training of staff
		Transportation means for staff
		Program planning and priority setting Level Farmer organization representation Percentage of women farmer representatives
		Linkages and partnerships
		Type and number of farmer organizations

Source: Swanson, Farner, and Bahal (1991); IFPRI and FAO (2010).

for all categories, particularly financial information. However, the GFRAS dataset is the most comprehensive global set of information about extension and advisory services since the FAO survey. Working through various partners and personal contacts for the GFRAS study yielded fewer responses from countries (that is, fewer countries but more organizations responding in those countries). Some 66 organizations never responded to the GFRAS survey. The GFRAS webpage has downloadable Excel sheets for those countries who completed the questionnaires. IFPRI has made the master database available online too.

The FAO study was meant as a benchmark to measure future progress and trends in extension. However, the GFRAS data were never compared with the earlier effort by FAO. This section examines and compares the general extension status of 1988–1989 and that of the 2012 study. The additional data collected, the inclusion of more nonpublic extension providers, and the limited centralized data collection resources combine to leave substantial gaps in coverage. The University of Illinois, FAO, and GFRAS study all collected sex-disaggregated data. From 1981, sex-disaggregated data from 57 countries showed 19 percent women staff—41 percent of whom were in home economics extension programs (Swanson and Rassi 1981); 97 of the 115 FAO countries reported personnel data disaggregated by sex. The GFRAS study included sex-disaggregated data for 73 countries.

2.2 Findings from the Data Collection

There are many data results from the global assessments (Table 2.3). However, for the sake of comparison, we focus on the best-fit characteristics of governance structures and legal status of providers, capacity and management in terms of staffing, and extension methods and clientele.

Governance Structures and Legal Status of Extension Providers

There are many types of extension providers and governance models around the world. Most authors classify the models by provision and financing of extension services, with provision from the public and the private sectors (Hoffmann et al. 2009; Rivera n.d.; Birner et al. 2006). Hoffman and colleagues (2009) talk about the profit-oriented and the nonprofit organizations. Some organizations offer extension services to sell their products and others offer only the services (Hoffman et al. 2009). Rivera (n.d.) shows the different options for funding and delivering extension services (Figure 2.1); this figure is included with much more detail as to the different types of providers in

TABLE 2.3 Organizational responses received by region, Food and Agriculture Organization, and Global Forum for Rural Advisory Services

Region	FAO (number of organizations)	FAO (number of countries)	GFRAS (number of organizations)	GFRAS (number of countries)
Africa	40	33	155	26
Asia	34	28	28	23
Europe	25	16	18	14
Latin America and the Caribbean	23	23	145	17
Oceania	8	11	1	1
Northern America	2	2	0	0
Total	**132**	**113**	**347**	**81**

Source: GFRAS (2012); Swanson, Farner, and Bahal (1991).

Note: Countries follow UNSTATS (https://unstats.un.org/home/). There may be disparities between numbers due to changed country names; some data came in late for the 1981 survey; and not all data were usable in all the surveys even when reported.

Birner and colleagues' best-fit paper (2009). The upper-left quadrant is public funding and delivery, which is still very common today, although within a pluralistic environment where the private sector and other nonprofit organizations also provide or finance services. The bottom-right quadrant is purely private funding and delivery, in the case where the public sector withdraws, privatizes, or transfers responsibility for extension services to nongovernmental or farmer organizations.

Neither the FAO 1991 nor the GFRAS 2012 survey provides adequate data on sources of funding for extension to allow allocation of programs across the four quadrants of Figure 2.1. However, the nature of the extension organizations, as reflected in Table 2.4, provides a fair degree of confidence in this. The 1991 survey dealt almost entirely with extension in quadrant 1—government-funded, government-delivery extension. In 2012 this option continued to dominate with nearly 92 percent of extension agents in government, semiautonomous government, research, or university organizations. Some of these—though likely a limited number—may have been government-delivery services privately funded through user fees or commodity levies. Private-delivery, private-funded extension in nongovernmental organizations, producer, and other private organizations accounted for only slightly more than 8 percent of extension agents, although almost certainly some of these would have been government-funded through grants or contracts. The obvious conclusion is that government extension remains predominant. There remains the suspicion that private-delivery, private-funded extension may

FIGURE 2.1 Public sector agricultural extension funding and delivery alternatives

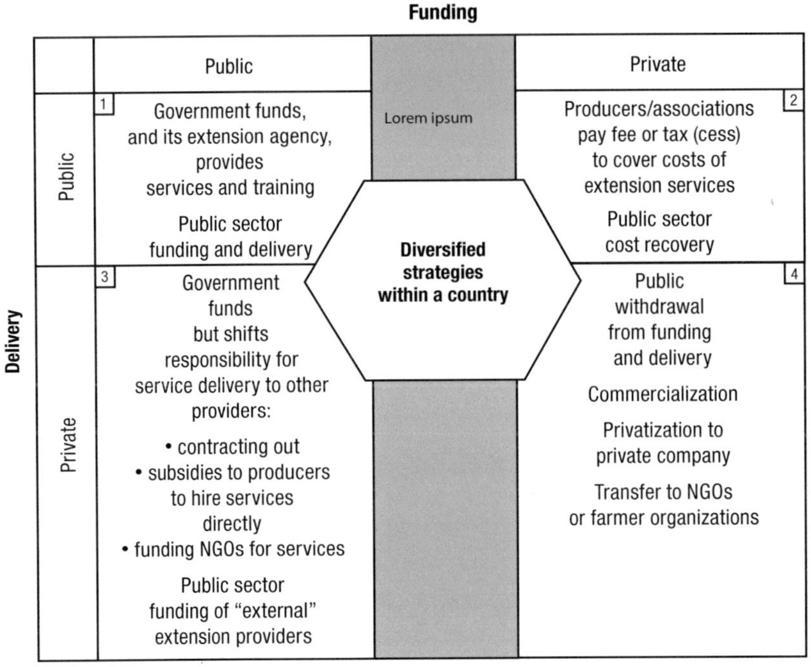

Source: Rivera (n.d.).

be grossly undercounted due to the number and dispersed nature of such activities.

We see a notable difference between the types of providers (indicated as "legal status of extension organizations") between the 1991 and the 2012 data (Table 2.4). The FAO study covered mainly government-based extension agencies, as these were at that time considered the only significant providers. University-based extension surveyed by FAO was also publicly provided. Only a small number of private, semiautonomous, nongovernmental, or farmer-based providers were reported. The GFRAS study shows a larger variation in types of extension providers, as the growing pluralism in extension was then recognized and is apparent in comparing these two points in time.

A number of factors account for the greater institutional pluralism. Most obviously, perhaps, is that the decline of communism led to private-service expansion in a number of countries (Hoffmann et al. 2009). In the late 1980s and in the 1990s, fiscal realities and pressures from international lending institutions led many countries to reduce public spending. Budget cuts to large

TABLE 2.4 Legal status of extension organizations, 1991 and 2012

	1991		2012	
Status	Number	%	Number	%
Governmental or ministry-based extension organization	108	82	95	27
Semiautonomous governmental extension organization	5	4	70	20
Public research institution with extension unit	0	0	5	1
University-based extension (public)	2	2	13	4
Subtotal (public sector)	*115*	*88*	*183*	*53*
Nongovernmental organization	7	5	90	26
Farmer-based organization	2	2	43	12
Private-sector organization or firm	8	6	32	9
Subtotal (private sector)	*17*	*13*	*165*	*47*
Total	**132**	**100**	**348**	**100**

Source: Authors.
Note: Because of rounding, totals might not add up precisely to 100.

civil service agencies such as extension were an attractive target, since recurrent costs were high, client demand was not well organized, benefits were not immediate, and many of the agencies were already criticized as inefficient and ineffective. In general, development strategies during this period shifted to deemphasize public-sector implementation in favor of a greater role for markets and the private sector. Another contributing factor was that government extension programs had expanded greatly during the 1970–1990 post–Green Revolution period and were at a peak in staffing. Critiques of programs, such as the training and visit system, noted problems of inefficiency, not reaching women, and most of all lack of financial sustainability (Anderson 2007). These criticisms and donor fatigue in funding public extension led international donors to a major shift toward funding private nonprofit and for-profit extension providers. Along with all of this, the growth of commercial agriculture and globalized markets enabled more farmers to finance needed services and led to development of many niche innovation needs that could best be addressed by private service providers. The resulting gap caused by weakened public services and continuing and diversifying need caused nongovernmental and private-sector actors to move in.

Although Table 2.4 indicates general trends, it is far from comprehensive and not fully indicative of overall percentages during the periods. Many of the extension providers are missing because at the time of data collection there was no mechanism to register all the extension providers in a country. The FAO study focused on public-sector providers. For the GFRAS study, the organizations used snowball sampling and personal connections to obtain much of the data. Thus it may have covered a wider range of providers but was less systematic.

Despite the growing pluralism, the total number of staff working for public institutions was still the majority in 2012 (Table 2.5), thus indicating the continued importance of public extension. The considerable difficulty in identifying and getting numbers from the many small nonpublic institutions providing extension services, such as farmer-based organizations and nongovernmental organizations, means that these are chronically underreported. Caution is necessary in interpreting findings for public research and university institutions with extension units due to likely distortions based on the small sample sizes. For smaller organizations and all those whose mandate is not primarily extension, attribution of time for staff working part-time on extension becomes a problem.

The original analysis of the 2012 survey dropped outliers in large institutions from Azerbaijan, Denmark, Norway, Poland, and the Republic of Moldova, as the situation in these industrialized countries appeared to differ from most developing countries and the very large institutions tended to distort findings. Poland, for example, reported 3,805 extension staff in one semiautonomous organization. Denmark and Norway had 3,465 extension staff in two farmer-based organizations. And Azerbaijan and the Republic of Moldova had 1,192 extension staff in two nongovernmental organizations. These do differ but may be indicative of a trend away from government extension in industrialized countries. Both the United Kingdom and the United States may be other outliers indicative of this trend with their drops in reported extension agents (assumed to be government extension) of 99 percent and 81 percent, respectively.

Along with the growing pluralism, governments have been decentralizing extension services, in part to enhance involvement of rural people in these services. This entails transferring control of specific program planning and management functions to the local system levels where extension programs are actually implemented (Swanson and Samy 2004). Many countries have now decentralized their services; however, the jury is still out as to whether this has made the services more relevant, efficient, or demand-driven.

Staffing

Table 2A.1 in the annex gives the number of extension agents in 1981 (290,246), 1991 (438,695), and 2012 (1,050,861). Additional work was done to collect data by Swanson and Davis in 2014, and Davis and

TABLE 2.5 **Total staff by type of organization, 2012**

Legal status	Total staff	Share (%)
Governmental or ministry-based extension organization	127,342	86.76
Semiautonomous governmental extension organization	6,848	4.67
Public research institution with extension unit	298	0.20
University-based extension	285	0.19
Nongovernmental organization	3,890	2.65
Farmer-based organization	7,007	4.77
Private sector organization or firm	1,112	0.76
Total	**146,782**	**100**

Source: GFRAS 2012.

Note: Outliers (organizations from Azerbaijan, Denmark, Moldova, Norway, and Poland) were removed; therefore 304 organizational responses were included here of the total 347 organizational responses received.

Spielman in 2016, thus some of the numbers from the GFRAS study have been updated or gaps have been filled (for example, Zimbabwe and some Central Asian countries were added; numbers were updated for Brazil, India, and South Africa). The updated information included data for China (some 600,000 staff). Again, these should be interpreted cautiously as there are many caveats to the data on number of extension agents; many private-sector institutions are missing, as are regional and departmental numbers. However, general trends can be observed.

The extension staff can be organized by type of organization, type of position, education level, and gender (Figures 2.2, 2.3, and 2.4); however, these data are only available for the GFRAS study (2012). Sixty-one countries have data from both the FAO and GFRAS surveys and, although the many caveats already noted on comparability of data apply, the data suggests continuing importance and commitment to extension. For these countries, extension agent numbers increased 136 percent to 976,393, due mainly to increase in China. Without China, the increase was 23 percent. Twenty-seven countries increased numbers of extension agents, but 34 decreased them. Many industrialized countries decreased extension agents (at least public extension agents), with the United States dropping by more than 12,000. Asian countries, Ethiopia, and Brazil had major increases. The number of extension agents is often expressed as a ratio to numbers of clients. Over the 1989–2012 period the global rural population grew about 12 percent from 2.98 billion to

FIGURE 2.2 Extension staff positions by type of organization

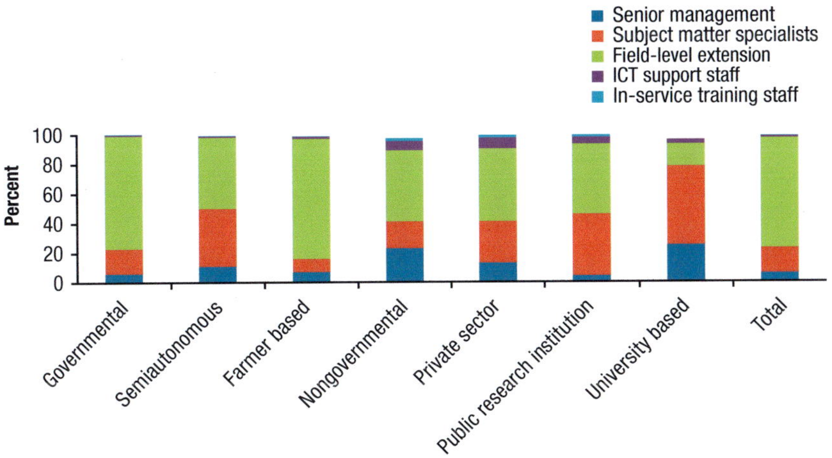

Source: GFRAS (2012).

Note: Outliers (organizations from Azerbaijan, Denmark, Moldova, Norway, and Poland) were removed; therefore, 304 organizational responses were included here of the total 347 organizational responses received.

3.36 billion (World Bank 2019). Many unanswerable questions emerge from these comparisons. For example, did China's commitment to extension contribute to its impressive economic performance? Have extension agent cuts by the United States and other Western countries had any impacts on agricultural performance?

Some clear differences can be observed. Governmental extension systems had a higher proportion (76 percent) of staff in field-level extension positions working with farmers. In this they were exceeded only by farmer organizations (81 percent), presumably because governance by clients themselves places a high priority on field-level services. Specialized agencies had a much smaller proportion of field-level agents. Universities were at the extreme with only 15 percent of their extension staff in field-level positions. There is a strong need for support service staff to ensure quality of messaging and efficiency of extension workers. These training, ICT support, and subject matter specialists may also contribute to innovation in extension. A high proportion of support service to field agent staffing may also suggest greater priority to agendas (for example, research, university programs, profits) other than extension services to clients. Subject matter specialists were found in all types of organizations, but the public research, semiautonomous, and university-based

FIGURE 2.3 Share by education level by organization type (percentage of total)

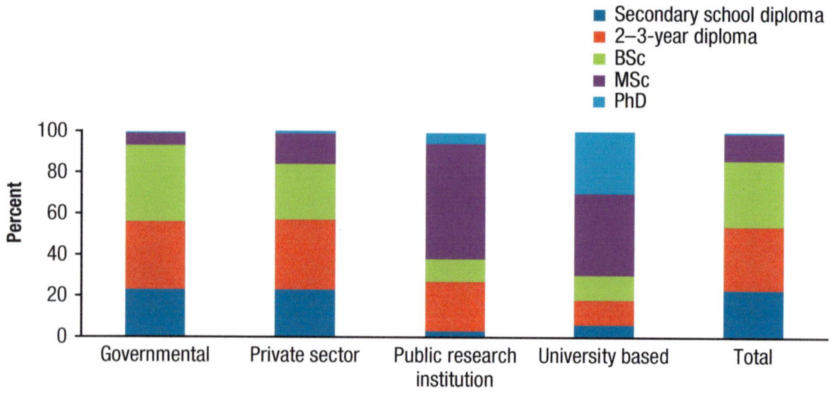

Source: GFRAS (2012).

Note: Outliers (semiautonomous, farmer-based, and nongovernmental organizations from Azerbaijan, Denmark, Moldova, Norway, and Poland) were removed.

organizations had the highest numbers of specialists for technical support. Nongovernmental and university-based organizations had the highest percentage of senior management. Most organizations did not have staff dedicated to information technology and communications support (indicated as ICT support staff). Governmental organizations and farmer-based organizations perhaps had the most realistic ratio of field staff relative to other staff. However, the other types of organizations tend to be more specialized in particular topics or client groups, which may account for high levels of nonfield staff.

Not surprisingly, education levels reflected institutional categories, with research institutions staffing predominantly with advanced degrees. Private-sector firms had a good range of staff between secondary up to MSc-level, with few PhDs. Governments had extension staffing that is predominantly with secondary or diploma qualifications. This also reflects the numbers of staff positions in categories requiring higher education. In general, the education levels of extension staff were surprisingly high. There were a good number of MSc holders in all types of organizations (fewest in government). Not surprisingly, the most PhDs were found in the university-based organizations.

A more relevant comparison may be the educational level of field-level staff. Figure 2.4 shows the educational qualifications of field-level staff disaggregated by gender. This was averaged out across all the reporting organizations

FIGURE 2.4 Average education across sectors of field-level extension staff by gender, 2009

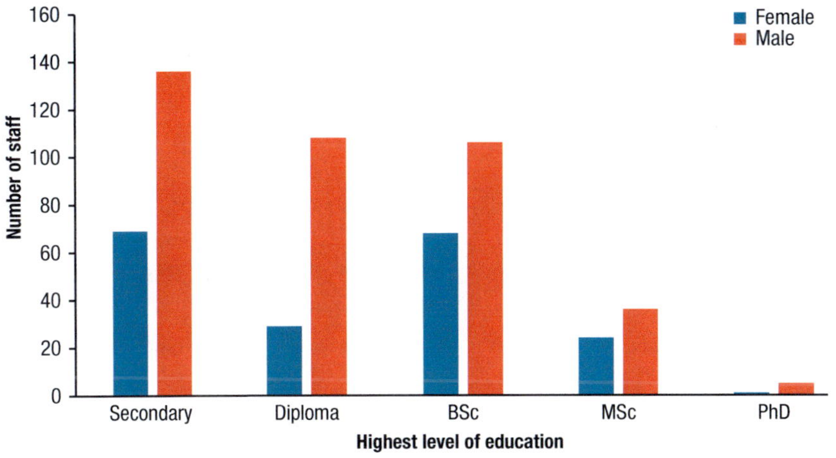

Source: GFRAS (2012).

across sectors. We find a high number of staff with secondary education, two- to three-year diplomas, or even BSc degrees. Very few numbers of field staff held MScs or PhDs, indicating that staff with the higher qualifications tended to be in cities. Women and men tended to have closer ratios at the BSc and MSc levels. Although these data were incomplete or unavailable, educational qualifications for extension staff has increased over time due to greater availability of educational opportunities and greater demands for extension staff with diverse skills to address increasingly complex client needs.

Clientele

The types of clientele depend very much upon the objectives of the extension organization. Many public extension organizations are trying to reach small-scale (often subsistence) farmers. However, farmers have become increasingly commercialized and linked to markets since 1991. Farmer-based extension organization services target their members, often producers of a specific commodity. Members can be small, medium, or large-scale. Some nonprofit NGOs target women or youth. The data are not directly comparable between the two global studies, because they were not asked the same way. The FAO study asked organizations what type of clientele they served. The GFRAS study asked respondents what percentage of their time they spent on precoded categories. Figure 2.5 depicts a word cloud of the extension organization clientele in 1991. Word clouds are an exploratory qualitative method to depict

FIGURE 2.5 Extension clientele word cloud

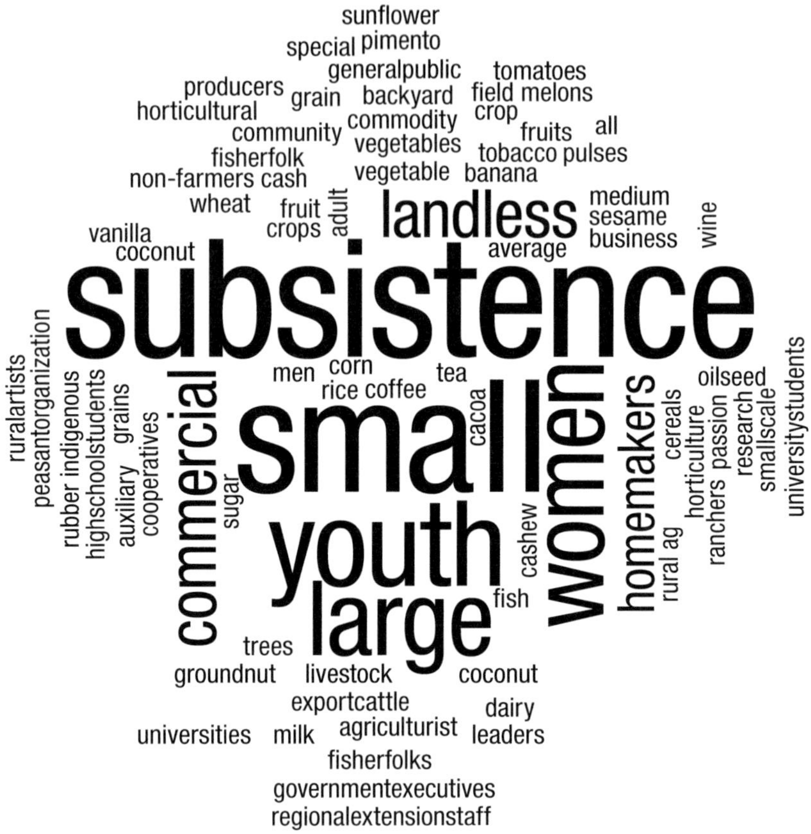

Source: Created by authors based on FAO (1991).

word frequency visually, giving greater emphasis on the words that appear most frequently (Cidell 2010).

A 1993 publication on agricultural extension and "farm women" in the 1980s is available from FAO. The use of the term "farm women" rather than "women farmers" shows the focus in that decade; indeed, much of the focus was on home economics and nutrition for "farmers' wives," indicating that women were not viewed as farmers by many extension services. According to the publication (FAO 1993: 3): "The 1989 FAO survey data show that about 5 percent of all agricultural extension resources worldwide are directed to female farmers and only 15 percent of the world's extension personnel are female." Percentage of time and resources allocated to women farmers was between 1 percent and 9 percent, according to the FAO (1993) publication, based on data from the FAO survey in 1989 (published in 1991). However, women and even "homemakers" featured somewhat prominently in the word cloud in Figure 2.5.

In 2012 we saw an increased percentage of female staff (34 percent in senior management and subject matter specialist positions, and 22 percent field-level staff) (GFRAS 2012). Figure 2.6 shows the average share of women across all types of organizations by position in the 2012 GFRAS study. Note that the highest percentages were in ICT support and senior management. In total, they averaged just under 25 percent of the share of total staff of the surveyed organizations.

In the GFRAS study, respondents indicated that they spent anywhere from 1 percent to 100 percent of time on women (the median was 14 percent and mode was 5 percent). Obviously, some of the respondents specifically targeted women farmers. However, many did not indicate that they spent any resources on women farmers. We can also look at data on time allocation by extension agents in 1991 and 2012. While not directly comparable, they do show some trends, especially with regard to women and youth clientele. Tables 2.6 and 2.7 show the time and resources allocated by extension to different clientele groups. FAO disaggregates by region and GFRAS by type of organization. As seen in Table 2.6, commercial farmers received at least 20 percent and up to 69 percent of extension's time in 1991. Africa had the lowest percentage at 20 percent. Subsistence farmers, not surprisingly, had high levels of time allocated in Africa, Asia and the Pacific, and to a certain extent in Latin America and the Near East, but not in Europe or North America. Landless producers received very little time allocation across the regions (the most in the Near East at 5 percent). Young farmers received quite a bit of attention in North America, no doubt due to well-known programs such as 4-H. Women farmers received quite low attention across the regions, as did home economics (usually focused on women).

The GFRAS study has data on how staff of different types of organizations allocated their time to different clientele groups (Table 2.7). Respondents were asked to specify the primary group or groups that their organization serves or targets and to indicate the relative importance of each group. If they targeted more than one group, they were asked to indicate the approximate amount of time and effort (as a percentage) that their organization devoted to each group. In general, there is quite even disbursement by the types of organizations (governmental, semiautonomous, farmer based, nongovernmental, private sector, public research institution, and university based)

FIGURE 2.6 Share of women in extension positions across organizations (%)

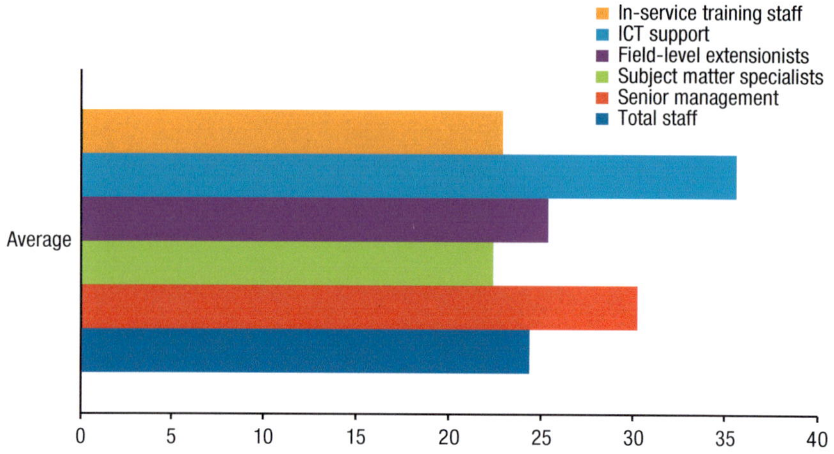

Source: GFRAS (2012).

by types of clientele (large commercial, small/medium commercial, small-scale subsistence, women, young [adult], landless, rural youth, and rural women). Clientele groups achieving the the most service across all providers included the small/medium commercial farmers (37 percent) and the small-scale subsistence farmers (38 percent). However, the university-based and the semiautonomous providers indicated that they spent the most time on this clientele.

Governmental providers were shown to spend the least time of any organization on small-scale subsistence farmers (24 percent). The private-sector respondents indicated they spent 62 percent of time on small/medium commercial farmers and 17 percent on large commercial farmers. They claimed to spend 24 percent of time working with rural women on nutrition, health, and hygiene. These findings are counterintuitive, not what would be expected of typical private-sector extension. Nongovernmental organization staff indicated that they spent 33 percent of their time on women farmers and 20 percent on rural youth. The public research institutes reported that they spent significant time (40 percent) on large commercial farmers. They reported spending the most time of any organization (45 percent) on rural youth. They spent 30 percent of their time on small/medium commercial farmers. Outreach to landless farmers ranged from 10 percent to 17 percent.

Figure 2.7 shows how extension agents across the different sectors and regions allocated time across three major activities in 2009. Note that "extension planning and support activities" included conducting needs assessments,

TABLE 2.6 Percentage of time and resources allocated to clientele groups by extension organizations by region in 1991 (%)

Clientele groups	Africa (n = 38)	Asia and the Pacific (n = 28)	Europe (n = 7)	Latin America (n = 39)	North America (n = 4)	Near East (n = 16)	Worldwide average (n = 132)
Commercial farmers	20	36	69	42	48	31	35
Commodity producers	26	17	5	24	1	34	23
Subsistence farmers	31	28	2	18	1	14	22
Landless producers	1	4	1	2	1	5	2
Young farmers/rural youth	10	8	7	5	16	2	7
Women farmers	7	3	3	5	1	9	5
Home economics	1	2	4	—	9	1	1
Other groups	4	2	9	4	23	4	5
Total	**100**	**100**	**100**	**100**	**100**	**100**	**100**

Source: Reproduced from FAO (1993), based on Table 4 in Swanson, Farner, and Bahal (1990).
Note: — = no data.

program planning, preparing performance reports, in-service training, program evaluation, and related activities; "educational and advisory service activities" included implementing educational programs such as farm visits, conducting on-farm demonstrations, training courses, workshops, and field days; and "noneducational activities" included carrying out noneducational activities, such as regulatory work, data collection (for example, agricultural census, crop forecasting), working on other government programs (for example, subsidies, credit, input supply), and assisting local government. Nearly 60 percent of time was spent on extension and advisory service activities.

In 2009 respondents were asked "What percentage of all farmers belong to some type of farmer or producer organization in your country?" The average response across all types of organizations was 45 percent.

Advisory Methods

Extension programs use a range of methodologies and approaches suited to differing situations. These have evolved over time, although most extension programs use a mix of varied methods, even when they may emphasize one particular approach. According to Birner and colleagues (2006: 43), methods can be classified according to the following:

- Types of training or technology transfer (demonstrations, field days, week-long courses, farmer-to-farmer exchanges);

TABLE 2.7 Clientele served (targeted) by type of organization in 2009 (%)

Type of organization	Large commercial farmers	Small/medium commercial farmers	Small-scale subsistence farmers	Women farmers	Young (adult) farmers	Landless farmers	Rural youth	Rural women	Women among total staff
Governmental	17	31	24	13	13	12	10	10	21
Semiautonomous	31	49	46	25	13	14	11	24	17
Farmer based	11	44	30	11	11	17	6	8	29
Nongovernmental	15	40	38	33	22	16	20	20	30
Private sector	17	62	34	15	15	10	6	24	29
Public research institution	40	30	38	10	—	—	45		20
University based	17	28	57	13	13	10	17	10	24
Weighted average (across all organizations)	21	37	38	15	15	13	12	14	25

Source: GFRAS 2012.

Note: Outliers (organizations from Azerbaijan, Denmark, Moldova, Norway, and Poland) were removed. Some 304 organizational responses were included. — = data not available.

FIGURE 2.7 Allocation of time by field extension or advisory staff, 2009

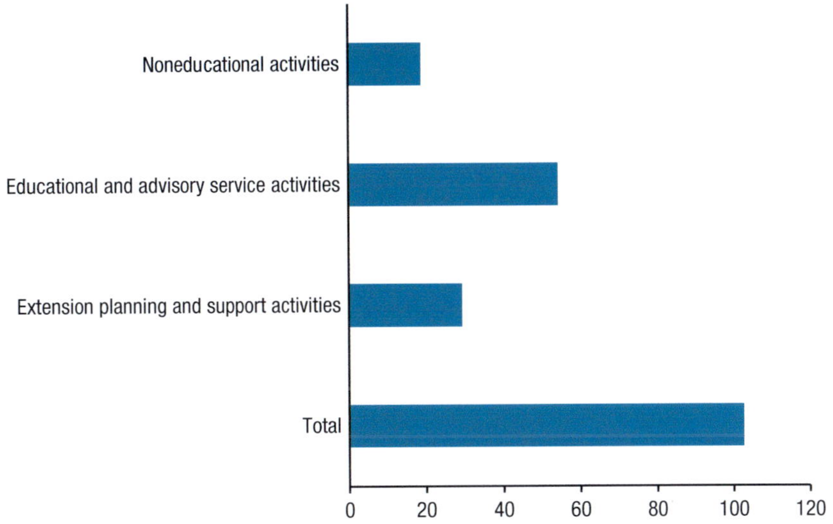

Source: GFRAS (2012).
Note: Percentages might not add up to exactly 100.

- Number of clientele (individual, group-based, mass approaches);
- Involvement of clients in planning and problem-solving (top-down methods, participatory methods);
- Specificity of content (limited to specific crops/livestock or dependent on needs identified by clients in different fields);
- Types of media used (information and communications technology or ICT, radio, drama, newspaper); and
- Adult education orientation (social learning, humanist, cognitive).

There is a substantial level of continuity in extension methodologies but some obvious evolution between the 1991 FAO study and the 2012 GFRAS study. The FAO study came toward the end of two important trends in extension. The training-and-visit system had been widely adopted by public extension agencies for technology transfer but was poised to begin a decline. The training-and-visit extension model was basically traditional field agent–based extension services with a codified regime of regular farmer training, field visits, contact farmers, farmer groups, and links to research to provide technology for dissemination. One might characterize it as traditional extension with sound management and discipline. It was criticized for being too prescriptive and top-down, being overly regimented, and having high recurrent costs. Impacts were positive but not up to expectations. With increasing government budget constraints, the high recurrent costs doomed many training-and-visit systems to cancellation or lingering decline. The approach has been somewhat discredited, but many national extension agencies still use a modified training-and-visit approach.

A second trend toward greater client participation in extension programs also undercut the training-and-visit system, even though that system was somewhat participatory in working through lead farmers and farmer groups. Nonprofit NGOs expanded activities through the 1980s and embraced varied mechanisms for increasing client participation in extension planning, implementation, and assessment. Farmer-to-farmer extension, use of lead farmers, farmer members of extension governance committees, and dissemination of traditional knowledge and technologies were common. Farming systems research and extension programs contributed to this by involving farmers in diagnostic analysis of farming systems, developing component technologies and system innovations to improve productivity. Social soundness analysis became important to program designs. All of these innovations in extension were somewhat labor-intensive and suited to NGO programs that usually operated on a relatively limited scale. Although participatory approaches are more difficult to incorporate into the larger national public extension programs, they remain an important emphasis across extension programs today.

Figure 2.8 shows how important the individual approach and the training-and-visit method was at the time of the FAO study. Farm visits, radio, and group approaches were also important. One can also see the integrated agriculture development programs showing strongly as well, as many of these were part of or an offshoot of integrated rural development programs that were common through the 1980s. Many of the traditional technology transfer methods remain common. Demonstrations, field days, farmer training courses, radio, and print media remain tools of most extension providers. Training courses may be less formal and shorter in duration. Group-based approaches may have increased but remain as part of an overall mix with individual and mass media approaches. Technology transfer remains the objective for most extension, and adult education strategies underpin some programs but do not get heavy emphasis.

The farmer field school approach had become widespread by the time of the GFRAS study. This approach—not unlike the training-and-visit system—is based on regular training (experiential learning sessions) for farmer groups

FIGURE 2.8 **Methods used by extension organizations, 1991**

Source: Created by authors based on FAO (1991).
Note: IRD = integrated rural development; TandV = training and visit. IADP = integrated agricultural development program.

facilitated by an extension agent or lead farmer. The experiential field learning is based on a fairly regimented methodology developed to promote integrated pest management. This draws on some of the strengths of the training-and-visit system but incorporates greater farmer participation and flexibility. Widespread adaptation of the farmer field school approach to different situations has led—for good and bad—to major deviations from the original model. The farmer field school approach aims for cost-efficient implementation, often emphasizing use of lead farmers for the experiential learning groups. Quality may depend heavily on the technical support and planning for learning sessions and the quality of the field school leader. The approach fits well with nonprofit NGO extension programs and remains influential as a model for extension generally.

Lead farmers, as in the case of farmer field schools, are important in many extension approaches. Their use serves to make up for lack of adequate numbers of extension agents, to expand reach of programs, and to limit costs. Issues of appropriate support and remuneration arrangements are worked out on a case-by-case basis. In countries where there is limited support from extension agents and the community, studies show limited coverage and outreach by lead farmers (see Chapter 7 on Malawi). Increasing evidence shows that lead farmers are complements of, and not substitutes for a well-functioning public extension system.

Another significant change is in extension content that now often emphasizes facilitation approaches. Under these programs, agents organize with clients to identify and address problems and opportunities, and develop linkages to other actors in the agricultural value chain to obtain needed support—financing, inputs, markets, technologies, or services. This shift to facilitation services stems from increased commercialization and specialization of farms, more diverse products and market demands, and greater capacity of farmers themselves. Such facilitation approaches are well suited to private NGOs and producer group extension services and can be effectively provided by consultant extensionists. Public extension agencies struggle to have the flexibility needed to provide such services but may be forced to move in that direction in the future.

A potential game changer for extension, as well as for the rest of the world, has been the emergence of new information and communications technologies (ICTs). Obviously, capabilities and use of ICT and digital data are tremendously different from 1991. Traditional ICT methods such as radio remain important and perhaps predominant. These and print media may reach more farmers, who may be better educated and more able to access such media. Many Internet, video, and cell phone–based applications are being developed and tested for disseminating extension messages, accessing specialist support, and analyzing data to improve recommendations to clients. Reaching resource-poor farmers, many of whom are illiterate, and those outside of ICT service areas is still a huge challenge.

Thus, although many traditional extension methods remain important, there are key changes. Participatory and facilitation approaches involve clients to a greater degree and help tailor services to their specific needs. Other changes with greater emphasis on group-based, mass media, farmer leaders, and ICTs replace a reliance on face-to-face extension outreach by agents. These may be motivated in part by need to reduce costs. Whether they are equally effective in serving clients may still be a relevant question. Trends in extension approaches contribute to greater diversity in methodologies and align well with the increasing institutional pluralism and private-sector provider roles in extension.

More and more field staff have access to the Internet. Figure 2.9 shows the relatively high access to the Internet by field staff in 2012. Although having distributed surveys by email may have introduced some sort of bias, the question was specifically about access by field-level staff. This cannot be compared to the FAO data since the Internet was not widely used at the time.

Other Trends and Differences

Other important trends are affecting the world of extension, related to the issues discussed above but with additional implications. Many of these relate to targeting clients and their differential needs.

Coordination mechanisms. These are needed more than ever due to the increased pluralism of extension providers. One major element in coordination is that between public and private extension. Beyond that the large numbers of private providers—many with relatively small programs—can benefit by increasing efficiencies with better coordination. Coordination committees and innovation platforms exist in many countries, but there is room for more innovation and introduction of new mechanisms for coordination. Commitment by public extension services to provide support, as "wholesalers" of extension services delivered by private providers, may be a key means of improving efficiencies. Quality control, as requiring registration or oversight, may also be an important public service for private extension (see Chapter 3 for further analysis regarding coordination).

Decentralization reforms. Reforms are in place or being implemented in many countries, generally shifting responsibility for extension to subnational government units at the provincial, state, or district levels. These reforms greatly complicated data collection for the GFRAS study, as information on public extension had to come from multiple decentralized offices, rather than one national agency. Decentralization reforms are initially quite disruptive of extension planning, financing, and implementation. They tend to take years to implement and adapt operating procedures and policies. Decentralization offers benefits in moving extension closer to the farmer clients, but extension can get lost in the overall local government agenda, and funding provisions can be particularly problematic.

Gender considerations. Gender is increasingly considered in most extension programs. Extension methodologies are adapted to better serve women farmers, and the number of women extension staff has apparently increased

FIGURE 2.9 Share of field extension workers with Internet access in their offices, 2009

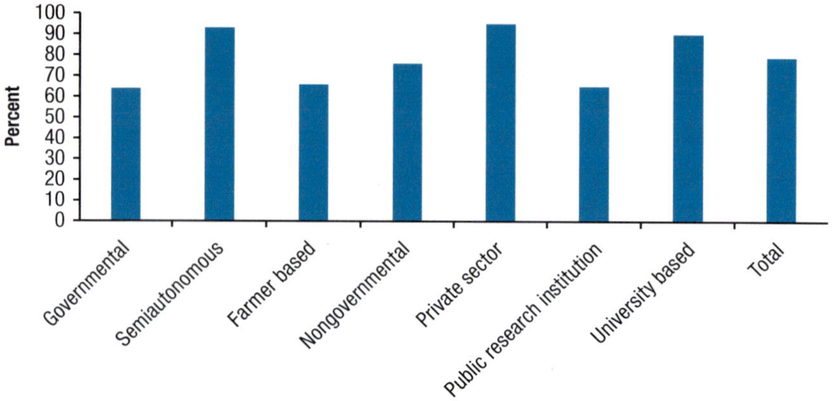

Source: GFRAS (2012).

from 15 percent to 22 percent (at the field level; management and subject matter specialists were both 34 percent in 2012). Despite these efforts, it is not clear that changes have been nor will be been fully effective, and women continue to be chronically underserved by extension. More attention is needed to assess and implement extension program reforms to reach women clients with relevant services.

Youth. Younger folks are increasingly recognized as an important target group for extension, although most extension programs struggle to reach youth, who are generally not the major decisionmakers for farm households. This presents a serious challenge for extension over the coming decade. Extension programs promoting youth entrepreneurship may be an ideal means to engage youth in modernizing agriculture and to give them opportunities they desperately need for productive jobs.

Markets. Markets are now frequently and appropriately the focus for extension activities. However, farmer clients bifurcate into two groups: commercial farmers engaged in and wanting to expand production for more demanding markets versus resource-poor farmers concerned with more diverse livelihood issues, production for home consumption, and building household resilience. This is nothing new. Both need markets, but capabilities differ; extension needs differ; and relevant extension methodologies, providers, and mechanisms may differ.

Urbanization. This continues, especially in transforming middle-income countries. Urbanization reduces the constituency for rural extension services and raises the question as to the need for urban and peri-urban extension. Urban and peri-urban agriculture is important in some places and, because of environmental and health issues, if not farmer productivity, warrants attention.

Agricultural research. Research is also changing, with new technologies and globalized research networks. Research-extension linkages are needed to give research feedback on farmer needs and to facilitate uptake of research findings. These linkages have not always been very effective. There is now need for such research linkages to both public and private extension.

The common thread running through these issues and those noted throughout the comparison of the 1991 global study versus the 2012 one is that of increasing extension complexity, as a diverse system. This can be a real strength, with different providers serving different client groups with different services using multiple approaches and methodologies. All this functions within the overall national agricultural innovation systems or agricultural sector value chain. But benefits are most likely with strategic partnering and interlinkages among providers and clients. Coordination of such systems is thus becoming both more challenging and more necessary.

2.3 Conclusion

This chapter has provided a landscape view of extension globally, comparing the recent status to the situation 20 to 30 years ago, when similar global assessments took place. The analysis uses components of the best-fit framework, including governance structure (and legal status) of extension providers, capacity (focusing on staffing), and advisory methods and clientele. It shows that extension and advisory services have become much more pluralistic. There are greater numbers of extension agents today (more than one million); however, they belong to a variety of different types of organizations from the private, nongovernmental, and public sectors, and even include farmers groups and volunteer farmer extension agents. More methods are available today, especially digital ones. Private-sector extension and the use of volunteer farmers is advocated and used worldwide as a complement for the traditional methods.

References

Ali, M. 2019. Personal communication via email. June 21.

ALINE (Agricultural Learning and Impacts Network). 2019. "Improving Agricultural Advisory Systems Metrics and Learning," meeting summary, April 8–9, 2019, Addis Ababa, Ethiopia.

Anderson, J. R. 2007. "Agricultural Advisory Services," background paper for *The World Development Report 2008*. Washington, DC: World Bank.

Birner, R., K. Davis, J. Pender, E. Nkonya, P. Anandajayasekeram, J. Ekboir, A. Mbabu, D. Spielman, D. Horna, and S. Benin. 2006. *From Best Practice to Best Fit: A Framework for Analyzing Agricultural Advisory Services Worldwide.* Inter-Divisional Discussion Paper. Washington, DC: International Food Policy Research Institute.

Cidell, J. 2010. "Content Clouds As Exploratory Qualitative Data Analysis." *Area* 42 (4): 514–523.

Cuangara, B., and T. Thompson. 2018. *Mozambique: Desk Study of Extension and Advisory Services.* Washington, DC: Developing Local Extension Capacity (DLEC) Project.

Davis, K., and D. J. Spielman. 2016. "A Global Update on the Status of Agricultural Extension." Unpublished manuscript. International Food Policy Research Institute (IFPRI), Washington, DC.

FAO (Food and Agriculture Organization of the United Nations). 1991. *International Directory of Agriculture Extension Organizations.* Rome: FAO.

———. 1993. *Agricultural Extension and Farm Women in the 1980s.* Rome: FAO.

Franzel, S., A. Ndiaye, and J. S. Tata. 2018. *Senegal: In-Depth Assessment of Extension and Advisory Services.* Washington, DC: DLEC Project.

GFRAS (Global Forum for Rural Advisory Services). 2012. Master Worldwide Extension Study Database. Excel File. Lindau, Switzerland: GFRAS.

Hoffmann, V., M. Gerster-Bentaya, A. Christinck, and M. Lemma, eds. 2009. *Handbook: Rural Extension Volume 1. Basic Issues and Concepts.* Weikersheim, Germany: Margraf Publishers.

Huber, S., and K. Davis. 2017. *Bangladesh: Desk Study of Extension and Advisory Services.* Washington, DC: DLEC Project.

IFAD (International Fund for Agricultural Development). 2018. *Delivering Extension Services to the Last-Mile: Improving Smallholders' Access to Innovation and Pluralistic, Demand-Driven Extension Services.* Grant Design Document. Rome: IFAD.

IFPRI. 2012. *Women's Empowerment in Agriculture Index*. Washington, DC: IFPRI.

IFPRI and FAO. 2010. *Survey Instrument: Agricultural Extension Service Providers.* Lindau, Switzerland: GFRAS.

Ngan, P. H., and S. C. Babu. 2018. "Viet Nam." Unpublished manuscript. IFPRI, Washington, DC.

Rivera, W. R. n.d. "Agricultural Extension System Reform and the Challenges Ahead." Unpublished manuscript. University of Maryland, College Park, MD, US.

Swanson, B. E., and K. Davis. 2014. *Status of Agricultural Extension and Rural Advisory Services Worldwide: Summary Report.* Lindau, Switzerland: GFRAS.

Swanson, B. E., B. J. Farner, and R. Bahal. 1990. "The Current Status of Agricultural Extension Worldwide." In *Report of the Global Consultation on Agricultural Extension*, 43–76. Rome: FAO.

———. 1991. *International Directory of Agricultural Extension Organizations.* Rome: FAO.

Swanson, B. E., and J. Rassi. 1981. *International Directory of National Extension Systems.* Urbana-Champaign, IL, US: Bureau of Educational Research, College of Education, University of Illinois at Urbana-Champaign. https://pdf.usaid.gov/pdf_docs/PNAAQ985.pdf.

Swanson, B. E., and M. M. Samy. 2004. "Introduction to Decentralization of Agricultural Extension Systems: Key Elements for Success." In *Decentralized Systems Case Studies of International Initiatives,* edited by W. Rivera and G. Alex, 1–5. World Bank Extension Reform for Rural Development Series, Volume 1. Washington, DC: World Bank and USAID.

United Nations, Statistics Division. 2009. UNSTATS database. Accessed July 8, 2019. https://unstats.un.org/home/.

World Bank. 2019. World Bank Data "Rural Population—World." Accessed July 8, 2019. https://data.worldbank.org/indicator/SP.RUR.TOTL?locations=1W.

Annex 2A. Additional Data

TABLE 2A.1 Numbers of extension staff by country, 1981, 1991, and 2012

Country	University of Illinois (1981)	FAO (1991)	GFRAS (2012)
Afghanistan		750	915
Algeria		1,400	835
American Samoa	14		
Antigua	28		
Argentina	555		1,500
Armenia			15
Australia	1,422		
Austria	414		
Azerbaijan			350
Bahamas	29		10
Bahrain	9	11	
Bangladesh	7,816	16,397	14,035
Barbados	11		6
Belarus			8
Belgium	461	160	
Belize	48		40
Benin		2,565	517
Bermuda	11		
Bhutan		250	500
Bolivia	158	118	
Botswana	388		
Brazil	11,567	4,740	24,000
Burkina Faso	1,210	1,803	684
Cambodia			1,244
Cameroon		5,218	2,389
Canada	978	380	
Central African Republic	922		66
Chad			3
Chile		799	493
China	121,865	617,706	
Colombia		1,512	3,191
Comoros	93		
Congo		261	

Country	University of Illinois (1981)	FAO (1991)	GFRAS (2012)
Cook Islands		46	
Costa Rica	310	601	500
Côte d'Ivoire		1,427	
Cyprus	192	354	
Democratic People's Republic of Korea	7,648	7,979	4,584
Democratic Republic of the Congo	11	769	
Denmark	954	1,750	3,198
Dominica		45	
Dominican Republic	14	850	913
Ecuador	264	274	
Egypt		4,926	7,421
El Salvador	480	446	28
Estonia			144
Ethiopia	826	6,584	45,812
Fiji	213	169	
Finland	662	710	
Gabon	93		
Gambia	823	215	
Germany	5,440**	5,810	
Ghana		2,752	1,244
Greece	1,504		17
Grenada	52		
Guadeloupe	58		
Guatemala	551	4,889	
Guinea	2,754		1,538
Guinea-Bissau		350	
Guyana	124	252	80
Haiti	224	670	
Honduras	399	504	35
Hungary		415	
Iceland	57	57	
India	99,395	65,957	90,000
Indonesia		29,957	53,949
Iran			6,497

(continued)

TABLE 2A.1 Continued

Country	University of Illinois (1981)	FAO (1991)	GFRAS (2012)
Iraq	1,085	195	
Ireland	813	650	
Israel	520	342	150
Jamaica		428	231
Japan	16,328	12,400	8,228
Jordan	188	95	84
Kazakhstan			55
Kenya			5,488
Kuwait	10		
Kyrgyzstan			38
Lao People's Democratic Republic	157	38	752
Latvia			330
Lebanon	74	54	67
Lesotho	298		7
Liberia	726		134
Lithuania			307
Madagascar			104
Malawi	1,882	2,304	3,054
Malaysia	4,804	3,391	1,216
Maldives	13		
Mali		1,757	1,129
Malta	39		
Mauritania		97	
Mauritius	78	68	
Mexico	2,094	7,916	5,836
Mongolia			1,100
Montserrat	12		
Morocco	5,340		7
Mozambique		351	1,304*
Myanmar		5,668	4,534
Namibia	24		
Nepal	5,368	2,207	2,606
Netherlands	1,194	650	
New Caledonia	126		

Country	University of Illinois (1981)	FAO (1991)	GFRAS (2012)
New Zealand	280	250	
Nicaragua	85	682	
Niger		682	847
Nigeria	2,295*		7,000*
Norway	1,289	1,500	267
Oman		117	
Pakistan	2,621	3,962	9,749*
Panama		693	
Papua New Guinea	1,785	1,788	34
Paraguay	136		16
Peru		1,623	253
Philippines	19,020	15,444	25,000
Poland			3,803
Portugal	633	1,480	
Qatar	10	12	
Republic of Korea			4,584
Republic of Moldova			918
Romania			860
Rwanda	417	2,537	1,244
Saint Kitts and Nevis	5	23	24
Saint Lucia		64	45
Saint Vincent and the Grenadines	46		24
Samoa		33	
Saudi Arabia	699	392	
Senegal	1,163	427	500*
Seychelles	18		
Sierra Leone		1,369	702
Singapore	0		
Solomon Islands		250	
South Africa	1,672		2,210
Spain	2,808		
Sri Lanka	3,022	4,528	568
Sudan		456	656
Suriname		137	
Swaziland	120		

(continued)

TABLE 2A.1 Continued

Country	University of Illinois (1981)	FAO (1991)	GFRAS (2012)
Sweden	1,573	710	12
Switzerland	650		104
Syria	211	2,208	
Taiwan, China	17,825		16
Tajikistan			420
Thailand	7,486	20,622	16,986
Timor-Leste			452
Togo	212		16
Tonga	77	44	
Trinidad and Tobago	146	190	100
Tunisia	15	2,431	854
Turkey	5,523	16,067	14,644
Ukraine		89	
United Arab Emirates	63	89	
United Kingdom	0	1,703	19
United Republic of Tanzania		5,752	10,891
United States of America	24,102	15,141	2,900
Uruguay	290	157	183
Vanuatu	49	47	
Venezuela	1,271		118
Viet Nam	68	30	13,185*
Yemen	50	126	1,438
Yemen	123***		
Former Yugoslav Republic of Macedonia			155
Zambia			908
Zimbabwe	1984		6,159
Total	**290,246**	**438,695**	**1,050,861**

Source: FAO (1991); GFRAS (2012); Davis and Spielman (2016); Swanson and Davis (2014).

Note: * Some of these figures have been updated via the following sources: Mozambique (Cuangara and Thompson [2018]); Nigeria (Huber and Davis 2017); Pakistan (M. Ali, personal communication); Senegal (Franzel, Ndiaye, and Tata [2018]); Viet Nam (Ngan and Babu [2018]). In 1981, Nigeria data were partial for a few states. ** West Germany. *** People's Democratic Republic of Yemen.

Chapter 3

COMPARISON OF NATIONAL EXTENSION SYSTEMS: APPLICATION OF THE BEST-FIT FRAMEWORK

Kristin Davis

3.1 Introduction

As seen in Chapter 2, extension has evolved over the decades in various ways: in terms of its governance structure, in terms of who provides extension, in terms of capacity (staffing) and management, and in terms of advisory methods. However, the analysis in Chapter 2 was centered on the global level, focusing on the big picture and trends over time. This did not include more detailed assessments of national and regional extension systems using the best-fit characteristics. This chapter offers a synthesis of information from in-depth country and regional assessments that were conducted mainly since 2015. These assessments took place with various partners and projects. To pull out comparisons of the national systems, this chapter applies the best-fit approach (see Chapter 1). Note that this level of analysis still does not look at the performance or the impact of extension services (Chapters 4–8 cover performance assessment issues). Rather, this chapter presents a snapshot of extension services zooming in at the country or regional level.

The best-fit approach analyzes extension and advisory services from a systems perspective. It uses the four best-fit characteristics: governance structures, capacity, management, and advisory methods used. The "governance structures" refer to the institutional setup of extension services (often based on national policies). They include the role of the public sector in extension services, the level of privatization, the degree of decentralization, the funding mechanisms, and the coordination and linkages. "Capacities" refer to the human and organizational competencies, motivation, and financial and physical assets. "Management style" is the way in which extension is managed within the respective governance structures. This also includes training and retraining efforts for extension staff at various levels; organizational management procedures carried out; incentives and methods of performance assessment of individual agents; and monitoring and evaluation of the services. "Advisory methods" are the means used by field extension staff in interactions with farmers. "Methods" can be classified in various (and overlapping) ways, such as number of clientele involved (individuals, groups); types of decisions on which advice is provided (specific to the production of certain crops or livestock, managerial decisions, group activities, etc.); and media used (radio, television, Internet, print, video, group meeting, face-to-face visits, etc.).

The rest of the chapter is organized as follows. Section 3.2 looks at governance structures, including funding, pulling out findings of interest from various countries. Section 3.3 looks at capacity, including human resources and training across the set of countries, while section 3.4 looks at management issues. Section 3.5 covers advisory methods, and section 3.6 contains cross-cutting issues. Section 3.7 concludes, and section 3.8 discusses some recommendations on potential indicators for extension and advisory services using the best-fit framework. Table 3.1 provides summary information by country, looking at similarities and differences according to elements of the best-fit framework. The table includes both qualitative and quantitative information. The text focuses on archetypical countries that typify certain elements of the four variables in the best-fit framework.

The source of information for these country and regional analyses is a number of in-depth reports, mainly out of extension projects described below or manuscripts prepared by individuals (for example, Cambodia—Ke and Babu 2018; Central Asia and the Caucasus—Dosov 2018; India—Babu and Shishodia 2018; Latin America—Preissing et al. 2018; Viet Nam—Ngan and Babu 2018; Zimbabwe—Mwakiwa 2017) as well as the five country cases described in Chapters 4–8 (Brazil [Chapter 4], Uganda [Chapter 5], Ethiopia [Chapter 6], Malawi [Chapter 7], and the Democratic Republic of the Congo [Chapter 8]). Secondary sources of information are used (except for the five country cases in Part 2, which use primary data). The reports were selected because they contain recent and in-depth analysis on agricultural extension and advisory services at the country or regional level.[1] Furthermore, the sources used had to include relevant data about the best-fit characteristics. More information on methods used for each study is included in Table 1.1 in Chapter 1. Note that Part 2 also includes in-depth country case studies; however, these country cases are focused on *performance* of extension services rather than the broad snapshot according to the best-fit framework that is covered in this chapter. They do, however, still analyze performance using the best-fit framework.

1 The oldest reports are from 2011; however, these country cases also have data from 2017 and later.

3.2 Governance Structures

As mentioned above, "governance structures" include the institutional setup and management of extension services and the roles of different types of providers, the degree of privatization and decentralization, the funding mechanisms, and the coordination and linkages.

Policies and Strategies

Policies and strategies determine governance structures, financing, and provision of extension services. Many countries have no policy at all, some have policies that are not well-implemented, and a few have strong policies that are well-implemented (see Table 3.1). Countries with no specific extension strategies include Azerbaijan, Kazakhstan, Mozambique, Viet Nam, and Zimbabwe. Such countries often root any strategies and activities in national agriculture policies. For instance, extension in Mozambique has been guided by a series of master plans and programs (Cuangara and Thompson 2018), although it is now developing a formal policy on extension.

Countries such as Ghana, Kenya, and Liberia have extension policies in place that are somewhat outdated and not implemented as well as they could be (in Kenya, due in part to the decentralization that occurred after policy development). Malawi (Chapter 7) is currently updating its policy. Postconflict countries such as the Democratic Republic of the Congo (Chapter 8) are considering a national extension policy as part of their recent agricultural development programs. A few countries stand out as having strong extension policies coupled with government and other types of support for implementation of these policies. These include Brazil (Chapter 4), Rwanda, and Ethiopia (Chapter 6).

Providers

The public sector is often the chief actor in extension provision in most of the countries examined. For example, the Brazilian System for Technical Assistance and Rural Extension mandated the Ministry for Agrarian Development as the official federal organ responsible for the smallholder sector to transfer funds to extension organizations in the country through their state and municipality counterparts.[2] In Cambodia the Department of Extension was the central body responsible for extension within the General Directorate of Agriculture (Ke and Babu 2018). *(text continues on page 66)*

2 As of 2017, the Ministry for Agrarian Reforms has been dismantled and located within a Secretariat of the President's Office.

TABLE 3.1 Summary of country extension status based on best-fit framework characteristics

	Governance structures		
Country or region	**Providers and policies**	**Financing and budget**	**Coordination**
Central Asia and Caucasus (Armenia, Azerbaijan, Georgia, Kazakhstan, Kyrgyzstan, Tajikistan, Uzbekistan) (Dosov [2018])	Many nongovernmental; weak public and private. No policies; extension usually falls under broader agrarian policies (for example, in Azerbaijan and Kazakhstan).	Little government funding; many donors and nongovernmental organizations (NGOs).	Very weak coordination among public, donors, and NGOs.
Latin America (Preissing et al. [2018])	Wide variety of providers: government, private, producer organizations. Policies related to extension usually part of broader sector policies.	Diverse including national, departmental, and municipal governments; producer payments; local donations; development aid; cofinancing schemes in Peru, Chile, Nicaragua, and Colombia.	Moderate to strong linkages in most cases; Argentina and Brazil have incentive systems and mechanisms for exchange with researchers; formal links in El Salvador, Guatemala, Honduras, and Nicaragua.
Bangladesh (Huber and Davis [2017]; Rahman et al. [2017]; Swanson [2011])	Very pluralistic with strong public provision and many nongovernmental providers. Policy was updated in 2012.	Most government funding goes to salaries and capital costs.	Have difficulties due to pluralism and lack of mechanism to coordinate.
Brazil (see Chapter 4)	Strong government role but in coordinating and financing many pluralistic players from private and nongovernmental sectors. Extension policy enacted in law.	Federal—universal coverage attempted but not yet reached.	System of public calls for services to smallholders and state-level functionaries for large farms.
Burundi (Ludgate and Tata [2015])	Public mainly. Plan National d'Investissement Agricole for 2012 to 2017.	Public funding reduced in 2005 due to reforms aimed to boost the private sector.	No information.

Capacity (staffing)	Management style (including monitoring and evaluation, incentives, training)	Advisory methods	
		Delivery tools	Inclusion (equity)
Various international projects and NGOs hire their own staff (numbers are not recorded), without much coordination.	Varies by project focus; not standardized; no continuing education; absent or inadequate regulatory incentives.	More traditional, less ICTs (information and communications technologies).	Donor-funded projects often target youth and women.
100–831 farmers per extension worker in public sector and 42–642 small producers per extension worker in private sector.	Significant efforts to train and upskill extension staff; Chile, Nicaragua, El Salvador, and the Dominican Republic had continuing education; Guatemala none. Salaries range from around US$9,000 in Paraguay to just under US$35,000 in Argentina. General lack of baseline information, monitoring and evaluation, and feedback systems on performance of extension.	Increasing diversity, especially ICT.	Some efforts to explicitly include marginalized groups.
Extension department is largest organization employing 14,092 field-level extension agents, with each responsible for 900 to 2,000 farm families.	Typically hold diplomas from one of the 11 agricultural training institutions. Training tends to be technical with focus on cropping systems. Public salaries relatively high.	Wide variety but informal channels and information sources such as agrodealers important.	Training on gender usually only in projects.
State-level extension has mandate to serve at the municipality level while civil society is encouraged to seek funding competitively; 20,000 agents.	Strong training on participatory methodologies and service delivery, such as providing advice and support on short-term credit; ability to reach different ages and genders weak.	Focus on participatory methods, agroecology, sustainable farming, organic farming, and high-value agriculture.	Policy focuses on marginalized groups and pays attention to age and gender.
Unknown.	Has Directorate of Agricultural Training and Animation.	Unknown.	National gender policy addresses inequalities.

(continued)

TABLE 3.1 Continued

	Governance structures		
Country or region	**Providers and policies**	**Financing and budget**	**Coordination**
Cambodia (Ke and Babu [2018]; INGENAES [2016k])	Pluralistic with strong public sector. Policies include some attempts to modernize and update extension.	Centralized funding but often inadequate to cover the needs.	Department of Agricultural Extension is responsible.
Democratic Republic of the Congo (see Chapter 8)	Pluralistic with public provision and many nongovernmental providers. No policy.	Low government funding; major donor-funded projects on extension.	Few organizations have incentive systems for good performance.
Ethiopia (INGENAES [2016j]; see Chapter 6)	Strongly state-led provision with some NGOs, producer organizations, and donor projects. Policy exists.	Major government investments; largely public extension; some projects, NGOs, private sector.	Incentives for good performance of staff but limited facilities demotivate.
Ghana (INGENAES [2016i])	Pluralistic with public provision and many nongovernmental providers and projects. Policy exists (2001).	Almost entire budget comes from government of Canada (personal communication, April 8, 2019).	Need for coordination due to number of providers.
Guatemala (Chaisson-Cardenas [2019]; INGENAES [2016h])	Pluralistic. No policy.	Most funds for public extension come from the National General Fund; decentralization efforts are reviving creation of 340 municipal-level extension agencies.	Unknown.
Guinea (MacNairn [2017])	Pluralistic with public provision and many nongovernmental providers and projects. No policy.	Government and donors.	Variable depending on external resources.
Haiti (Goertz [2016])	Weak public sector with many projects and nongovernmental actors. No policy.	Donors and government.	Unknown.

Capacity (staffing)	Management style (including monitoring and evaluation, incentives, training)	Advisory methods	
		Delivery tools	Inclusion (equity)
Majority of staff at the provincial level with limited number at district level.	Training extension staff recognized as imminent need, but has not supported with adequate funding.	Focus on ICT slowly increasing.	No clear policy on inclusiveness.
Large number of extension agents, but no operating funds.	Largely no training; postconflict country; public salaries extremely low.	More traditional, less ICTs.	No.
Good extension agent ratio; still face major capacity challenges; supply-driven extension.	Training mainly technical; financial rewards, educational opportunities, certificates, and promotions for good performance.	Variety including group approach, visual methods.	Trainings organized for women and youth.
Pluralistic from all sectors.	Government is doing reskilling with help of Canadian universities.	Multiple.	Unknown.
Extension has services in 19 of 22 departments, and Ministry of Agriculture estimates that half of municipalities have extension coverage.	Eight-month Rural Extension Certificate Program designed to support extension in strategy, production technology, environment, socioeconomics, food security, and nutrition.	Unknown.	Poor gender equality, some donor projects focus on indigenous people; previous projects focused on youth.
800 public agents.	Five national agricultural education and training centers give diplomas. No incentives for public agent performance.	Multiple.	Several projects target women and youth.
Mainly NGO-provided but government also.	World Bank recommends rehabilitation of vocational schools and farmer field education to drive demand and supply of extension.	Unknown.	Poor on gender equity.

(continued)

TABLE 3.1 Continued

	Governance structures		
Country or region	Providers and policies	Financing and budget	Coordination
Honduras (Valenzuela and Saavedra [2017]; Williams [2016])	Pluralistic. No extension policy; agricultural policy guided by 2004–2021 State Policy for the Agri-Food Sector and Rural Areas and 2014–2018 Government Strategic Plan Everyone's Plan for a Better Life.	Government and donors.	Pluralism results in fragmented delivery with little coordination and weak leadership.
India (Babu and Shishodia [2018])	Very pluralistic and decentralized with strong public provision, producer organizations, private sector, and many nongovernmental providers. Policy exists.	Increased funding to human resource development but likely to devolve to a state level.	Coordination very poor due to multiple roles extension agents play.
Kenya (INGENAES [2016g])	Strongly pluralistic and decentralized with many public and private players. National Agriculture and Livestock Extension Policy (2000).	Government (now devolved to the county level) and donors; many NGOs and private sector.	Agricultural Sector Coordination Unit.
Liberia (Sigman [2017]; INGENAES [2016f]; McNamara, Swanson, and Simpson [2011])	Weak government provision as well as many projects and some private companies. Policy exists (2012).	Government but very constrained.	Though not functional, policy stipulates coordination platform at national, county, and district levels, managed through Agricultural Coordinating Committee.
Malawi (Cai [2017]; INGENAES [2016e]; Simpson, Heinrich, and Malindi [2012]; see Chapter 7)	Very pluralistic and decentralized with strong public provision, producer organizations, private sector and many nongovernmental providers and projects. Currently revising policy.	Low government funding; major donor-funded projects on extension; large farmer-based organizations with own extension agents.	Highly developed structures to coordinate and plan.

		Advisory methods	
Capacity (staffing)	Management style (including monitoring and evaluation, incentives, training)	Delivery tools	Inclusion (equity)
Mainly NGOs; also education and research, unions and associations, public sector, and private sector.	Mainly university graduates, generally agricultural engineers; most organizations provide continuing education; incentives used by 66% of organizations; government salaries low.	Farmer field schools most common.	Some organizations focus on gender, youth, and marginalized populations.
Limited levels of human resources, no new recruitment since 1998, although this is changing with reforms in some states.	Regular training is part of the extension system, but the curriculum has not kept up with changing needs of the farmers. A new national curriculum is being developed.	Large focus on ICTs combined with traditional methods and farmer field schools.	Gender-sensitive programs at the local level; state extension system focuses on smallholders, but richer and better educated farmers seem to capture most of the benefits.
5,470 in 2013 (before devolution).	Unknown.	Various.	Policy identifies several approaches and methods that target rural women.
Estimated extension to farmer ratios from 1:1,000 to 1:5,000	Two-year National Agriculture Diploma Program; nongovernment workers have higher salaries and more incentives such as training, equipment, and transportation.	Various.	Gender-sensitive practices and culture within extension and wider agricultural community.
Large number of government extension agents, although with some vacancies.	Some trainings mainly from donor projects; farmer-to-agent and farmer-to-lead farmer ratios used as main extension performance indicators.	Variety of methods, including radio programming, group approaches, lead farmer, model village, among others.	Policies on gender exist; strong donor and NGO efforts for gender equality.

(continued)

TABLE 3.1 Continued

	Governance structures		
Country or region	**Providers and policies**	**Financing and budget**	**Coordination**
Mali (DLEC [2018]; INGENAES [2016d])	Pluralistic and government led. No extension policy, but various related policies in the agricultural sector.	Government and donors.	National Directorate of Agriculture responsible to coordinate.
Mozambique (Cuangara and Thompson [2018]; INGENAES [2016l])	Very pluralistic and decentralized with public provision, producer organizations, private sector and many nongovernmental providers and projects. New Extension Master Plan 2018–2027 was under development (V. Parkinson, personal communication, 2017).	National budget and donors through basket funding.	Limited; research-extension coordination takes place through periodic reviews.
Nepal (Suvedi and McNamara [2012])	Government-led but with many projects and nongovernmental organizations. No policy.	Government and donors.	Poor links with research and education.
Nigeria (Huber, Davis, and Lion [2017])	Decentralized with a number of government and nongovernment providers. Policy in development.	Federal government and some states.	Had strong system when supported by the World Bank.
Rwanda (MacNairn [2018]; INGENAES [2016a])	Strongly government led but with many projects and nongovernmental providers. Policy exists (2009).	Government led with donor support, many NGOs.	Research and extension are merged.
Senegal (Franzel, Ndiaye, and Tata [2018]; INGENAES [2015])	Government, including national development agencies; farmer organizations and projects. No extension policy.	Parastatal led by national government with producer organizations, private sector, and local government.	Weak, especially at the local level; in theory take place through research-development committees under extension in each region.

Capacity (staffing)	Management style (including monitoring and evaluation, incentives, training)	Advisory methods	
		Delivery tools	Inclusion (equity)
Public staff were 839 in 2009 and are major providers by numbers and geographic and technical coverage (women were 10–25 percent); staff are rapidly aging.	Rural Polytechnic Institute of Training and Applied Research—certificates up to MSc level; insufficient resources to perform work (transport, equipment).	Various delivery tools.	Some projects target women and youth.
1,304 public agents (1 public agent per 3,000 households); some incentives.	Very weak foundational and continuing education; public sector cannot compete with nonstate salary levels; public sector has performance indicators.	Various.	Female agents number 16% of public agents.
4,974 technical agricultural and 4,050 livestock staff.	Adequate infrastructure but need to strengthen market, group development skills; M&E weak; 60–70% district budget to salaries.	Various including mass media.	Mainly generalist approach; frontline workers mainly men.
7,000 public agents, 28% female	Diploma or BSc; many agricultural institutes exist; no performance-based management system for extension staff.	Various including one-stop shops.	Women have low access to extension information and technologies and limited contact.
2,500 farmer field school facilitators and 14,200 farmer promoters.	Field school facilitators have up to secondary level; performance contracts ensure accountability.	Farmer field schools and farmer promoters predominate.	High gender equality and groups promote engagement.
500 but 24% vacancy rate in 2017.	Field agents have diplomas; many NGO staff have BSc or MSc; access to vehicles; performance management systems absent in the public sector.	Various.	Women appear underrepresented among staff and farmer trainers.

(continued)

TABLE 3.1 Continued

	Governance structures		
Country or region	**Providers and policies**	**Financing and budget**	**Coordination**
Sierra Leone (INGENAES [2017])	Weak government provision with many projects. No policy.	Various with high donor involvement.	Extension division coordinates delivery.
Tajikistan (INGENAES [2018]; Dosov [2018])	Many nongovernmental; weak public and private. No policy, only National Development Strategy.	Donor-dependent; majority NGOs; also government, research, academia.	Limited.
Tanzania (INGENAES [2016b])	Very pluralistic and decentralized with public provision, producer organizations, private sector and many nongovernmental providers and projects. No extension policy; National Agriculture Policy (2011).	Strong public role with donor funding; also NGOs.	Unknown.
Uganda (INGENAES [2016c]; see Chapter 5)	Very pluralistic and decentralized with public provision, producer organizations, private sector and many nongovernmental providers and projects. Policy exists.	Pluralistic with donor funding, many NGOs and private providers.	New Single Spine Agricultural Extension System has strong collaboration with nongovernment actors.
Viet Nam (Ngan and Babu [2018])	Pluralistic and government led. Attempting to restructure agriculture.	Following decentralization, increased local funding.	Coordination structures exist and function reasonably well.

Capacity (staffing)	Management style (including monitoring and evaluation, incentives, training)	Advisory methods	
		Delivery tools	Inclusion (equity)
Limited; district offices have subject matter specialists, supervisors, and field-level agents.	Inadequate number of technically trained and qualified personnel; donor project promotes staff development.	Training and visit, innovation platforms, farmer field schools, agricultural business centers, farmer-based organizations.	Attempts to mainstream gender equity and inclusion, but limited voice for women and information access. Agenda for Prosperity five-year (2010–2015) strategy promoted empowerment of girls and women.
Officers typically subject matter specialists with limited access to resources.	Centralized system of professional training and capacity building.	Various.	Cultural beliefs value men over women; no specific strategies; some programs focused on women.
In 2012, 10,891 extension officers, 6,925 for crops and 3,966 for livestock; 75% men.	Unknown.	Various; individual and group contact, farmer field schools important.	Women have limited access to land and credit to purchase modern farming inputs.
Some 5,000 new staff just trained; preexisting number unknown.	In 2017 the government launched Guidelines, Standards, Code of Ethics and Process for Registration and Accreditation of Extension Service Providers.	Village Agent Model, farmer field schools, others.	Many programs target women and youth.
High turnover of staff in the extension system.	Attempt to enhance training from 1995 to 2015; low incentives for public agents.	Use of ICT limited. Still heavy reliance on field demonstrations.	No; women have less access to extension.

(continued)

TABLE 3.1 Continued

Country or region	Governance structures: Providers and policies	Financing and budget	Coordination
Zambia (Burrows, Bell, and Rutamu [2017])	Pluralistic with public provision, producer organizations, private sector, and many nongovernmental providers and projects. No; development strategy based on Vision 2030 and 6th National Development Plan (revised in 2013).	Pluralistic; major problem of underfunding (up to one-third of positions go unfilled as a result).	Coordination meetings to harmonize extension services with other actors.
Zimbabwe (Mwakiwa [2017])	Public, private, and nongovernmental projects provide services. No policy exists.	Donor support with limited government.	Previous mechanisms not operating and no funds.

Source: Author's compilation based on the various chapters in this book and their earlier, longer versions.

Viet Nam's Ministry of Agriculture and Rural Development led all activities of the Department of Agriculture and Rural Development and the National Center for Extension (Ngan and Babu 2018), guiding provincial extension centers and District People's Committee, which manages district extension stations, commune extension staff, and village extension collaborators. In Africa extension in Ethiopia continues to be heavily government run, although the country is highly committed to financing and reforming its extension system as a cornerstone of its agricultural transformation agenda. Rwandan extension is also heavily government run yet pluralistic.

Pluralism (the presence of many types of extension providers from different sectors), however, is the norm in all the countries to varying degrees, with government, private sector, and civil society all providing extension services. Some countries in Latin America that did away with public-sector extension several decades ago (hoping that the private sector would fill the gap) are beginning to fund public services once more (for example, Colombia, Ecuador, and Honduras). Countries in Latin America now have a mix of public and private extension providers: local municipalities, nongovernmental organizations, and universities. Although more organizations are providing extension services, the country cases show that pluralism of service delivery is far from being complete, with elements of competition across different

Capacity (staffing)	Management style (including monitoring and evaluation, incentives, training)	Advisory methods: Delivery tools	Advisory methods: Inclusion (equity)
1,700 public agents.	University of Zambia's School of Agricultural Sciences and agriculture training institutes train staff as general agricultural practitioners.	Various, including study circles, lead farmers.	Women often do not own or control productive resources, are excluded from decision-making processes, and are less likely to benefit from services.
4,700 front-line staff employed by government.	0.4% have MSc; 3.8% have BSc; 11.3% are diploma holders; 84.5% have certificates; fast-tracking training resulted in suboptimal skill sets; high turnover.	Training of trainers; training and visit; lead farmer; group approach; demonstrations; field days; media.	No framework to deal with gender; women underrepresented in extension force.

providers and opportunity for clients to demand and choose among them largely missing. Service providers are dominated by international NGOs, often with limited human resources and relying heavily on government extension workers. They fill the gaps and limited resources from the government by providing mobility and operating funds to government extension agents. A common scenario is government providing the human resources while international NGOs provide the financial resources, but hardly any competition of service providers or expansion of options or choices for farmers is happening.

Moreover, pluralism must be coordinated to avoid duplication of services and conflicting messages and methods. The national extension policy in Malawi that advocates for pluralism has led to many nongovernmental extension providers that have been largely uncoordinated. However, Malawi has developed highly sophisticated coordination mechanisms within the Ministry of Agriculture, Irrigation and Water Development (MoAIWD) and district-level agricultural extension system, although it faces several challenges due to limited funding and clear mandates and implementing guidelines (see Chapter 7 for analysis of Malawi's performance). Nonetheless, the National Agriculture Content Development Committee (NACDC) that has been harmonizing technology and extension messages particularly on its radio programming has been shown to be effective in improving access to quality extension services. These could be intensified for other technologies and advisory methods and could be expanded to other countries.

Extension in India has become increasingly pluralistic over the past two decades (Babu and Shishodia 2018), with a deliberate move by government in 2011 toward decentralized provision of extension with the complete rollout of the Agricultural Technology Management Agency and the rise in the role of private-sector and nongovernmental organizations in extension. Private-sector extension provision in India faces many challenges. The reach of private extension remains limited to wealthier commercial farmers who can afford to pay for it. And, although productivity gains through private extension services received by smallholder farmers are not uncommon, there are many cases of exploitation of farmers by private companies providing extension in India (Zhou and Babu 2015). Nonetheless, in other settings private extension, often bundled with other input provision and marketing, frequently through contract farming schemes, has been effective in delivering relevant and useful knowhow to farmers and increasing their production and incomes (Ragasa, Lambrecht, and Kufoalor 2018; Ruml, Ragasa, and Qaim 2019; see Chapters 4 and 7). These findings highlight the need to monitor and harmonize extension messages and to develop farmers' collective negotiation skills and bargaining power from them to work as equal partners with private companies.

Financing

Countries finance their extension services from national or subnational budgets, levies and taxes, and development aid as well as by paid services from the private sector. Countries that struggle to fund their extension systems include most of those in Africa and Central Asia. According to Dosov (2018), the lack of policies or regulations that specifically provide guidance or define the rules of a national extension system means there are limited options for national governments to invest in extension. Donors often step in, which, if not well-regulated, may lead to the lack of cohesion and a duplication of services. What government funds do exist go mainly to pay salaries, leaving little for operational expenses. Although donors and international NGOs cover operating funds for extension, often working with government extension agents, their projects are short-lived and largely uncoordinated. Thus government funding is required for sustainability and for much needed coordination and monitoring for these various efforts.

The exception is Ethiopia and Rwanda, which have strong government commitment to agriculture and to extension to achieve their development objectives. These two countries strongly fund the sector as well. This does not preclude donor funding, but Ethiopia and Rwanda lead the donor working groups to coordinate financing and provision of services. Greater diversity in funding sources was seen in Argentina, Chile, Colombia, and Ecuador as the Latin American region has experimented with new models and orientations in extension implementation, matched with investments in the reform of extension systems. In Colombia the sources of financing were diverse: 29 percent of the organizations received resources from the national government; 19 percent from departmental governments; 35 percent from municipal governments; 14 percent from producer payments; 11 percent from local donations; and 14 percent from development aid. In Chile's Technical Assistance Service the national government contributed 70 percent, while the remaining 30 percent is financed by the producers (Preissing et al. 2018). Argentina had considerable self-financing directly or indirectly from producers, but occasionally they receive external financing from national programs (Preissing et al. 2018).

Ecuador had a diverse funding mix, based on the types of programs and initiatives. Public programs depended 100 percent on the national budget. Decentralized governments differed; for example, the Provincial Government of Tungurahua depended 50 percent on government resources and 50 percent on international cooperation for extension provision (Preissing et al. 2018). India also exhibits some funding diversity. However, gaps still exist. Nearly 45 percent of total funds under the Agricultural Technology Management Agency (ATMA)—the major public-sector approach to extension, implemented through state-level funded extension personnel—was spent on farmer-oriented activities such as organizing training programs, conducting demonstrations, exposure visits for farmers, and mobilizing farmers to form farmer interest groups or self-help groups (Babu and Shishodia 2018). At the same time, the traditional state extension agents continued to be employed to implement state-level subsidy schemes. The sustainability of funding for ATMA in the long run is still debated, as there are signs that the central government will transfer the full responsibility of ATMA to the states at some time in the future (Babu and Shishodia 2018).

Decentralization has been tried in almost all of the countries studied in this book as a way to increase accountability by bringing the services closer to the clientele. However, the legal transfer of responsibilities for agricultural services is often not matched by sufficient fiscal decentralization, therefore continuing dependence on the transfer of funds from the central government undermines the autonomy of local governments. Where such transfers have been reduced and taxes and revenues by the local government remain meager, the promises of decentralization often are not realized. For example, attempted reforms toward a district-level agricultural extension services system in Malawi lack the resources and capacity to coordinate different service providers and engage with farmers (see Chapter 7 on Malawi). This is also consistent with public expenditure studies in other countries—for example, in Ghana analysis of district budgets over time indicates that total average agricultural expenditures at the subnational level decreased from 11 percent to 6 percent of spending between the onset of decentralization in 2012 and 2015 (Resnick 2018).

Institutional Linkages and Coordination

The coordination issues tend to follow the policies and financing trends just discussed. Those with no policies or weak policies and many different types of funding, especially donor and project funding, tend to struggle with coordination and linkages. Those with strong policies and strong government financing tend to have better coordination and linkages. A major challenge has been the poor institutional linkage between the formal research, education, and extension systems as well as coordination within extension (Babu and Shishodia 2018; Dosov 2018; Preissing et al. 2018). Coordination has been a major issue given that many different organizations are providing extension today and there is weak institutional capacity to coordinate them. In Colombia, for example, 38 percent of the organizations surveyed had weak links with other organizations; 30 percent had moderate links; and 20 percent had close links. However, stronger links occurred with local government agencies and microcredit banks and institutions (31 percent) (Preissing et al. 2018). Except for a few cases, such as in Argentina and Brazil, which had formal institutional cooperation agreements and assistance, incentive and mechanisms for exchanges among researchers and extension services are weak and nonexistent in Latin America (Preissing et al. 2018).

Malawi is well known as a country with a multitude of projects, donors, and nongovernmental organizations as well as government commitment to extension. Extension in Malawi (see Chapter 7) is dominated by the public extension system, although there is a rich diversity of civil society, nongovernmental agencies, and private-sector companies implementing extension projects as well. The Ministry of Agriculture, Irrigation and Water Development (MoAIWD) is decentralized, yet translation of the research outputs into tangible benefits for farmers will require improved engagement of district-level actors. An *Agricultural Extension Policy Implementation Guide* was adopted in 2004 (Malawi, MoAIWD 2004) to implement the national extension policy. It also aims to guide the District Agricultural Extension Services System (DAESS) in coordination among the many actors. However, considerable investment in institutional capacity is needed before coordination can be improved. The National Agriculture Content Development Committee (NACDC), instituted in 2014, is a good start; if continued and funded, the NACDC should strengthen coordination.

In India's extension system, coordination of extension professionals in various entities has been a long-standing challenge (Babu and Shishodia 2018). Staff of farm science centers report to central government, state government, and NGOs, which manage the centers. In the same state some of the farm science centers are run directly by central government staff, while others are run by state agricultural university staff, and yet others by NGOs. There is hardly any coordination among them, even though they may be in the same agroecological zones. Furthermore, there is a high level of disconnect between the farm science centers and the state extension functionaries. Linkages between the Indian Council for Agricultural Research—an important information source for extension workers—and the state departments of extension remain weak (Babu and Shishodia 2018). Although farm science centers exist for each district to translate research into adoptable technologies, coordination of state extension personnel in this effort remains weak, even after the introduction of the decentralized Agricultural Technology Management Agency (ATMA) model.

Some stronger links were seen in Argentina and Uruguay. Argentina showed a high degree of linkages and institutional agreements (Preissing et al. 2018), with most links formalized by cooperation agreements. The strongest links were with universities, local or municipal government agencies, and research centers; more moderate links were with technical training services and NGOs; and the weakest links were with cooperative organizations, banks and credit institutions, and input suppliers. In Uruguay there were strong links between actors, especially at the local level and occasionally in an informal way. These links developed because of mutual knowledge of issues at the local level. Chile's Local Development Program is linked to few entities, and the perception is that the links are mostly moderate to weak with the strongest established with government agencies and municipalities.

To promote intersectoral collaboration and checks, multistakeholder panels and innovation platforms have been formed in various countries to help address information gaps on preferences and demands for extension services and on performance of providers. Although many of these platforms function poorly due to lack of motivation and limited capacity of service providers to respond to demands of rural communities, newer forms of innovation platforms that focus on market access and center on specific value chains have been successful in some cases (Ragasa, Badibanga, and Ulimwengu 2016).

Other promising mechanisms are available that can contribute to improving coordination, transparency, and accountability in extension service provision. In Ethiopia, Digital Green's Connect Online—Connect Offline platform, which regularly collects data and automatically computes relevant statistics on adoption and yields, provides relevant and timely data for decision-making and programming (see Chapter 6). In Rwanda an approach combining elements of performance-based contracts for various actors, consultative platforms for greater accountability, earmarked funding, and capacity strengthening at all levels is successfully improving adoption and yields (MacNairn 2018). Three-party performance contracts between Rwandan district officials, the Rwanda Agricultural Board, and farmer field school facilitators' cooperatives are helping monitor the progress of field school facilitators. This approach is combined with the *imihigo* system, a traditional consultative process for demand articulation and accountability. Combining these different elements, by earmarking funding and creating metrics for successful service delivery that go beyond budgeted expenditures or input indicators, coupled with receipt of frequent updates on these priority metrics through ICT, service providers and decisionmakers at all levels can address problems in real time.

3.3 Capacity

Human Resources

Most of the case studies report human resources as one of the constraints to increasing effectiveness of the extension system. In general, capacities include staff skills, the infrastructure and equipment to carry out the job, as well as foundational and continuing education for extension personnel. Extension capacity is not only at the level of the extension officials or frontline workers; it also applies to extension organizations at different levels and the role they play in the agricultural innovation system. Extension capacity at the personnel level includes quantity and quality of those officials and staff providing extension. This section looks first at the human resources (number or coverage and type of staff) and then their foundational and continuing education.

In Latin America extension coverage ranged from 100 to 831 farmers per extension worker in the public sector and between 42 and 642 small producers per extension worker in private-sector programs (Preissing et al. 2018). According to Preissing and colleagues (2018), 55 percent of the Central American extension staff were technical and professional extension workers in the organizations interviewed, and 45 percent were support staff. The public sector had the largest number of technicians and extension professionals as well as support staff (Guatemala had a 75:25 ratio and Nicaragua 70:30) (Preissing et al. 2018).

In Africa south of the Sahara, the farmer-to-extension agent ratios are much higher than in Latin America, and there are also high vacancy rates of government extension agents' positions. In Senegal a high vacancy rate (24 percent in 2017) has been a chronic problem due to lack of funds to pay salaries of new staff (Franzel, Sinja, and Simpson 2018). The total number of extension agents across sectors in Senegal was approximately 500, not counting all the managers.

Ethiopia has invested significantly in its human and physical capital in recent years. The country has a significant extension force and infrastructure in the form of farmer training centers at many of the lowest administrative units (Chapter 6). In addition to the workforce of some 47,500 extension agents, there are roughly 7,000 subject matter specialists and 4,000 supervisors employed in the public extension system in district and regional offices. Extension agents in Ethiopia are generally trained in crops, livestock, or natural resource management at one of the Agricultural Technical and Vocational Education and Training (ATVET) colleges. All public extension agents have postsecondary education, and they generally have a good mix of expertise on crops and livestock production and natural resources management.

In some countries in Asia coverage is also an issue. In India most state departments function with limited levels of capacity in terms of human resources, with no new recruitment since 1998 (Babu and Shishodia 2018). This overburdens the existing staff and prevents them from increasing their capacity and updating their knowledge. Only 20 percent of staff are university graduates. The ratio of staff to farmers varies widely across the states and is particularly low in remote areas in the north and east of the country. Increased system, individual, and institutional capacity are needed to increase the productivity of crops and the sustainability of production systems. In Viet Nam about 20 percent of extension workers retire or resign each year, equivalent to a loss of 5,600 staff a year. Therefore, the intake needs for the extension system is 5,000 people a year. However, Viet Nam is experiencing a shortage of human resources as the number of students enrolled in agricultural extension schools is declining (Ngan and Babu 2018).

Capacity of existing staff to be effective on the job is a further problem. Along with the quantity of extension workers, there is also a shortage of quality. Training levels for extension staff range from diplomas received from agricultural training institutes (Bangladesh, Ethiopia, Senegal) up to MSc or BSc degrees in agriculture for managerial staff in Senegal (Franzel, Ndiaye, and Tata 2018). Some countries provide continuous education to extension agents, particularly within donor projects, but these are largely uncoordinated (see Chapter 7 on Malawi and Chapter 8 on the Democratic Republic of the Congo). An exception, Chile has a well-supported process for continuous training of service providers, and in Colombia, 25 percent to 35 percent of the Municipal Units of Agricultural Technical Assistance technical staff had access to continuous training activities, while 60 percent to 75 percent of consulting firm and NGO staff had access to training programs in universities and other courses promoted by the Ministry of Agriculture and Rural Development (Preissing et al. 2018).

Furthermore, training is not always in all the relevant areas. Extension staff may have adequate technical backgrounds (albeit with limited hands-on applications during training) but are weak on the so-called functional or "soft" skills. In Tajikistan, for instance, agricultural advisers were often technically trained and well-oriented, but they needed commercial expertise—that is, skills and attitudes that are necessary in a market economy (Dosov 2018). These advisers needed knowledge and skills of how small farmers can enhance and diversify their farming systems. They lacked skills for organization and operation of self-help activities and group work and cannot help groups with marketing. Agricultural advisers also needed skills and equipment to work with information and communications technologies to work more cost-effectively, had limited access to relevant market information, and lacked the ability to use modern tools for decision-making (Dosov 2018). Assessments in Honduras and Malawi found limited focus on functional skills, such as group development, communication, or entrepreneurship (Cai 2017; Valenzuela and Saavedra 2017).

Many countries use local "farmer extensionists" to bolster extension services and help to increase the reach of extension more sustainably. For instance, to deal with Cambodia's extension officers' low capacity, village-level para-extension workers are employed who work with the government extension providers (Ke and Babu 2018). Viet Nam has about 20,000 grassroots volunteer extension workers at the commune and village levels (Ngan and Babu 2018). Similarly, Senegal uses farmer trainers, who tally about 9,100 (including community nutrition volunteers) (Franzel, Ndiaye, and Tata 2018). Almost all extension agents in Ethiopia (Chapter 6) used model farmers who, in theory, each reach five more farmers, to expand their reach. In Malawi official figures show a ratio of one lead farmer to six farmers (see Ragasa et al. 2019; Chapter 7 on Malawi).

There is, of course, a trade-off between quality and quantity of these efforts. Many short-term projects may temporarily train farmer extensionists, but it is not always institutionalized for the longer term as in the case of Peru, where training efforts are focused on building capacities of the rural promoters (Preissing et al. 2018). In addition to adequate training and capacity building, community ownership, recognition and support, as well as government extension agents' support, are often lacking, which continue to demotivate lead farmers, reducing their effectiveness and impact in the community (see Ragasa 2019; see Chapter 7 on Malawi). In Uganda even simple signposts to signal the presence of lead farmers in the community could make a difference in the impact of the program (Behaghel et al. 2018). Farmers reported that the farmer trainers with signposts were more knowledgeable about dairy farming than those without, indicating that this variation increased their credibility; it may also have enhanced their motivation (Behaghel et al. 2018). Registration and certification efforts are just getting under way in many countries. Uganda has been undergoing a process of registration and accreditation of agricultural extension and advisory service providers, along with guidelines and standards for agricultural extension services (see Chapter 5). Except for efforts in Peru, little has been done to certify the skills of extension workers or producer leaders at the community level in other South American countries.

3.4 Management

Organizational management in extension includes the management style of extension organizations, procedures carried out, incentives and methods of performance assessment of individual agents, and monitoring and evaluation of the services. Staff salaries for extension agents are usually similar to those of other different public-sector agencies but considerably lower than those in research (for example, in Senegal; see Franzel, Ndiaye, and Tata 2018). Nongovernmental and private-sector extension agent salaries are generally higher. The variation between public- and private-sector salaries affects many African countries and tends to lead to high staff turnover in the public sector. The public sector in Mozambique, for instance, is unable to compete with the private sector and NGOs for salaries, and many extension staff leave the government for better paid salaries in the private sector. The Ministry of Agriculture must constantly train new staff to fill the gap. Incentive policies to retain staff in the public sector do not exist or are nonfunctional (Cuangara and Thompson 2018).

However, public extension agents in Bangladesh were relatively well paid, receiving between US$450 and US$500 per month on joining and US$630 per month with five years of service (Huber and Davis 2017). Similarly, in Central America salaries of extension workers fluctuated between US$5,400 and US$12,000 per year in public institutions and were lower than private and nongovernmental organizations (Preissing et al. 2018). Salaries for South American public-sector staff are shown in Figure 3.1.

Beyond salaries, incentive systems also affect the performance of public and private extension staff. Many have no incentives for public extension agent performance (for example, Guinea; see MacNairn 2017). The Ethiopian extension system provided rewards and prizes for good performance of their staff, such as financial rewards, educational opportunities, certificates, and promotions. In fact, 40 percent of extension agents said they received some reward for good performance in the past three years (see Chapter 6). However, promotion, benefits, and allowances seem to be not the same for all. Despite recent efforts to clarify the extension agents' specific roles and tasks, the amount and type of work they are expected to perform is extensive and continually increasing, leaving agents overburdened and poorly prepared to carry out regular extension activities. Despite a well-staffed extension system with incentives, the organizational management of extension workers in Ethiopia remains a challenge (Chapter 6). Extension agents in Ethiopia have been overburdened by activities beyond their regular mandates, leaving them little time to search for additional knowledge and information. The farmer training centers where field agents are based are generally underresourced.

Aside from salaries and organizational funding, a study in Nepal shows that local knowledge and motivation of extension staff at various levels, and not the traditional indicators of organizational capacity (for example, resources, professionalism, and autonomy), are more effective in effecting access to extension services (Kyle and Resnick 2019). Monitoring and evaluation systems for extension were also weak. In the Latin American case, there was a widespread lack of baseline information, monitoring and evaluation, and feedback systems on the delivery and performance of extension systems. Cofinancing schemes were introduced in many countries (Chile, Colombia, Nicaragua, and Peru) to increase small farmer ownership of outputs and to enable greater coverage (Preissing et al. 2018). The lack of extension

FIGURE 3.1 Net annual salary of agricultural extension service professionals in the public sector in South America (US$)

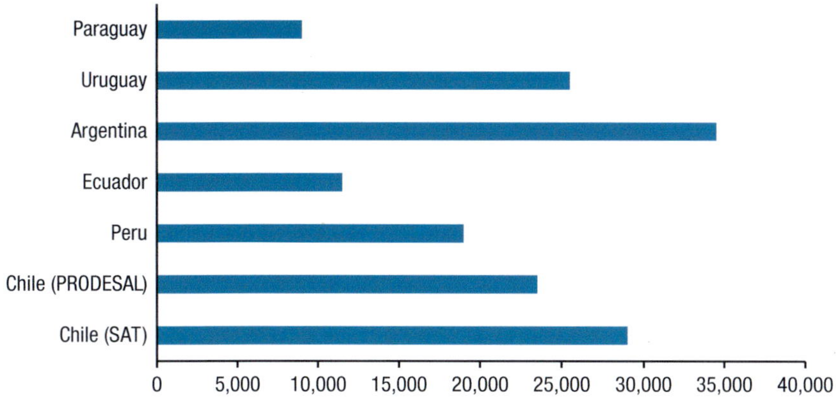

Source: Preissing et al. (2018).

Note: PRODESAL = Programa de Desarollo Local (Local Development Program); SAT = Servicios Agricolas Technicos (Technical Advisory Services).

performance indicators is a critical vacuum in Latin American extension systems. In Malawi farmer-to-agent and farmer-to-lead farmer ratios are persistently used as its main extension performance indicators at the national level. Monitoring should also go beyond measuring just access to extension services, to the level of adoption of improved technologies and practices as well as changes in productivity and incomes. There is potential for information technology and data analytics to collect data and provide relevant and frequent updates to different stakeholders at various levels (see Chapter 6 on Ethiopia) on climate and soil conditions; status of extension services, input use, and technology adoption; and how all these drive development outcome indicators.

3.5 Advisory Methods

Advisory methods include the different means to serve clients and provide information, empower farmers, and otherwise fulfill the extension mandate in each country. Many countries are moving from top-down to more participatory methods and to market-oriented approaches rather than a sole focus on production. There is also a trend to use more modern information and communications technologies in extension, as well as the tried-and-tested technologies such as radio. A number of approaches intended to enhance inclusion of disadvantaged groups (such as women, youth, and pastoralists) also exist. One point that is clear is that all countries use a wide variety of methods (for example, Figure 3.2). The sections below discuss traditional, participatory, information and communication technology-based methods, and farmer-focused and inclusive approaches.

Traditional methods such as workshops, field days, meetings, and media (for example, brochures, newsletters, and websites) were used by public and private extension systems of Central Asia and the Caucasus (Dosov 2018). Armenian extension mainly used trainings (workshops, field days), consultancies (meetings, discussions), research, introduction of new technologies and best practices, and media such as brochures, newsletters, and websites. Extension methods in Kyrgyzstan included training, lectures, presentations, demonstrations, field visit days, mass media, movies, articles, group activities, practical work, experience exchange, seminars, and round tables (Dosov 2018). In the Democratic Republic of the Congo the main approach to extension was still heavily based on the training-and-visit system, remnants of the World Bank–supported program of the 1980s (see Chapter 8). In Ethiopia several extension methods were often used in combination, including door-to-door extension, farm-to-farm extension, development group meetings, community meetings, field demonstrations, and training at farmer training centers (see Chapter 6).

The breadth and depth of participatory extension methodologies have been increasing (rural promoters, learning groups, local innovation networks, etc.). These provide small producers with greater opportunities for coexecution of extension and inclusion of greater local knowledge. In Argentina, Chile, Colombia, and Uruguay, more participatory strategies and capacity development also are being used (Preissing et al. 2018). In the case of Peru and Ecuador, participatory approaches such as farmer-to-farmer, farmer field schools, and learning-by-doing were used. Participatory methodologies dominated in Brazil (see Chapter 4), where extension workers employed facilitation approaches in connecting farmers to government programs and the financial sector through their endorsement of the farmers for creditworthiness.

In Central America, 55 percent assigned a medium priority and 26 percent a high priority in the use of information and communications technologies in extension strategies (Preissing et al. 2018). In the private sector, 75 percent stated that these technologies should have a greater priority in the future. In Central America, 84 percent of the extensionists noted that beneficiaries had access to the technicians through cell phones. However, only 42 percent stated that their beneficiaries had access to email.

FIGURE 3.2 Agriculture extension service methodologies used in Argentina, Chile, and Colombia (%)

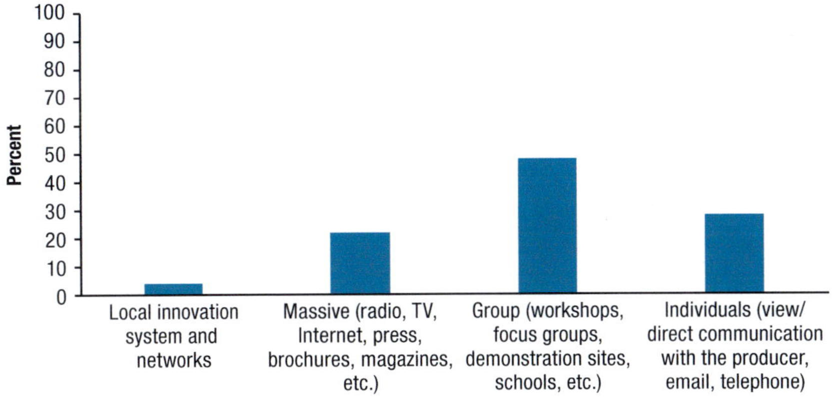

Source: Preissing et al. (2018).

ICT and Digital Technologies

There also is growing recognition of the value of information and communication technologies. The use of cell phones is almost universal even in rural communities. However, there was still very little use of digital technologies by extension workers in many of the countries, beyond pilot projects (Franzel, Ndiaye, and Tata 2018; Preissing et al. 2018). Radio has been and continues to be an important technology to reach farmers with information (Rao 2015; Ragasa et al. 2019; see Chapter 7 on Malawi and Chapter 8 on the Democratic Republic of the Congo). However, extension programs are now using television (Kiptot et al. 2016), video (Bentley, Chowdhury, and David 2015), apps (Payne and Makh 2018), and mobile and smartphones (van Campenhout, Spielman, and Lecoutere 2018) to share information. Programs such as WhatsApp and Facebook Live are used by service providers to reach their clientele.

The use of ICT applications for extension has expanded greatly in India due to the high penetration of mobile phones in rural areas (Babu and Shishodia 2018). Provision of technological solutions, especially from private-sector providers through mobile phones, has been increasing rapidly. Initiatives are transforming traditional agricultural systems by using such methods as community radio, short message services, voice-based messaging, and Internet kiosks. Yet the context specificity of the messages is lacking as the farming systems are becoming highly heterogenous and more and more farmers are going in for diversified farm enterprises and moving away from traditional farming practices. Farmer associations tend to use participatory methods for sharing latest information on specific commodity value chains.

There has been a large increase in the number of telephone users and Internet subscribers since 2008 in Cambodia as well (Ke and Babu 2018). However, lack of capacity and shortage of staff at the national and local levels constrain information sharing among extension workers and farmers. The Department of Extension promoted information and communications technologies for message delivery as well as digitally-enabled rural information centers and community-based farmer radio and television programs. However, there is a need to increase the relevance of the content shared through mobile applications. The typical methods used in Viet Nam include technical training, extension model building, study tours, communication, and providing farmers with updated knowledge and new farming techniques (Ngan and Babu 2018). Farmer associations have played a critical role in sharing knowledge among their members as well.

Another trend in terms of advisory approaches is the use of volunteer farmers, also known as lead farmers, farmer-trainers, volunteer farmers, village agents, or community advisers. This can be seen as a form of private-sector extension where there is sometimes commercialization of the services provided by local farmers and other rural actors. Many lead farmers are selected by communities to work closely with government, private, or nongovernmental extension. Lead farmers are, of course, locally-based and speak local languages; they thus have the trust of fellow villagers. A survey of 80 organizations using lead farmers in Cameroon, Kenya, and Malawi found that the approach was used in part due to its low cost, expansion of reach, and improved accountability to the community (Franzel et al. 2015). Lead farmers may or may not receive salary or incentives in the form of operating cost payments or subsidized or free inputs for demonstrations. Studies have shown that nonmonetary factors such as recognition and altruism play a role in motivating lead farmers (Franzel, Sinja, and Simpson 2014). Such local extension support people need coaching and technical backstopping to be effective, and they need community recognition and support; without these they may perform poorly (Franzel et al. 2015; Behaghel et al. 2018; Ragasa 2019).

A similar model is the "village agent model" started by Sasakawa Global 2000 and promoted by the United States Agency for International Development (USAID) in Uganda. Similar models are used in Tanzania (DLEC 2019) and Rwanda (the latter with the so-called farmer promoters) (MacNairn 2018). Village agents may be linked up to transport companies, agro-input companies, or offtakers, and may provide such services as advice, spraying, or bulking. They may receive a fee for their services or a commission from sales. Particularly in Africa, there has been a focus on reaching and engaging youth in extension and advisory services (defining "youth" broadly, normally someone between 15 and 35 years of age). This is a big switch from the 1989 study. The Food and Agriculture Organization of the United Nations (FAO) reported in 1993 that rural youth was one of the program areas of least importance. Percentage of time and resources spent to reach young farmers or rural youth varied between 2 percent and 16 percent (FAO 1993). In 2012 this varied between 4 percent and 14 percent (GFRAS 2012). Ghana has been experimenting on training and deploying youth extension agents. Latest estimates (April 2019) supplied by Ghana's Ministry of Food and Agriculture (MoFA) suggests there are around 1,700 agricultural extension agents, 950 district (agricultural) development officers, and 5,100 youth extension agents deployed under the Nation Builders Corps (NABCO) and Youth Employment Agency (YEA) programs to address (graduate) youth unemployment.[3] A planned assessment of this approach is under way.

Today there are special programs and projects to engage youth in extension and advisory services. In Guinea a donor-funded program emphasizes human capacity development of youth entrepreneurs and equips them to become private-sector agents after training (DLEC 2019; MacNairn 2017). In Niger youth advisory services organizations provide services mostly to youth (more than 50 percent of the clientele). These services appear free of charge but are actually remunerated through the provision of other services including sales inputs (seeds, fertilizers, etc.) (Djamen 2019). In Guatemala, Rural Development Learning Centers are training centers comprised of organized community members coordinated by rural agricultural promoters. The rural agricultural promoter is a community member who has demonstrated leadership in guiding rural development and serves as a liaison between the community and the extension agents (Chaisson-Cardenas 2019).

3.6 Cross-Cutting Issues

The country and regional case studies reported in this chapter brought out several cross-cutting issues such as gender and youth (and generally social inclusion), nutrition, and climate change. Inclusiveness in extension systems

3 For more information, see NABCO (https://nabco.gov.gh/about/) and YEA (https://www.yea.gov.gh/).

is often seen as a criterion for social impact of extension since the majority of extension services is captured by the large progressive farmers who are on the lookout for new information. Given the limited resources for extension, extension agents in Latin America more often reach out to farmers along the major highways and places easily accessible. Farmers in remote areas are often left out in the process. Furthermore, women farmers are left out as most of the extension workers tend to be men and, due to cultural issues, women are often not connected to innovation systems. The farmers of the lower social castes, black farmers and farmers, in Native communities in Latin America, are more likely to be left out by the standard extension system.

In Latin America programs emphasized social inclusion, rural innovation, markets, and rural territorial development, and moved extension from vertical technical and productivity issues to broader roles. Guatemala had several programs and strategies targeting youth and indigenous communities (Chaisson-Cardenas 2019). There were many successful programs that targeted women farmers (for example, the Ecuadorian Tungurahua Agricultural Strategy, which serves 70 percent of women, and the Zero Hunger program in Nicaragua, which targets women farmers and produced valuable lessons). The territorial development approach used in the region provided an opportunity for extension to be more receptive to local needs and to work with diversified approaches based on agroecological and socioeconomic conditions, ensuring that the service offered is made to measure (Preissing et al. 2018). The Brazilian extension policy attempted to be very inclusive and had action lines on gender, ethnic groups, and youth (see Chapter 4). However, farmers in several of the surveyed territories saw this indicator in the study as having the lowest performance among the different indicators of extension.

In the Democratic Republic of the Congo, the case study showed little participation of women in extension (5 percent to 9 percent), especially in leadership positions (see Chapter 8). Leading constraints to women's participation were social norms and mobility. However, female extension workers serve a higher proportion of female farmers than do male agents. Also related to gender of the extension staff is the issue of age, whereby a sizable portion of Congolese agents, especially from government, were near the retirement age (this was also the case in many African countries such as Guinea) (MacNairn 2017).

In India, district-level decentralized farm science centers known as Krishi Vigyan Kendras were shown to be effective at increasing income and self-confidence among women through gender-sensitive programs (Babu and

Shishodia 2018). However, there is considerable room for improvement in giving greater emphasis on adaptation to climate change to increase yields sustainably. Nutrition as a subject matter is not part of the extension system in India. Although nutrition was taught as part of the undergraduate curriculum two decades ago, slowly it was removed as separate colleges of home sciences were created as part of the agricultural universities. Yet the graduates of the home science colleges are not employed as extension professionals. There is room for collaboration at the community level between village nutrition assistants of the integrated nutrition programs and agricultural extension workers. This coordination will have to take a back seat, until the coordination among the institutions of the agricultural innovation system is improved.

In Cambodia women had low access to land, extension, financial services, markets, and technology (Ke and Babu 2018). Recently, climate change has become a main issue in Cambodia due to increasing droughts, floods, and changes in monsoon seasons. In Viet Nam women generally had low access to extension, production inputs, and capital compared to that of men (Ngan and Babu 2018). Young agricultural extension staff account for a large proportion of the total in Viet Nam. Since 2016, the National Center for Extension has organized many training courses to encourage young farmers to come into agriculture.

Special topics such as climate change or nutrition are generally only covered in projects rather than in mainstreamed extension services (for example, in Guinea [MacNairn 2017] and Liberia [Sigman 2017]). In Liberia the Ministry of Agriculture had a capacity development plan for Climate Change Management in Agriculture developed through a climate change adaptation project that trained public extension staff (Sigman 2017). Climate change adaptation is a subtopic (with limited allocated funding) of the Comprehensive Africa Agriculture Development Programme (CAADP) and the National Agricultural Investment Plan (NAIP) in the Democratic Republic of the Congo. Extension agents were not generally trained on the topic of nutrition (see Chapter 8).

Climate change is projected to have a negative impact on the Central Asia and Caucasus region, where countries are at different stages in terms of defining and implementing their climate change strategies. Currently, there is no regional intergovernmental body with an explicit mandate to support regional cooperation on climate change across a broad range of sectors (Dosov 2018). Although nutrition is a topic in many high-level meetings and documents, it was not seriously addressed by extension staff and their programs.

3.7 Conclusion

There are a wide number of studies looking at the status of extension and advisory services mainly since 2015. Although these studies had different purposes and methodologies, there is still useful information with regard to the best-fit characteristics of extension: governance structures, capacity, management, and advisory methods used. Cross-cutting issues such as gender, youth, nutrition, and climate change were also examined. The majority of countries have pluralistic extension systems with a large number of public, private, and civil society providers (see Table 3.1). The public sector, usually the ministries of agriculture, is responsible for the overall coordination and regulation of extension. Many countries are starting to decentralize the services to lower levels of governance.

A number of countries do not have a specific policy for extension and advisory services; these are usually rooted in broader agricultural sector development policies. Some countries have a policy that is outdated or not well implemented. There are a few examples of countries with updated, well-financed, and well-implemented extension policies. Coordination between players and with other actors in the agricultural innovation system remains a challenge in most countries, although we see some mechanisms to deal with this in Malawi and Brazil. Financing of extension services is quite diverse in Latin America; in Africa and Central Asia and the Caucasus, it is very donor-dependent and public sector–led (while Rwanda and Ethiopia do use donor funds, their extension programs are driven by their extension and agriculture strategies rather than by donor priorities).

Numbers of extension agents from the public sector are seldom sufficient for the job at hand, and there are high vacancy rates and turnover in some countries. The foundational training for extension agents is usually focused on technical topics and may miss out on functional skills that are also needed for extension. One exception is several Central and South American countries that appear to have well-established foundational and continuing education systems for extension providers. A few countries provide incentives for extension agents. In most cases extension agents are demotivated and lack basic transportation and equipment. Salaries are normally low, especially in the public sector, and there are few rewards or recognition. Lead farmers and rural promoters are often used, but they also struggle with community recognition and support as well as support from the public extension system.

Methods used for advising are quite diverse. Most of the countries tend to use traditional methods such as individual and group communication approaches; however, there is a trend toward more participatory approaches, such as farmer field schools, and toward the use of more ICTs beyond radio and television. Most countries use some type of multiplier approach such as farmer extensionists or rural promoters. A number of methods exist to strengthen social inclusion. While there is increasing emphasis on reaching women, youth, or other disadvantaged populations, this is largely in certain projects and not necessarily mainstreamed, nor is much known on the effectiveness of such approaches. Focus on climate change and climate-smart approaches are rapidly growing and being taken up in many programs and promoted with a number of methods. Incorporation of nutrition issues in the extension content is only in the discussion stage in several countries.

3.8 Policy Recommendations

Part 1 of this book has given indications of the difficulties but also the necessity of collecting indicators on extension and advisory services. Although this book uses the best-fit framework to guide analysis of extension services, the framework falls short in establishing a common set of indicators against which to measure outcomes and impacts. Since the 2012 GFRAS survey of extension systems in the developing countries, there have been several initiatives to continue to collect global (and internationally comparable) data and to establish common indicators for extension. The World Bank, USAID, IFPRI, and FAO have been discussing the issue since the early 2000s. Around 2016, the Bill and Melinda Gates Foundation initiated work on a number of "dashboards," sets of indicators that could be compared across countries. On the extension topic, this was done with regard to budget distribution, numbers of extension workers, relevance of content, reach, adoption rate, and other, more system-level characteristics. The extension dashboard was initially trialed in three places: Tanzania, Ethiopia, and Bihar in India. This work is still in development and has struggled with stakeholder buy-in (DLEC 2018).

At the same time, the Global Forum for Rural Advisory Services, together with its regional networks and national platforms, proposed the establishment of "extension observatories" as part of a new funding proposal (IFAD 2018). The observatories are meant to continuously collect and make information on extension indicators accessible through national platforms, which will contribute to the information base of the African Observatory of Science, Technology and Innovation that is being developed by the African Union to champion evidence-based science, technology, and innovation policy-making by backstopping African countries to manage and use statistical information

(AOSTI 2013). These observatories seemingly echo the Agricultural Science and Technology Indicators (ASTI) initiative, which provides open-source data on agricultural research systems across the developing world. ASTI, however, has likely gained more traction because research data are easier to obtain than extension data due in part to the existence of agreed-upon definitions and conventions on what constitutes research and how to measure it—the Frascati Manual, which is based on work dating back to 1963 (OECD 2015). This further reinforces the need for agreed-upon definitions and conventions for extension and advisory services (DLEC 2018).

The organizations mentioned above that are grappling with the issue of collecting extension data see value in having an established set of metrics to consistently assess the performance of extension systems across many contexts. This can help to share and adapt learning across multiple settings and guide policymakers, funders, and extension implementers regarding what innovations and activities to invest in to maximize impact (DLEC 2018). The earlier efforts (FAO 1991; GFRAS 2012) to collect extension metrics focused on what might be described as the "input" side of the equation. These indicators included, for example, the number of extension agents in service, public expenditures allocated to extension, farmer-to-extension agent ratios, and number and type of extension providers across the public, private, and civil society sectors. Input metrics are useful in determining where to allocate scarce public resources within the broad area of agriculture and rural development but tend to reveal only part of the story and are not as useful in determining how to use these resources within existing or planned programs. First, and by definition, they are generally unable to capture the short- or long-term outcomes associated with extension: improvements in productivity, sustainability, income, or social equity. Second, they do not reflect system-level performance, such as the performance of structures, processes, and relationships between and among extension providers and other actors in the agricultural sector. Thus higher order metrics are needed to supplement the input metrics and help extension stakeholders design for and measure across a range of outcomes (DLEC 2018).

Any common set of extension metrics will still run into the challenge that data are hard to come by given that, among other reasons, many countries do not explicitly budget extension services as a separate or distinct line item so the lines of what defines extension is often blurry and different across contexts (DLEC 2018). In addition, the combination of pluralistic actors providing extension, multiple channels being used for information sharing, and the decentralization of advisory services in many countries provide challenges in identifying from whom to collect extension metrics, and how to identify their financial, technical, and human resource investments in extension in a comparable manner. Where extension and advisory services are associated with proprietary information and supply-chain contracts or commercial technologies, data are often even more difficult to obtain.

As governments and development partners recognize the importance of farmer-based learning processes, last-mile delivery, and other elements that are critical to advancing rural advisory services, there is a recognized need to invest more strategically in extension. This requires good data and analysis to adequately understand the most effective extension strategies. The data need to serve the learning needs of researchers, policymakers, funders, and extension providers on how to design and implement an advisory service system that draws from global successes while taking into consideration contextual specificities. However, country- and project-specific metrics and data as they are currently designed do not necessarily enable a global conversation on what national extension systems are getting right and what other countries can learn from them (DLEC 2018).

In 2018 a set of organizations again took up the challenge of "common extension metrics" to propose a common set of indicators that not only focus on the "input" side of extension, but system, delivery, outputs, outcomes, and even impact domains (DLEC 2018). Spearheaded by the Developing Local Extension Capacity (DLEC) project, they formed a task force consisting of members from the World Bank, USAID, and IFPRI, and started work on common definitions and indicators for extension. The task force started from the Bill and Melinda Gates Foundation initial dashboard but also used a literature review to identify other sources of extension metrics across the various domains.[4] While still a work in progress, indicators that were identified as promising by the task force are shared in Table 3.2.

The table relates the indicators to the dimensions from the best-fit framework (see Figure 1.2 in Chapter 1). There are several caveats. Many of these indicators are aspirational and do not currently exist. Indicator numbers 1–19 are descriptive data; indicators 20–25 should be evaluated rigorously as to how extension services contribute to these outcome indicators. Despite being a work-in-progress, based on the continued *(text continues on page 91)*

4 For Ethiopia specifically the Gates Foundation developed 10 indicators: (1) share of public budget on extension; (2) extension budget per household; (3) financial incentive of an extension agent; (4) continuous improvement processes; (5) households reached by public extension; (6) households reached by all extension; (7) ICT as a main source of extension; (8) quality of public extension; (9) services to female farmers; and (10) inclusivity.

TABLE 3.2 Potential indicators, data sources, availability by best-fit dimension

Indicator number	Best-fit dimension	Indicator name	Data source (year)	Availability
1	Governance—financing	**Share of public budget on extension.** Total government (public) extension budget divided by total agriculture budget (%)	Monitoring and Analyzing Food and Agricultural Policies; country governments; national statistics offices	Available data
2	Governance—financing	**Government budget per household.** Total government (public) extension budget per 1,000 rural farm households (purchasing power parity dollar per farming household per year)	Monitoring and Analyzing Food and Agricultural Policies; country governments; national statistics offices; World Bank, World Development Indicators	Available data
3	Governance—financing	**Cost of extension.** Total government (public) extension expenditure divided by per unit value of production	Country governments; FAOSTAT	Aspirational
4	Frame conditions	**Systems index.** Country score on extension policy, linkages, pluralism, and digital management	Country governments	Aspirational
5	Performance	**Relevance/farmer demand index.** Country score on whether extension system incorporates farmer voice and is relevant to their needs	Country governments	Aspirational
6	Capacity	**Farming households reached by all extension.** Share of households receiving extension services (public or otherwise) (%)	Living Standards in Measurement Study—Integrated Surveys on Agriculture for some countries	Available data
7	Capacity	**Farming households directly reached by public extension.** Share of households receiving advice from a public extension agent (%)	Living Standards in Measurement Study—Integrated Surveys on Agriculture for some countries	Available data

Indicator number	Best-fit dimension	Indicator name	Data source (year)	Availability
8	Advisory methods	**Households reached by information and communications technologies.** Share of households identifying information and communications technologies as main source of extension (%)	Living Standards in Measurement Study—Integrated Surveys on Agriculture for some countries	Available data
9	Capacity	**Farming households per public extension agents.** Number of farming households per public extension agent	World Bank, World Development Indicators; GFRAS Worldwide extension survey (GFRAS 2012)	Available data
10	Capacity	**Age of public extension agents.** Average age of public extension agents	Country governments	Aspirational
11	Management	**Financial incentive of an extension agent.** Ratio of an extension agent's daily wage rate to a rural teacher's daily wage rate	Country governments	Aspirational
12	Capacity	**Continuous improvement process.** Proportion of extension personnel annually trained in specialized courses	Country governments	Aspirational
13	Performance	**Service to female farmers.** Share of females in smallholder farming households reporting receiving extension services (%)	Living Standards in Measurement Study Integrated Surveys on Agriculture for some countries	Available data
14	Performance	**Frequency of visits.** Number of times public extension agents visit the plot in the past 12 months	Living Standards in Measurement Study—Integrated Surveys on Agriculture	Available data
15	Performance	**Inclusivity.** Index (0–1) of inclusivity of extension services relative to a defined disenfranchised group	Country governments for some countries	Aspirational

(continued)

TABLE 3.2 Continued

Indicator number	Best-fit dimension	Indicator name	Data source (year)	Availability
16	Performance	**Farming households that receive marketing information.** Share of households receiving advice from public extension on marketing (%)	Living Standards in Measurement Study—Integrated Surveys on Agriculture for some countries	Available data
17	Performance	**Farming households that receive livestock information.** Share of households with livestock receiving advice from public extension on livestock (%)	Living Standards in Measurement Study—Integrated Surveys on Agriculture for some countries	Available data
18	Performance	**Quality of public extension.** Share of households reporting public extension services as at least "somewhat useful" (%)	Living Standards in Measurement Study—Integrated Surveys on Agriculture for some countries	Available data
19	Farm households	**Adopters of practices and technologies.** Share of total number of adopters divided by number of households receiving extension advice from public services	Country governments	Aspirational
20	Farm households	**Adoptions of practices and technologies.** Share of total number of adoptions divided by total number of adopters	Country governments	Available household survey panel (secondary or primary), analyzed using rigorous impact evaluation methods
21	Impact	**Women's empowerment.** Women's Empowerment in Agriculture Index	Women's Empowerment in Agriculture Index (IFPRI 2012)	Available household survey panel (secondary or primary), analyzed using rigorous impact evaluation methods
22	Impact	**Yield.** Average agricultural production per hectare of land	Living Standards in Measurement Study—Integrated Surveys on Agriculture for some countries	Available household survey panel (secondary or primary), analyzed using rigorous impact evaluation methods

Indicator number	Best-fit dimension	Indicator name	Data source (year)	Availability
23	Impact	**Income (net profit).** Average income (that is, net profit) from farming activities per hectare of land	Living Standards in Measurement Study—Integrated Surveys on Agriculture for some countries	Available household survey panel (secondary or primary), analyzed using rigorous impact evaluation methods
24	Impact	**Poverty.** Percentage of the rural population living below the national poverty lines	World Bank, Global Poverty Working Group	Available household survey panel (secondary or primary), analyzed using rigorous impact evaluation methods
25	Impact	**Environmental sustain ability.** Emissions from agriculture	FAOSTAT, http://www.fao.org/faostat	Available household survey panel (secondary or primary), analyzed using rigorous impact evaluation methods

Source: Authors' compilation, adapted from ALINE (2019).

importance of extension and advisory services, it will be valuable for stakeholders—funders, implementers, and evaluators of the services—to continue thinking about the best measures for extension.

References

ALINE (Agricultural Learning and Impacts Network). 2019. "Improving Agricultural Advisory Systems Metrics and Learning." Meeting summary, April 8–9, 2019, Addis Ababa, Ethiopia.

AOSTI (African Observatory of Science, Technology, and Innovation). 2013. *Science, Technology and Innovation Policy-Making in Africa: An Assessment of Capacity Needs and Priorities.* AOSTI Working Papers 2. Malabo, Equatorial Guinea: AOSTI.

Babu, S. C., and M. Shishodia. 2018. "India." Unpublished report. International Food Policy Research Institute (IFPRI), Washington, DC.

Behaghel, L., J. Gignoux, M. K. Beraho, J. Kugonza, K. Macours, M. N. Mangheni, and J. Oduol. 2018. *Dissemination of New Agricultural Technologies in Africa: Making Extension Work: Impact Evaluation of a Farmer Trainer Program in Uganda.* Final project report. New Delhi; Cambridge, MA, US; Paris: 3ie/J-PAL/Paris School of Economics.

Bentley, J., A. Chowdhury, and S. David. 2015. *Videos for Agricultural Extension.* Note 6. GFRAS Good Practice Notes for Extension and Advisory Services. Lindau, Switzerland: Global Forum for Rural Advisory Services (GFRAS).

Burrows, E., M. Bell, and N. G. Rutamu. 2017. *Extension and Advisory Services in Zambia: Understanding Structures, Services, Roles and Incentives for Reaching Farmer Households As a Basis for Discussing Potential for Scale.* Integrating Gender and Nutrition within Agricultural Extension Services (INGENAES). Urbana-Champaign, IL, US: University of Illinois.

Cai, T. 2017. *Malawi: Desk Study of Extension and Advisory Services.* Washington, DC: Developing Local Extension Capacity (DLEC) Project.

Chaisson-Cardenas, J. P. 2019. *Youth in Extension and Advisory Services: Guatemala.* Washington, DC: DLEC Project.

Cuangara, B., and T. Thompson. 2018. *Mozambique: Desk Study of Extension and Advisory Services.* Washington, DC: DLEC Project.

Djamen, P. N. 2019. "Youth in Agricultural Extension in Niger." Unpublished report. Washington, DC, DLEC Project.

DLEC (Developing Local Extension Capacity Project). 2018. *Mali: In-Depth Assessment of Extension and Advisory Services.* Washington, DC: DLEC Project.

———. 2019. *Strengthening Private Sector Extension and Advisory Services—Portfolio Review.* Washington, DC: DLEC Project.

Dosov, B. 2018. "Central Asia and the Caucasus." Unpublished report, Washington, DC, IFPRI.

FAO (Food and Agriculture Organization of the United Nations). 1991. *International Directory of Agriculture Extension Organizations.* Rome: FAO.

———. 1993. *Agricultural Extension and Farm Women in the 1980s.* Rome: FAO.

———. 2010. FAOSTAT database. Accessed October 9, 2018. http://www.fao.org/faostat.

Franzel, S., A. Degrande, E. Kiptot, J. Kirui, J. Kugonza, J. Preissing, and B. Simpson. 2015. *Farmer-to-Farmer Extension.* Note 7. GFRAS Good Practice Notes for Extension and Advisory Services. GFRAS: Lindau, Switzerland.

Franzel, S., A. Ndiaye, and J. S. Tata. 2018. *Senegal: In-Depth Assessment of Extension and Advisory Services.* DLEC Project. Washington, DC: USAID.

Franzel, S., J. Sinja, and B. Simpson. 2014. *Farmer-to-Farmer Extension in Kenya: The Perspectives of Organizations Using the Approach.* ICRAF Working Paper 181. Nairobi: World Agroforestry Centre. DOI: http://dx.doi.org/10.5716/WP14380.PDF.

GFRAS (Global Forum for Rural Advisory Services). 2012. "Master Worldwide Extension Study Database." Excel File, GFRAS, Lindau, Switzerland. https://www.ifpri.org/publication/worldwide-extension-study.

Goertz, H. 2016. *Haiti Landscape Analysis*. Integrating Gender and Nutrition within Agricultural Extension Services (INGENAES). Urbana-Champaign, IL, US: University of Illinois.

Huber, S., and K. Davis. 2017. *Bangladesh: Desk Study of Extension and Advisory Services*. Washington, DC: DLEC Project.

Huber, S., K. Davis, and K. Lion. 2017. *Nigeria: In-Depth Assessment of Extension and Advisory Services*. Washington, DC: DLEC Project.

IFAD (International Fund for Agricultural Development). 2018. *Delivering Extension Services to the Last-Mile: Improving Smallholders' Access to Innovation and Pluralistic, Demand-Driven Extension Services*. Grant Design Document. Rome: IFAD.

IFPRI. 2012. *Women's Empowerment in Agriculture Index*. Washington, DC: IFPRI.

INGENAES. 2015. *Senegal Landscape Analysis*. Urbana-Champaign, IL, US: University of Illinois.

———. 2016a. *Rwanda Landscape Analysis*. Urbana-Champaign, IL, US: University of Illinois.

———. 2016b. *Tanzania Landscape Analysis*. Urbana-Champaign, IL, US: University of Illinois.

———. 2016c. *Uganda Landscape Analysis*. Urbana-Champaign, IL, US: University of Illinois.

———. 2016d. *Mali Landscape Analysis*. Urbana-Champaign, IL, US: University of Illinois.

———. 2016e. *Malawi Landscape Analysis*. Urbana-Champaign, IL, US: University of Illinois.

———. 2016f. *Liberia Landscape Analysis*. Urbana-Champaign, IL, US: University of Illinois.

———. 2016g. *Kenya Landscape Analysis*. Urbana-Champaign, IL, US: University of Illinois.

———. 2016h. *Guatemala Landscape Analysis*. Urbana-Champaign, IL, US: University of Illinois.

———. 2016i. *Ghana Landscape Analysis*. Urbana-Champaign, IL, US: University of Illinois.

———. 2016j. *Ethiopia Landscape Analysis*. Urbana-Champaign, IL, US: University of Illinois.

———. 2016k. *Cambodia Landscape Analysis*. Urbana-Champaign, IL, US: University of Illinois.

———. 2016l. *Mozambique Landscape Analysis*. Urbana-Champaign, IL, US: University of Illinois.

———. 2017. *Sierra Leone Landscape Analysis*. Urbana-Champaign, IL, US: University of Illinois.

———. 2018. *Tajikistan Landscape Analysis*. Urbana-Champaign, IL, US: University of Illinois.

Ke, S. O., and S. C. Babu. 2018. "Cambodia." Unpublished report. Washington, DC, IFPRI.

Kiptot, E., S. Franzel, C. Nora, and A.-M. Steyn. 2016. *Edutainment TV for Disseminating Information about Agriculture*. Note 22. GFRAS Good Practice Notes for Extension and Advisory Services. GFRAS: Lausanne, Switzerland.

Kyle, J., and D. Resnick. 2019. "Delivering More with Less: Subnational Service Provision in Low Capacity States." *Studies in Comparative International Development* 54 (1): 133–163.

Ludgate, N., and S. J. Tata. 2015. *Burundi Landscape Analysis*. INGENAES. Urbana-Champaign, IL, US: University of Illinois.

MacNairn, I. 2017. *Guinea: Desk Study of Extension and Advisory Services*. Washington, DC: DLEC Project.

———. 2018. *Rwanda: Desk Study of Extension and Advisory Services*. Washington, DC: DLEC Project.

Malawi, MoAIWD (Ministry of Agriculture, Irrigation, and Water Development). 2004. *Agricultural Extension Policy Implementation Guide*. Lilongwe.

McNamara, P., B. E. Swanson, and B. M. Simpson. 2011. *Rebuilding and Strengthening the Pluralistic Extension System in Liberia*. Modernizing Extension and Advisory Services (MEAS) Project. Urbana-Champaign, IL, US: University of Illinois.

Mwakiwa, E. 2017. *Zimbabwe Agricultural Extension Issues Paper*. Harare, Zimbabwe: Ministry of Agriculture, Mechanization and Irrigation Development, and the United Nations Development Program.

Ngan, H. P., and S. C. Babu. 2018. "Viet Nam." Unpublished report, IFPRI, Washington, DC.

OECD (Organisation for Economic Co-Operation and Development). 2015. *Frascati Manual 2015: Guidelines for Collecting and Reporting Data on Research and Experimental Development: The Measurement of Scientific, Technological and Innovation Activities*. Paris: OECD. https://doi.org/10.1787/9789264239012-en.

Parkinson, V. 2017. Personal communication via email. November 9.

Payne, J., and S. Makh. 2018. *Scaling Up Digital Technologies for Extension and Advisory Services Part 1: Digital Extension Typology (Blog)*. https://www.agrilinks.org/post/digital-extension-typology.

Preissing, J., S. Ardila, F. Aguirre, and J. Buitrón. 2018. "Latin America." Unpublished report, IFPRI, Washington, DC.

Ragasa, C. 2019. *Modeling the Effectiveness of the Lead Farmer Approach in Agricultural Extension Service Provision: Nationally Representative Panel Data Analysis in Malawi*. IFPRI Discussion Paper 1848. Washington, DC: IFPRI.

Ragasa, C., T. Badibanga, and J. Ulimwengu. 2016. "Effectiveness and Challenges of Participatory Governance: The Case of Agricultural and Rural Management Councils in the Western Democratic Republic of the Congo." *Food Security* 8: 827–854.

Ragasa, C., I. Lambrecht, and D. Kufoalor. 2018. "Limitations of Contract Farming as a Pro-poor Strategy: The Case of Maize Outgrower Schemes in Upper West Ghana." *World Development* 102: 30–56.

Ragasa, C., C. Mthinda, C. Chowa, D. Mzungu, K. Kalagho, and C. Kazembe. 2019. "Assessing and Strengthening Malawi's Pluralistic Agricultural Extension System: Evidence and Lessons from a Three-Year Research Study." Project Note, November 13. IFPRI, Washington, DC.

Rahman, M. W., M. S. Islam, L. Hassan, N. Z. Tanny, L. Parvin, and A. Bohn. 2017. *Bangladesh Landscape Analysis.* INGENAES. Urbana-Champaign, IL, US: University of Illinois.

Rao, S. 2015. *Using Radio in Agricultural Extension.* Note 18. GFRAS Good Practice Notes for Extension and Advisory Services. GFRAS: Lindau, Switzerland.

Resnick, D. 2018. *The Devolution Revolution: Implications for Agricultural Service Delivery in Ghana.* IFPRI Discussion Paper 1714. Washington, DC: IFPRI.

Ruml, A., C. Ragasa, and M. Qaim. 2019. "Heterogeneous Effects of Marketing Contracts and Resource-Providing Contracts on Household Income." Global Food paper series draft, University of Goettingen, Germany.

Sigman, V. 2017. *Liberia: Desk Study of Extension and Advisory Services.* Washington, DC: DLEC Project.

Simpson, B. E., G. Heinrich, and G. Malindi. 2012. *Strengthening Pluralistic Agricultural Extension in Malawi.* Modernizing Extension and Advisory Services (MEAS) Project. Urbana-Champaign, IL, US: University of Illinois.

Suvedi, M., and P. McNamara. 2012. *Strengthening the Pluralistic Agricultural Extension System in Nepal.* Modernizing Extension and Advisory Services (MEAS) Project. Urbana-Champaign, IL, US: University of Illinois.

Swanson, B. E. 2011. *Assessment of Bangladesh's Pluralistic Extension System.* Modernizing Extension and Advisory Services (MEAS) Project. Urbana-Champaign, IL, US: University of Illinois.

Valenzuela, M. A. B., and D. Saavedra. 2017. *In-Depth Assessment of Extension and Advisory Services.* Washington, DC: DLEC Project.

Van Campenhout, B., D. J. Spielman, and E. Lecoutere. 2018. *Information and Communication Technologies (ICTs) to Provide Agricultural Advice to Smallholder Farmers: Experimental Evidence from Uganda.* IFPRI Discussion Paper 1778. Washington, DC: IFPRI. http://ebrary.ifpri.org/cdm/singleitem/collection/p15738coll2/id/133022.

Williams, R. 2016. *Honduras Landscape Analysis.* INGENAES. Urbana-Champaign, IL, US: University of Illinois.

Zhou, Y., and S. C. Babu. 2015. *Knowledge Driven Development.* New York: Elsevier.

PART 2

Performance of Extension Systems

Complementing and extending the global landscaping and comparison of national extension systems in Part 1 of this book, Part 2 provides country cases where in-depth assessments of national extension system performance were undertaken. In the best-fit framework (see Figure 1.2) performance measures include content, targeting, feedback, timeliness, relevance, effectiveness, efficiency, and sustainability (Box I). However, these terminologies were vaguely defined, and no indicators were included. This makes operationalization of the framework quite challenging for researchers, and as a result these performance indicators were inconsistently applied in various assessments. In Part 2, five country cases provide an in-depth examination of the framing conditions of the countries; the governance, structure, and management of the national extension systems (and subnational systems in some cases), including human and financial resources; and the delivery mechanisms and approaches (Boxes A–H). The country cases extend the assessment to include performance and impact indicators.

First, country cases explored in these chapters extend the assessment of performance at the systems level, including content, targeting, feedback, timeliness, relevance, effectiveness, efficiency, and sustainability (Box I), with some country cases including more comprehensive measures of performance than others. Second, the country cases extend the assessment of performance to the level of target households and individuals, including access to extension services and some indicators of the quality and relevance of extension services (all country cases) (Box J). All country cases, except Brazil, include indicators of changes in behavior and practices manifested in technology-adoption patterns (Box J). Third, some country cases extend the assessments to changes in yields or productivity and food security (Malawi, the Democratic Republic of the Congo, and Uganda) and agricultural transformation and economic development (in the case of Ethiopia) (Box K). Some country cases cover the whole country with available national representative datasets (Malawi, Uganda, and Ethiopia); while others cover selected zones or regions (Brazil and the Democratic Republic of the Congo), and therefore care and caution are made in the analysis and interpretation of the results.

Due to the diversity of research teams and available data, different performance and impact indicators were used in the different country cases. The concluding chapter (Chapter 9) summarizes these indicators with some limitations and suggestions for future assessments.

Chapter 4

BRAZIL

Sonia Maria Pessoa Pereira Bergamasco, Ricardo Serra Borsatto, Carolina Rios Thomson, Kristin Davis, and Suresh C. Babu

4.1 Introduction

Over the past two decades, many developing countries reformed their rural extension services in response to emerging issues and challenges (Benson and Jafry 2013; Chowdury, Odame, and Leeuwis 2014; Qamar 2005; Rivera and Alex 2005). Some of these emerging issues and challenges include "globalization, market liberalization, privatization, pluralism, decentralization and devolution, client participation in decision-making, natural and man-made disasters, rural poverty, food insecurity, HIV/AIDS epidemic, and emphasis on integrated, multidisciplinary, holistic and sustainable development" (Qamar 2005: viii). In response to these issues and challenges, the modernization of rural extension services included broadening extension's technical mandate, enacting a national extension policy, professionalization of extension services, use of information and communications technologies, and participatory and agroecological approaches (Qamar 2005).

Reviving Latin American extension systems has been back on the policy agenda the past decade or so (Klerkx, Landini, and Santoyo-Cortés 2016), indicating a reversal of decades of neglect following the neoliberal wave of fiscal and economic crises that led to the removal of public funding for extension services in many countries in the region. This renewed attention in the region has led to several reforms of their extension systems. The case of Brazil is typical. Yet despite much investment in Brazil's National Policy for Technical Assistance and Rural Extension (PNATER), there is still a lack of research assessing the impacts of the extension policy reforms from the perspective of smallholder farmers and extensionists. This chapter presents results of a study that aims to examine the perceived impacts of the Brazilian federal government's extension services from the perspective of the smallholder farmers and the extensionists in reaching the policy goals and objectives.

An interdisciplinary approach drawing upon institutional economics, neoinstitutional theory, and public policy-making was employed for this study. Using an indicator system based on essential elements of the extension policy guiding the services, farmers, extension technicians, and extension managers were interviewed to obtain their perspectives on the functioning and the effectiveness of extension services. Data were collected in five territories in three states in Brazil during 2014 and 2015. The states and territories were purposively selected to represent the reality of the family farming system in Brazil.

The findings show that the extension services have been relatively successful in terms of reinforcing the role of government policies regarding food sovereignty and food security issues, spreading agroecological principles, making use of innovative pedagogical approaches by extensionists, encouraging social and community mobilization in individual states, and improving farmer income. However, the extension services were less effective in terms of addressing issues of gender, age, and ethnic inequality, which were rated low in all states by both the smallholder farmers and the extensionists. Furthermore, the effectiveness of the extension services to deal with environmental issues and access to natural resources and the frequency of services were also rated low by the respondents.

4.2 Data Sources and Methodology

Five territories were selected in three Brazilian states to evaluate the perceived impacts of the agricultural extension policy. The study areas were part of the Brazilian Territory Citizenship Program created in 2008 by the federal government. This program identified the poorest and least developed municipalities in different administrative regions as part of establishing such territories as specific targets of interventions for human development. The final choice of territories was made in consultation with the representatives from the Ministry of Agrarian Development (MDA). The study territories were selected due to high concentration of family farmers, rural settlements of black farmer communities (*quilombos*), and indigenous populations. The following territories were included in the study: Alto Jequitinhonha (Minas Gerais state), Cantuquiriguaçu (Paraná state), Pontal do Paranapanema (São Paulo state), São Paulo's Southwestern (São Paulo state), and Vale do Ribeira (Paraná state) (Figure 4.1).

Data-collection instruments included questionnaires focusing on objective questions, allowing answers to identify the interviewee's perception of his

FIGURE 4.1 Geographic locations of the study territories in Brazil, 2017

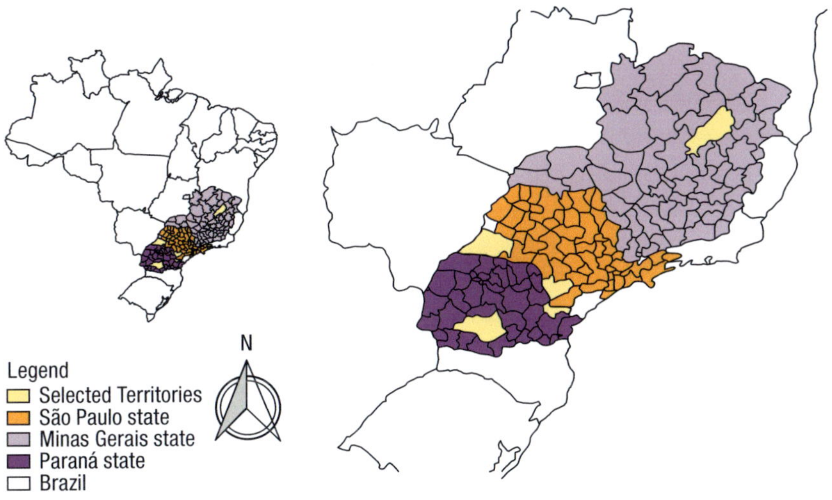

Source: Authors.

or her reality. Responses were based on a five-point Likert scale—from the least to the greatest—asking respondents to indicate how much they agreed or disagreed, approved or disapproved, or believed to be true or false (Albaum 1997; Allen and Seaman 2007). For instance, one question was "Do the extension technicians conduct specific actions related to youth and the elderly?" Respondents could answer "No, never; Rarely; Regularly; Sometimes; Always." In addition to closed questions, the questionnaires for farmers and extensionists also had some open-ended questions, aiming to add a qualitative perspective to the interview and increasing the explanatory capacity of the phenomena observed in the quantitative information.

Two different data-collection instruments were used; the first questionnaire collected information from family farmers; and a second questionnaire elicited information from field technicians implementing extension services in the territory. The family farmers' questionnaire contained 56 questions, encompassing different indicators, among which three were specific for black rural and indigenous communities. The questionnaire developed for field technicians consisted of 78 questions. It took about 40 minutes to 50 minutes for each interview. Questionnaires were pilot-tested before starting the data collection. Following the preliminary tests, some questions were altered for various reasons (difficulty in understanding terms, ambiguities, insufficient options for answers, grammatical errors, and others). After finishing the interviews, all the scores from a given question from all the interviewees within a territory were summed. The resulting score was used to evaluate extension in each territory.

An indicator system was developed to collect and analyze data from farmers and extension agents in each territory. The indicators represent a measure with substantial social significance. This measure, with quantitative and qualitative values, is used to replace or operationalize an abstract social concept, usually with theoretical (for academic research) or programmatic (to formulate, analyze, and evaluate policies) interest (Januzzi 2009). Defining an indicator, or a set of them, is a cognitive, abstract task that aims to identify the essential characteristics of a given reality and express it in quantitative or qualitative values. Therefore, it is a reductionist process, which means no indicator system, even the more sophisticated and complex ones, will be able to express reality as it is. Some elements of this reality will always be lost; but this loss was compensated by the elaboration of an indicator system capable of showing the determinant elements of social, economic, cultural, and political configurations of this reality, contributing with subsidies to desired changes (Jannuzzi 2009; Bellen 2010).

Eleven indicators were selected to represent abstract social constructs reflecting key aspects of PNATER values, principles, and objectives. (Table 4.1). The indicators were also validated at a meeting organized by the Ministry of Agrarian Development. Regular consultations with public managers were held to ensure that the results of this research will have both academic and policy-making relevance. Although all the indicators were relevant to evaluate the extension services, there were different levels of importance defined by the researchers and the MDA staff. The criteria used in this study followed the same logic of importance given by the Municipal Human Development Index, developed by the United Nations Development Program for Brazil (Índice de Desenvolvimento Humano Municipal Brasileiro 2013) for the dimension "Access to Knowledge." As this research seeks to evaluate a recent policy, different weights (multiplier factors) were given according to the importance of the indicator to achieve PNATER's most essential principles (higher), ranging from 1 (relevant), 2 (somewhat relevant), or 3 (highly relevant).

The goal for the sample survey was to interview at least 200 smallholder family farmers and 20 agricultural extension technicians in each territory. Farmers were purposively selected to ensure a diversity of representatives of family farmers, including indigenous, black rural communities, and farmers

TABLE 4.1 Weights and descriptions for the different indicators

Indicator	Weights	Brief description
Extension frequency	3	Expresses the frequency that farmers get services from extension agents
Social and community organization	3	Assesses the dimension related to the farmers' organizational aspects (associations, cooperatives, trade unions, informal groups)
Income	2	Determines the economic impact of extension services in the rural community (monetary and nonmonetary earnings, origins of earned income)
Life quality	2	Includes material and nonmaterial conditions, such as access to essential material means, physical and psychological well-being, social relationships, housing conditions, recreational access, education, freedom of expression and organization, among others
Food sovereignty and food security	3	Assesses the access to adequate nutrient-rich food, in sufficient quantity, without compromising access to other basic needs, based on healthy food practices, respecting cultural diversity and being socially, economically, and environmentally sustainable
Environmental issues	3	Evaluates the actions related to promoting effective changes to ensure life continuity and life quality for a long-term period, to manage and ensure vital and finite resources in a social system
Gender, age, and ethnicity	2	Assesses if the extension is developing specific actions that consider these dimensions in their scope and if other general activities consider the specificities of these social groups
Pedagogical conception	3	Evaluates if the extension activities are following the PNATER guideline that suggests the use of participative, humanistic, and constructivist approaches
Access to natural resources	1	Assesses if extension activities that enhance the access to natural resources (land, water, forestry resources) by the farmers are being developed
Extension concerning other public policies	2	Evaluates how extension activities contribute to access to other policies.
Technological and management resources	1	Assesses if extension activities are being developed to enhance the use of management tools and technologies

Source: Compiled by the authors.

on agrarian reform settlements. Extension workers were purposively selected to come from the territories and also from both governmental and nongovernmental organizations. In total, 1,000 interviews with farmers and 87 with extensionists were conducted in the 5 territories between August 2014 and January 2015.

Data tabulation and analysis were conducted using Microsoft Office Excel 2007 spreadsheets. The given score for a particular question of all interviewees from the same territory was summed, obtaining a single evaluation for a specific question within a territory. This first step allowed each question to have a single score for each territory. After that, the score from each question was multiplied by the assigned weight for the respondent. The sum of these operations for each question resulted in the indicator score. This score was then divided by the maximum possible score for each category. The result of this operation was converted into a percentage, achieving a measure of evaluation for each indicator. Although several recent studies have addressed the structures of extension systems and their consequences (Compagnone and Simon 2012; Faure, Desjeux, and Gasselin 2012; Klerkx, Landini, and Santoyo-Cortés 2016; Klerkx and Proctor 2013; Knierim et al. 2017; Nettle, Crawford, and Brightling 2018; Prager et al. 2016), there is still a lack of full understanding of rural extension systems from a political economy perspective. Thus this research focuses on analyzing the processes that determine the formats and directions of agricultural extension systems.

In this study we use the framework proposed by Howlett, Ramesh, and Pearl (2009) in which the political process is seen as a series of interrelated and interactive sequential stages. This framework allows the user to understand the development of the political process over time; consequently, unveiling the key variables that influence the implementation and impact of the policy. For instance, the framework facilitates identification of bottlenecks in the political process, allowing the identification of the actors and actions of each stage and the consequences of decisions taken (Howlett, Ramesh, and Pearl 2009; Howlett and Giest 2015). We assess PNATER using the following four steps proposed by this approach: (1) agenda-setting—when a social problem is perceived as relevant and becomes part of the government agenda; (2) policy formulation—when some possibilities for addressing the issue are discussed and selected; (3) decision-making—when formal government actors adopt a particular course of action; and (4) policy implementation—when decisions are put into effect using public structures to change the distribution of goods and services in the society.

Considering that the process of assessing a policy is an attempt to understand its merit, worth, and utility, we use the evaluation criteria proposed by the Organization for Economic Co-Operation and Development (OECD 1991), and recommended by the Global Forum for Rural Advisory Services (Christoplos, Sandison, and Chipeta 2012). Both suggest the following

TABLE 4.2 Description of the evaluation criteria used to discuss the PNATER assessment

Criteria	Evaluation questions
Relevance "The extent to which the objectives of a development intervention are consistent with beneficiaries' requirements, country needs, global priorities, and partners' and donors' policies."	• Are extension priorities appropriate from the perspective of the clients? • Are the interventions relevant to national and local agricultural policy goals? • Are intervention plans adapted to changing market and climate conditions?
Efficiency "A measure of how economically resources/inputs (funds, expertise, time, etc.) converted to results."	• Have the target groups of clients received the planned services at an "appropriate" cost? • Have the capacity of extension service providers to reach intended clients changed, and at what cost? • What are the alternative systems for providing (quality) services or capacity development? • How do the costs of services compare with the alternatives?
Effectiveness "The extent to which the development intervention's objectives were achieved, or are expected to be achieved, taking into account their relative importance."	• Has the intervention improved access to services and inputs? • Has the intervention facilitated market access and marketing methods? • Has the intervention facilitated the formation of sustainable farmer groups?
Impact "Positive and negative, primary, and secondary long-term effects produced by a development intervention, directly or indirectly, intended or unintended."	• What has been the result of the intervention in terms of greater food security, better nutrition, or improved profitability for different groups of extension clients? • Has the intervention had unintended negative impacts on the environment, on the workload of women, or increased the risks that smallholders face?
Sustainability "The continuation of benefits from a development intervention after major development assistance has been completed. The probability of continued long-term benefits. The resilience to risk of the net benefit flows over time."	• Have (or will) extension service providers been able to cover the recurrent costs of the intervention's approach after external funding is phased out? • What are the long-term impacts on soil fertility, access to water, and maintenance of common property natural resources resulting from the promoted technologies?

Source: Compiled by the authors, adapted from Christoplos, Sandison, and Chipeta (2012).
Note: PNATER = National Policy for Technical Assistance and Rural Extension.

criteria to evaluate agricultural extension policies/programs/projects—relevance, efficiency, effectiveness, impact, and sustainability. Despite their limits, the criteria proposed by the OECD are widely recognized as a useful tool for evaluation in the field of development. They allow users to extract interesting feedback from the lessons learned, enabling their incorporation into the decision-making processes, guiding future improvements by identifying success factors, and explaining failures (Chianca 2008; Terrapon-Pfaff et al. 2014). A brief description of each evaluation criterion is in Table 4.2.

4.3 Governance Structures and Policies

Context

Brazil has gone through a long history of agricultural policies starting in 1831 (Bergamasco, Borsatto, and Thomson 2016; Bergamasco, Thomson, and Borsatto 2017; Thomson, Bergamasco, and Borsatto 2017). The inclusion of family farmers as beneficiaries of such policies is quite recent but gaining space on the political agenda since the 2000s (Diesel and Miná Dias 2016). Until the implementation of the National Policy for Technical Assistance and Rural Extension (PNATER by its acronym in Portuguese) in 2004, the history of Brazilian agricultural extension followed the same trajectory of several other Latin American countries. Brazil, in the late 1940s and early 1950s, in an international context dominated by the Cold War, started the implementation of agricultural extension systems based on the North American model. However, rural extension gained political importance and consequently received substantial investments from the mid-1970s, when Brazil assumed, as part of its national development strategy, the mission of providing rural extension services focused on the objective of modernization of its agriculture (Bergamasco, Borsatto, and Thomson 2016; Bergamasco, Thomson, and Borsatto 2017; Klerkx, Landini, and Santoyo-Cortés 2016; Thomson, Bergamasco, and Borsatto 2017).

In the late 1980s and early 1990s, the advancement of a political agenda with neoliberal bias was promoted throughout Latin America as a solution to the national economic crises (Liverman and Vilas 2006; Portes and Hoffman 2003). This resulted in a drastic reduction in federal funding for agricultural extension policies. As a consequence, the Brazilian agricultural extension system became simultaneously uncoordinated and lacking national finances (Bergamasco, Borsatto, and Thomson 2016; Bergamasco, Thomson, and Borsatto 2017; Brazil, MDA and SAF 2004; Peixoto 2008; Pettan 2010; Thomson, Bergamasco, and Borsatto 2017). Hence, there was progressive scrapping of state-level public organizations; some closed and others merged with other public enterprises such as research institutes. Private organizations were closed down or began to provide services to companies interested in selling agricultural inputs. In the 1990s, in the absence of a national policy for the agricultural sector, Brazil saw a number of local initiatives to fill in the vacuum. For example, initiatives funded by municipalities, nongovernmental organizations, and farmer-based organizations sprang up but without coordination (Bergamasco, Borsatto, and Thomson 2016; Bergamasco, Thomson, and Borsatto 2017; Caporal and Costabeber 2002; Miná Dias 2008).

Absence of small family farms, indigenous people, black rural households, and fishers' communities from public policies marks Brazilian history in rural policy-making. The roots of such neglect go back to the country's colonization, which followed a development model that pursued large-scale production of commodities for exportation. National governments have followed such a model over the past five centuries as one of the most critical strategies for Brazil's economic growth. In addition, recent policies have focused on agribusiness-related activities that represent the significant part of Brazilian exports and contribute to the gross domestic product. However, as a direct consequence, such policies have marginalized smallholder farmers who lack access to capital and land from productive value chains. Especially since the 1960s, this process has become even more profound as the access to machinery and other technologies were highly promoted and subsided in favor of large-scale producers.

Besides, as there was never an effective land redistribution policy in the country, land concentration has been increasing. Over the past fifty years, most family farmers have abandoned their original activities and migrated to urban areas as the above-mentioned development model advanced in the countryside. However, the remaining rural families are mostly living under economic insecurity resulting in rural poverty in Brazil as an urgent and modern problem. A study carried on by Soares et al. (2016) revealed that rural poverty in Brazil is directly linked to the lack of access to land, as 81 percent of those living in rural households occupy an area considered insufficient for a family's survival. The numbers from the most recent agricultural census from 2006 also reveal the level of land and income concentration in Brazil as a result of such skewed land policies. Family farming is responsible for 84 percent of rural properties, yet it occupies only 24.3 percent of lands suitable for human exploitation (Brazil, IBGE 2006). Furthermore, family farming is responsible for employing 74 percent of people in rural activities, which means 26 percent of the country's economically active population (Brazil, IBGE 2006). However, when it comes to their revenues, family farmers generate only one-third of the country's income from agricultural activities, an average of 13,000 reals per year per property.

The farming system of the smallholder farmers in Brazil is often characterized by multicrops and livestock operated under primitive technological and marketing systems. According to Wanderley (1999), one of the strategies that allow family farming to persist in rural areas is its type of production. Family farmers that do not have access to modern technologies or choose not to specialize in conventional production commonly produce different kinds of products for human feeding. Besides their main activity, such as cropping or dairying, they often have animals for family consumption of protein as well as vegetables and fruit gardens around the household. Such systems of production have also contributed to family farming, which typically produces for domestic supply and has been highly exploited by intermediaries, including the government and the food industry throughout history. This study attempts to understand the role and effectiveness of policies relating to extension and rural services in increasing the productivity and food security of the smallholder farmers.

Past, Present, and Future Policies

The year 2004 marked a turning point when the then new government determined family farmers as the primary beneficiaries of agricultural extension policies, based on a model of sustainable rural development (Bergamasco, Borsatto, and Thomson 2016; Bergamasco, Thomson, and Borsatto 2017; Correa da Silva 2011; Diesel and Miná Dias 2016). During 2003, policymakers of the new government, in partnership with representatives of civil society and with various stakeholders, began to formulate PNATER and launched it in 2004. PNATER outlines a list of principles that should be followed for the provision of agricultural extension services supported by public resources and presents a series of innovative guidelines. The guidelines established that participatory methodologies should guide service delivery, that agroecological principles should be the technical guideline, and that family farming is the primary beneficiary, among others (Brazil, MDA, and SAF 2004).

The document states that policy management must include different actors involved in the policy arena—the federal government, state-level public institutions, family farming organizations, and other civil society organizations (including service providers). It also encourages the participation of beneficiaries (family farmers) in the planning, monitoring, and evaluation of its actions. Moreover, PNATER determines that rural extension services are pluralistic, and providers can come from a broad gamut of organizations, including the private and public sectors, nongovernmental organizations, farmer-based organizations, professional cooperatives, and educational institutions, among others. To implement the policy, in 2005 the government launched the National Program for Technical Assistance and Rural Extension in Family Farming and Agrarian Reform. The document established goals and actions to stimulate public extension programs, train family farmers, develop sectoral extension activities (working with indigenous communities, black rural communities, riverside communities, fisheries, young and female agricultural workers), and improve and extend the rural extension services in the country (Bergamasco, Borsatto, and Thomson 2016; Bergamasco, Thomson, and Borsatto 2017; Diesel, Miná Dias, and Neumann 2015).

In 2006, due to the demand to move forward in the process of coordinating the implementation of the policy, the new Brazilian Decentralized System of Rural Extension was formalized, which, while being coordinated by the Ministry of Agrarian Development, stipulated social control and participation in its processes. In 2010 the Congress gave PNATER the status of law, informally known as the Rural Extension Law. Its publication represented a major political breakthrough in terms of ensuring public rural extension services for family farming with some independence from changes in government.

In addition to the PNATER, other government policies aimed at smallholder farmers, with the goals of fostering food and nutritional security, social and economic inclusion, the fight against hunger, family benefits, public procurement, credit and insurance, and income generation, had essential interactions with extension services. It is important to emphasize three of these policies that strictly relate to PNATER and depend directly on the success of its execution. The main one is the National Program for Family Agriculture (PRONAF), created in 1996 but only receiving consistent funds after 2003. PRONAF is credit policy specifically for family farmers. The second one is the Food Purchase Program, through which the government purchases products from family farmers on fair and constant prices and mediates its distribution for public institutions in which the population at food insecurity risk has regular access to food. Finally, since 2009 there was also an essential new government rule added to a previous program called the National School Feeding Program. The federal government started to require that at least 30 percent of the total resources passed on by the National Education Development Fund to the states and municipalities be designated to acquire products for school meals from family farming (Brazil, MDA 2010). It is important to mention that if the delivered products are organic, both programs pay 30 percent more than the value paid for conventional products in the regional markets.

Despite these advances and investments for a more sustainable and socially responsible rural development model in Brazil, there is still a lack of studies assessing the impacts of PNATER specifically, especially from the perspective of its main stakeholders—smallholder farmers and extensionists. The research presented here examines the views of these key stakeholders on the extension services concerning the PNATER policy goals and objectives. The researchers attempt to understand and explain whether and how these efforts have succeeded in promoting the improvements intended by the extension policy in its implementation.

Approaches to Implementing PNATER through Extension and Rural Services

The extension and rural services that were implemented based on PNATER took various modes of operation. Although the major goal of the extension and rural services at the state and municipality level is to translate the rural development goals set for improving the quality of life of the marginalized and smallholder communities, it took several forms of intervention depending on the demand for extension services. Major approaches of extension include the following:

- Social mobilization of family farms.
- Training of the extension staff in PNATER guidelines, formation of farmer associations and cooperatives to deal with specific problems of the farming communities.
- Organization of the key stakeholder for identifying the problems and developing locality-based solutions.
- Advising famers on the crop and enterprise choices and methods of crop production and protection.
- Introduction of modern agricultural practices.
- Advice on existing government programs such as Bolsa Família.
- Help in access to short-term and medium-term credit for development of the commercial aspects of farming.
- Increasing food security though crop diversification and home gardening.
- Improving equality though gender and ethnicity programs.
- Increasing access to natural resources.

4.4 Organizational Capacity and Management

"Family farmer" emerged as a political category in Brazil during the 1900s during a crisis scenario for agricultural extension and other rural public policies. The consolidation of such a specific social group is a direct response from civil society to the neglect of poor smallholders in rural development policies by neoliberal governments. As a result, family farming is a plural reference, related both to the traditional European notion of the peasantry and also to indigenous, black rural, and fishers' communities, among others. The Brazilian interpretation of family farming resides in the social and economic link between the rural family and the activities developed in the land in which they live. It must be a place of both work and living. It is a notion opposed to entrepreneur agriculture, characterized by large areas of monoculture and paid employment.

The MDA was the official federal organ responsible for contracting and providing funds to the extension organizations (public or private) in the country. Figure 4.2 details the Brazilian Decentralized System of Rural Extension, including the extension services, the policy advising and evaluation organs, the policy-making bodies, and the policy execution agencies.

Public policy deals with issues faced by citizens of different countries (Dewey 1927 in Davies, Nutley, and Smith 1997). Researchers studying public policies examine what the government does and why, and how this affects intended (or unintended) outcomes (Dye 1976 in Davies, Nutley, and Smith 1997). Extension policies are often part of broader agricultural systems, and many countries do not have specific extension policies for different reasons (Oladele 2011). However, those that do have been gearing the strategies to strengthen agriculture as an engine of pro-poor growth and meet broad goals such as reduction of poverty (Birner and Anderson 2007; Sulaiman and Hall 2005). But overall, extension proponents struggle to ensure political commitment and financing for agricultural extension services to implement the policies (Feder, Willett, and Zijp 2001).

Part of the reason for lack of commitment is the fact that many extension policies—and their resulting programs and services—have not been adequately examined as to their outcomes and impacts. While there is some evidence showing high rates of return extension (Alston et al. 2000; Birkhauser, Evenson, and Feder 1991), in general, measurement and design challenges make it difficult to attribute impacts directly to extension programs and link cause and effect quantitatively (Anderson 2007; Purcell and Anderson 1997). In addition, the evolving nature and current status of the extension have contributed to difficulties in evaluating the impacts of extension systems at the national level. Extension today is characterized by both pluralism and privatization, making it difficult to define and to measure extension outcomes and impact (Knierim et al. 2017; Klerkx and Proctor 2013; Prager et al. 2016; Labarthe and Laurent 2013).

FIGURE 4.2 The decentralized system of extension services in Brazil

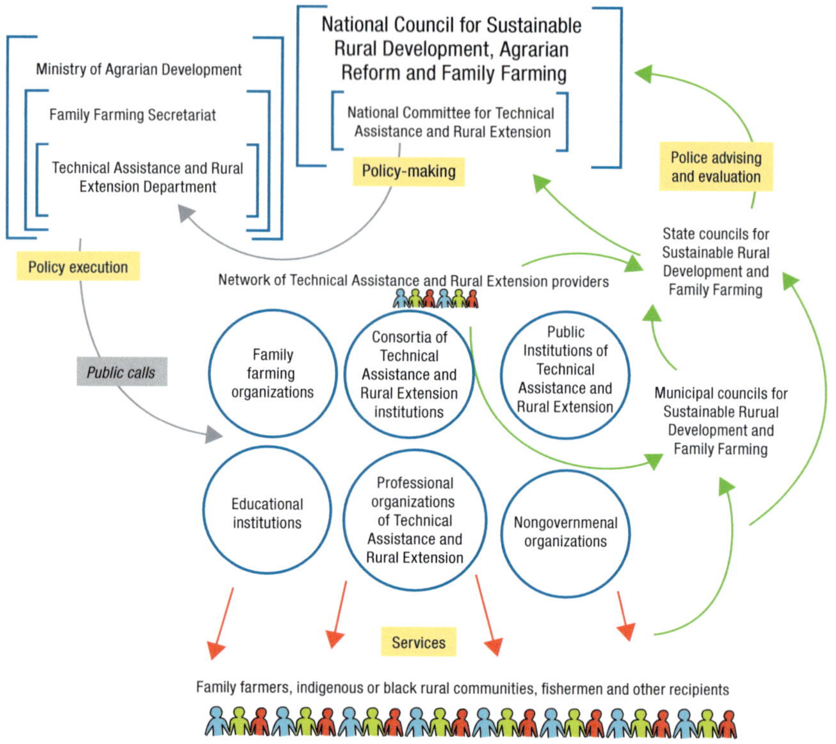

Source: Compiled by the authors.

Some researchers have examined extension policies and reforms, however. Chowdhury, Odame, and Leeuwis (2014) attempted to describe how a public-sector agricultural extension organization attempted to reform its roles in implementing a major agrarian extension project. Critical challenges encountered included the reluctance of actors to change, neglect or lack of capacity for broader intermediary functions for the extension, and a refusal to view extension as facilitating learning processes. Benson and Jafrey (2013) point out the difficulties many governments have faced in reforming conventional, top-down management systems to demand-driven and pluralistic ones. Based on a literature review, these authors examined the challenges faced by extension reforms. Their findings recommend using a framework that clarifies the aims of government policy to ensure that national extension strategies best fit a country's economic and social objectives.

In the case of Brazil, there is a clear policy linking the country's economic and social objectives to the extension strategies and programs. The researchers conducted this study in Brazil to understand the effects of the 2004 extension policy PNATER. They did this by examining the extension services provided under the PNATER principles and objectives. Specifically, the researchers collected primary data from producers and extension personnel regarding perceptions about the extension services. Research regarding perceptions of rural services has been done in other countries; researchers used opinions of extension staff and farming households in India, Ghana, and Ethiopia in a study on gender in rural services (World Bank 2010). In Nigeria researchers studied perceptions of farmers and extension professionals regarding cost-sharing of extension services (Ozor et al. 2008).

4.5 Advisory Methods

Brazil has a long history of extension paradigms that have determined methods and approaches used in serving clientele. These have included training and visit (used by the Brazilian Agricultural Research Corporation at one time), farmer field schools, farmer-to-farmer extension, research groups, field days, trainings, demonstrations, and individual visits. The current Brazilian extension policy prescribes participatory extension methodologies and dialogue to build new knowledge and develop technological innovations, taking into consideration both indigenous and formal knowledge (Corrêa da Silva 2011). The role of the extension agent is defined as facilitator, organizer, and educator. However, Bergamasco, Borsatto, and Thomson (2016) found that extension services activities for women, the young and the elderly, and ethnic groups received low ratings from farmers in all territories. Extension agents stated that few activities, if any, are not developed for this category.

Since 2006 under Brazilian law, the extension services are directly aimed at family farming, defined according to size of production unit, predominance of family work, and earned income. "Family farmers" includes a large diversity of clientele: people in agrarian reform settlements, indigenous communities, traditional family farmers, *quilombolas* (traditional populations of former slaves), forest people and artisanal fishermen, and technical family farmers in agro-industrial chains. There are more than four million farming families in the country, comprising 80 percent of farmers (Corrêa da Silva 2011). The policy guidelines also address including gender, racial equality, and youth—issues that were largely neglected in previous policies (Corrêa da Silva 2011). The policy sets up guidelines for promoting income generation and value addition and uses a value-chain approach. Agricultural research organizations still focus mainly on commercial farming and address monoculture commodities, which are of little relevance to family farming agriculture systems. They do not tend to serve family farming because they are not prepared to deliver technology for diverse production systems (Corrêa da Silva 2011).

There are more than 20,000 public extension agents in the country, occupying over 5,000 offices, which cover 95 percent of the country's municipalities. Fifteen NGO networks operate in different parts of the country and account for about 5,000 extension agents. This corps reaches about 1.5 million family farmers and has an annual budget of more than US$1 billion (Corrêa da Silva 2011). It is worth mentioning that extension agents from public entities usually had better working conditions that those from private entities. This is because the extensionists' salary, offices, vehicles, and travel costs were usually funded by state governments. In some cases, extension agents from entities with less capital must use their personal vehicle for field work.

4.6 Effectiveness of Extension Services from Farmers and Extensionists' Perspectives

The five territories from which the data were collected for this study are: Alto Jequitinhonha, Cantuquiriguaçu, Pontal do Paranapanema, São Paulo's Southwestern, and Vale do Ribeira. First, we present the results of each territory; and second, we present the results at the national level.

Alto Jequitinhonha Territory

This territory is located at the state of Minas Gerais and comprises 20 municipalities. In this territory the researchers did 200 interviews with family farmers and 13 interviews with extensionists, from 18 municipalities. A public institution named Minas Gerais Technical Assistance and Rural Extension Company (EMATER, by its acronym in Portuguese) provides the major part of agricultural extension services. However, there were also nongovernmental and private companies performing the service through public calls during the period of the interviews. Ten extensionists interviewed at the territory worked for EMATER and another three for different private entities. Figure 4.3 shows the territory results.

Results show that the main disagreements are regarding the indicators "pedagogical conception" and "social and community organization," which reached 91 percent and 92 percent, respectively, according to extensionists, but only 36 percent and 45 percent, respectively, according to farmers. Such

FIGURE 4.3 Alto Jequitinhonha's results for extension services, 2017

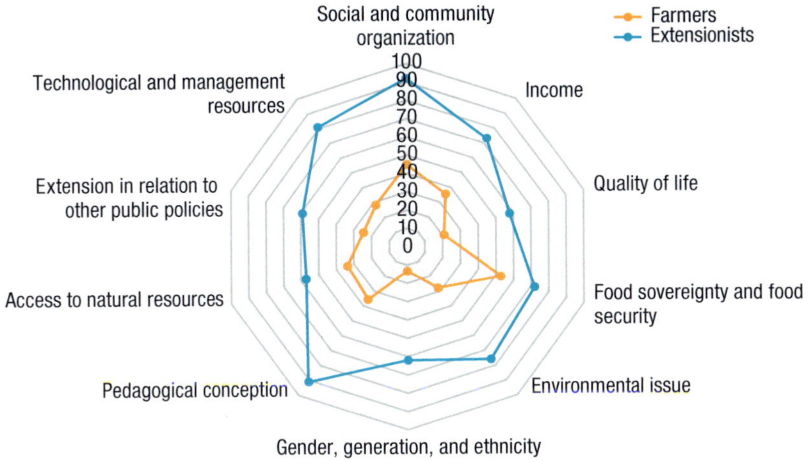

Source: Authors.

divergence is because most of the interviewed extensionists are employed by the public extension company of Minas Gerais, which adopted the PNATER guidelines and has trained its extensionists along those lines. Therefore, many extensionists reported the encouragement to establish farmers' associations and cooperatives in the communities, and the importance and efforts to accomplish their services in a participative way with the farmers. However, the farmers' evaluation might be explained by the low number of families visited by a single extension technician, even in collective formations, since it is a vast territory with difficult access.

According to the interviews, 13 interviewed extensionists were responsible for executing extension services to 5,448 families, an average of 419 families per extensionist. This average of farm families per extensionist exceeds the maximum stipulated by the contract with MDA, which specifies 100 families per extensionist. In Minas Gerais each extension staff from the extension public company is responsible for all the farmers from the same municipality, and in some cases, private entities can also perform the services in the same region. During the field interviews, researchers asked extensionists about the number of families that they believed they could serve while preserving service quality. The average reply was 105 families per extensionist. Therefore, even though the extension technicians recognize the quality of their services, it is unrealistic that the farmers' evaluation would be proportional. The indicator "frequency" revealed the insufficiency of extensionist services in the territory, achieving only 33 percent of the hypothetical ideal. This indicator is not included in the radar figures because we did not consider it as an "impact" of the policy for technical reasons.

We can observe that the indicator with the best performance evaluated by farmers from Alto Jequitinhonha was "food sovereignty and food security" (53 percent). It is essential to highlight the projection of national policies to eradicate starvation, such as the social welfare program Bolsa Família, allied with the improvement of social rights access, and rural retirement. Besides facilitating access to those policies, extension is also crucial in the food security indicator, since the farmers cited the encouragement from the extensionists (when farmers had access to the services) to diversify the production and valorization of plantation for self-consumption.

The indicator with the lowest performance, according to the farmers, was regarding the extension actions in their communities aimed at "gender, generation, and ethnicity," at 13 percent. This indicator was also the second-worst according to the extensionists, reaching 63 percent in the evaluation of their services. Therefore, it is essential to highlight the limitation of extensionists to develop services with specific groups of women, young and older people, indigenous, black rural, and other traditional communities. This indicator had a low percentage in all five territories studied.

In Alto Jequitinhonha the access of farmers to natural resources is limited, especially access to water. This indicator had a lower percentage (57 percent) when extensionists evaluated their services. This restriction is explained by the modernization of farming, which started during the 1970s in the region. Since then, the state has been authorizing the use of public lands for planting Eucalyptus trees by the paper and cellulose industry. Before the modernization process, these public lands, locally called *mangas*, were meant for shared use, such as cattle fattening and the gathering of native fruits from the tropical savannah region called *Cerrado*. Also, most of the springs from the regions' rivers are located at the *mangas*, but due to monoculture plantations of eucalyptus, the water resources are drying up and the water is not reaching the family farmers' properties.

Cantuquiriguaçu Territory

This territory is located in the state of Paraná and comprises 20 municipalities. The territory had a population of 232,551 inhabitants in 2010, with 52 percent living in rural areas (Brazil, IBGE 2010). Also, 83 percent of the farms are engaged in family farming, totaling 21,184 establishments. Another important characteristic of Cantuquiriguaçu Territory is the significant presence of rural settlements. Established as a result of pressure from social movements and agrarian reform policies, they consist of more than 4,000 farming units. Therefore, the extension services in this territory mainly catered to the farmers located in rural settlements. There are also four black rural communities and two indigenous communities, forming one of the largest indigenous reserves in the state of Paraná, constituted by Kaingang and Guarani Tribes (Brazil, IBGE 2010). A total of 200 interviews were conducted with family farmers and 14 with extensionists, in 10 different municipalities in Cantuquiriguaçu Territory.

The results for this territory (Figure 4.4) show a lower level of satisfaction by the extensionists here when compared with Alto Jequitinhonha Territory. This is because the extensionists reported low appreciation for their career development, insufficient professionals in their institutions, and the large numbers of family farmers to serve in the territory. With about 21,000 establishments of family farming in the region, there is an average of 204 families per extensionist. They stated that the ideal number should be about 60 families per extensionist.

We further observed in this territory that the extensionists' evaluation was mostly lower than the farmers' evaluation (Figure 4.4). The extensionists understand the principles of the extensionist work, but they state that the precarious working conditions are the main reason for poor service quality. This statement reinforces the evaluation of the indicators "pedagogical conception" (53 percent according to extensionists and 47 percent according to farmers) and "social and community organization" (46 percent according to extensionists and 55 percent according to farmers). Therefore, once the limitations of their institutions were pointed out, especially regarding lack of resources and number of families to be visited, the extensionists stressed that their primary strategy was to enable the development of family farming in the territory, encourage clients to organize associations and cooperatives, and develop participatory methodological tools. The farmers, in turn, highlighted the importance of platforms to share experiences and strengthen socioeconomic aspects in the community.

During the fieldwork, we observed that the primary extension strategy in the territory was to support groups linked to social movements. Since these farmers were more organized, the extensionists believed that their work has more impact. If we calculate the extension coverage in the territory, considering the number of family farming establishments and the number of families visited by the interviewed extensionists, one would see only 13 percent

FIGURE 4.4 **Cantuquiriguaçu's results for extension services, 2017**

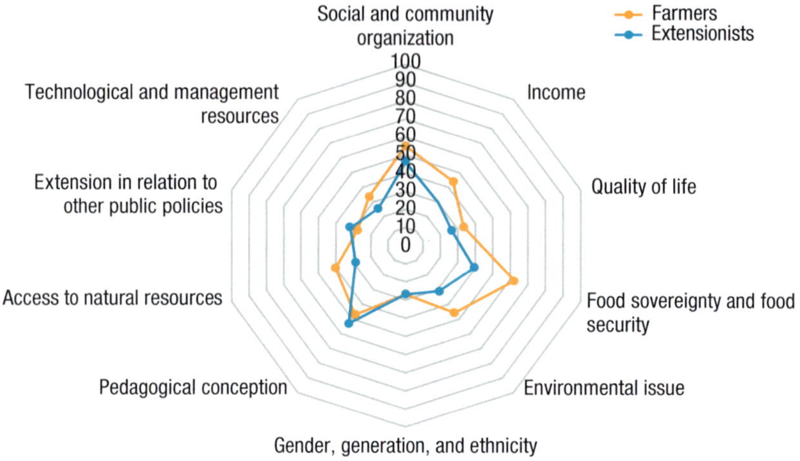

Source: Authors.

of family farmers were visited by extensionists in the territory. Therefore, we can judge that the extensionists' activities are inefficient, as the indicator "frequency" evaluated by the farmers was 40 percent of the hypothetical ideal, although considerably superior to the 13 percent calculated above based on extensionists' reporting.

It is essential to observe that the indicator "food sovereignty and food security" reached 60 percent of the hypothetical ideal by farmers, even though the "income" indicator reached only 44 percent of the hypothetical ideal. Once again, we note the importance of public policies to overcome extreme poverty and food sovereignty in the poor Brazilian territories since economic and social development is slower in family farming, especially where land and income are concentrated. In this territory, the social welfare program Bolsa Família was mainly responsible for the success of the indicator "food sovereignty and food security," allied to extension efforts to diversify production and self-consumption–oriented production for the families. The extensionists once again evaluated their actions as inferior to those reported by the farmers, reaching only 39 percent of the hypothetical ideal.

As in Alto Jequitinhonha Territory, the indicator with the lowest percentage (27 percent) relates to extension actions in their communities aimed at "gender, generation, and ethnicity" issues. The extensionists' evaluation reached about the same percentage (27 percent). The farmers indicated a few or inexistent actions in this field, and extensionists claimed that clear methodologies are lacking for addressing these issues through extension approaches. Together with the indicator "gender, generation, and ethnicity," the indicator "ATER in relation to other public policies" had a low percentage—28 percent according to the farmers and 32 percent according to the extensionists. These data are mainly related to access to credit and commercialization policies, and the National Program for School Feeding. Default nonpayment by farmers and bureaucratic difficulties place restrictions to reaching goals of such policies. The extensionists report farmers' problems to organize credit and finances. However, farmers claimed payment delays from the federal government, difficulties in regulating requested documentation, and resistance from the financial institutions to enable access to credit policies.

Pontal do Paranapanema Territory

This territory is located in the state of São Paulo and is composed of 32 municipalities, with a total population of 583,703. Of these, 90 percent live in urban areas, and 10 percent in rural areas (Brazil, IBGE 2010). As in Cantuquiriguaçu Territory, the Pontal do Paranapanema Territory has a significant presence of rural settlements established during the 1980s and 1990s, as a result of social movements pressure and agrarian reform policies. In total, we conducted 200 interviews with family farmers and 20 interviews with extensionists in 19 municipalities in the Pontal do Paranapanema Territory.

The indicators in this territory represent the most substantial divergence between the extensionists and the rural families' evaluation. As to the farmers' assessment, it presented the worst indexes among the five territories investigated for this study (Figure 4.5). However, in contrast to results in the other territories, despite the considerable workload for extensionists (an average of 357 families per extension technician, while the ideal should be about 163 families, according to the interviewed extensionists), the indicator "frequency" was 61 percent, according to farmers. In this territory, "frequency" achieved the highest percentage in the research.

Due to its history, the biggest challenge for extensionists in this territory is to enable the PNATER guidelines and to reestablish family farming organizations, which became nearly extinct during the 1990s. The farmers were encouraged to invest in commodities production and set up big cooperatives to compete in the market after the first settlements established in the region. For various reasons this model collapsed in the mid-1990s, and the farmers became indebted. In spite of this effort, however, the monoculture mentality still prevails among farmers. Another discrepancy that limits extensionists' work is the result of the indicator "access to natural resources," which

FIGURE 4.5 Pontal do Paranapanema's results for extension services, 2017

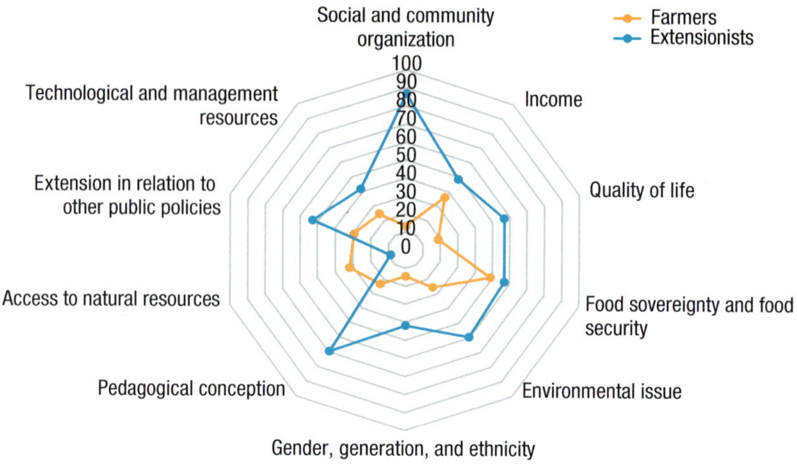

Source: Authors.

was 9 percent according to interviewed extensionists. The access to and preservation of natural resources is pointed out by PNATER as one of the main guidelines to promote agroecology, and therefore this result reveals that even though extensionists declared that they are committed to PNATER guidelines, they did not fully succeed in promoting it.

The main finding from the results of the indicator system in Pontal do Paranapanema Territory is that even though extensionists declare they are committed with PNATER guidelines, their practices do not reflect these guidelines in the territory. For instance, the indicator "social and community organization"—the worst according to farmers' evaluation (13 percent) and the best according to extensionists (87 percent)—highlights the observed discrepancy. As in other territories, the indicators with higher convergence were "food sovereignty and food security" and "income." Even though the percentages of these indicators were low, we can once again observe the importance of public policies to overcome extreme poverty and provide food security. However, this territory is the one where family farming has a higher social and economic vulnerability.

São Paulo State's Southwestern Territory

This territory comprises a group of 15 municipalities. A remarkable characteristic of this territory is the presence of a significant number of decapitalized

FIGURE 4.6 São Paulo's Southwestern's results for extension services, 2017

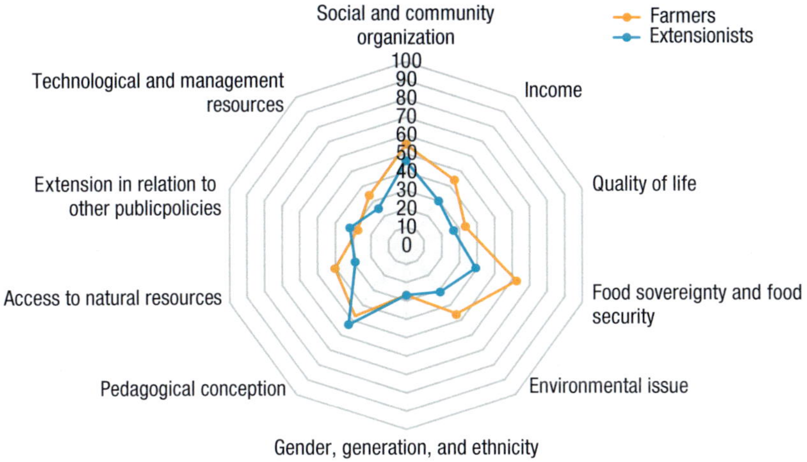

Source: Authors.

family farmers with diversified production systems (mainly grains, milk, vegetables), coexisting side by side with highly specialized industrial agriculture. According to the 2010 census, 312,063 people live in the territory, 21.6 percent in rural areas (Brazil, IBGE 2010). Furthermore, a significant percentage of this rural population lives under extreme poverty conditions, reaching in some municipalities up to 70 percent of their rural population. The territory has 7,208 units managed by family farmers, of which 415 were in rural settlements. In addition, there is one black farmers' community (*quilombo*) and two indigenous settlements in this territory. A total of 200 interviews with family farmers and 19 with extensionists in 16 municipalities were conducted in this territory.

In this territory the indicators studied revealed a significant convergence between farmers and extensionists, although extensionists' evaluation of indicators was inferior to farmers' assessments (Figure 4.6). Indicators that were better evaluated were "pedagogical conception" and "social and community organization." This is partly because this territory has several associations and cooperatives aimed at institutional market development, enabled by the federal government over the past decade. Therefore, farmers pointed out the importance of extension services in public organizations, especially dealing with bureaucracy. Thus the existence of better conditions to access markets and a superior logistics infrastructure resulted in the observed convergence.

Extension services depend on external conditions, so extensionists' activities can accomplish farmers' expectations and also enable dialogue about their objectives.

As in other territories, the percentage of the indicator "food sovereignty and food security" was high according to farmers' evaluation, reaching 65 percent of the hypothetical ideal. It is worth mentioning that this territory is the most impoverished region in the state of São Paulo. Once again this highlights the importance of public policies to overcome extreme poverty and promote food sovereignty, besides production for self-consumption. As seen in the other territories, the indicator with the lowest evaluation is related to "gender, age, and ethnicity." In this territory this result requires particular attention since there are a large number of black and indigenous communities and the need for specific extension services are much greater for these communities.

Vale do Ribeira Territory

In the state of Parana, this territory comprises 7 municipalities and has 100,821 inhabitants, 43 percent of which live in rural areas (Brazil, IBGE 2010). The region has 5,596 family farmers, with a significant presence of artisanal fishing communities (80 communities with 2,500 families), and black farmers' communities (11 communities). In total, the researcher carried out a total of 200 interviews with family farmers and 12 with extensionists in the 7 different municipalities.

The primary provider of agricultural extension services in the territory is EMATER-PR, a state-level public entity, responsible for assisting farmers and facilitating access to rural credit, food sovereignty, social benefits, and other public policies. We observed other organizations developing specific actions with family farmers, such as courses on organic agriculture or facilitating access to rural credit. In this territory the indicator "income" was better evaluated by farmers than by extensionists (Figure 4.7). Once we deeply analyzed the farmers' and extensionists' answers, we understood that farmers' income is mainly composed by Bolsa Família and rural retirement, which explains why extensionists do not consider their services essential to improve farmers' income.

Farmers from this territory commonly complement their income with temporary off-farm jobs. Only a small portion of the interviewed farmers' income is from their agricultural activities on-farm. Therefore, production is mainly destined to self-consumption, reflecting the convergence of results for the

FIGURE 4.7 Vale do Ribeira's results for extension services, 2017

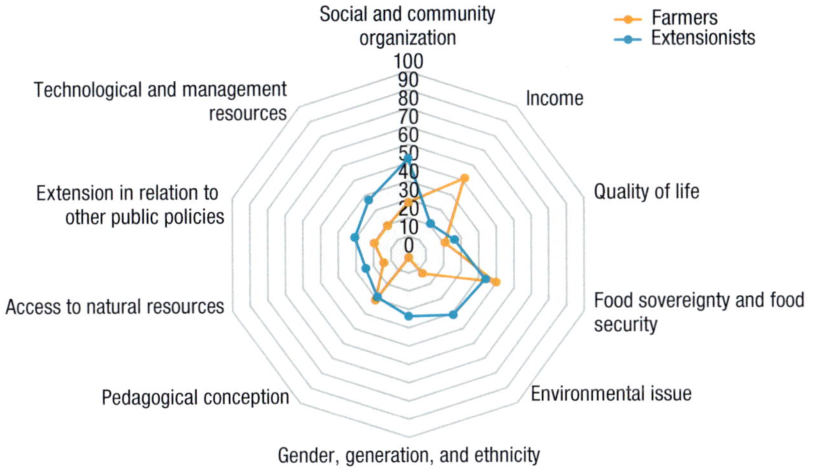

Source: Authors.

indicator "food sovereignty and food security" by farmers and extensionists. Field research revealed that the territory had difficulty in accessing the market and limited logistic infrastructure for product distribution. The rural communities visited were challenging to locate, with no permanent availability of health-care and education services.

The precarious local infrastructure explains why extensionists reported that their services were insufficient to organize farmers in the territory, creating associations or cooperatives. This fact influenced the evaluation of the indicator "social and community organization," which was 30 percent of the hypothetical ideal, according to farmers. Farmers from this territory presented the worst evaluation for the indicator "gender, generation, and ethnicity." This limitation is worth mentioning since this territory had the most significant number of black rural communities. Interviewed black farmers affirmed that local and state governments neglected their communities, and extensionists were not trained to deal with black farmers' specific demands.

Summary of Results at the National Level

Figure 4.8 and Figure 4.9 present national results regarding reach and assessment of agricultural extension services (Figure 4.8 for farmers and Figure 4.9 for extensionists).

FIGURE 4.8 Farmers' national results for extension services, 2017

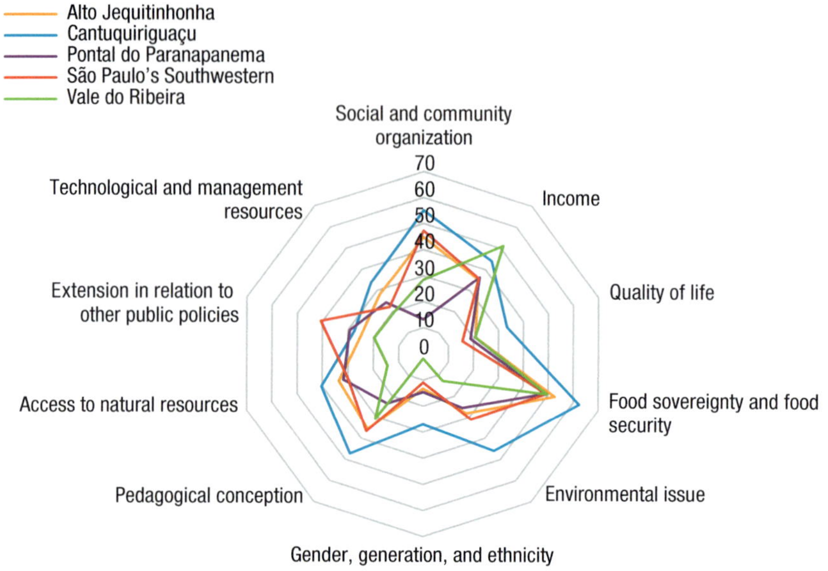

Source: Authors.

FIGURE 4.9 Extensionists' national results for extension services, 2017

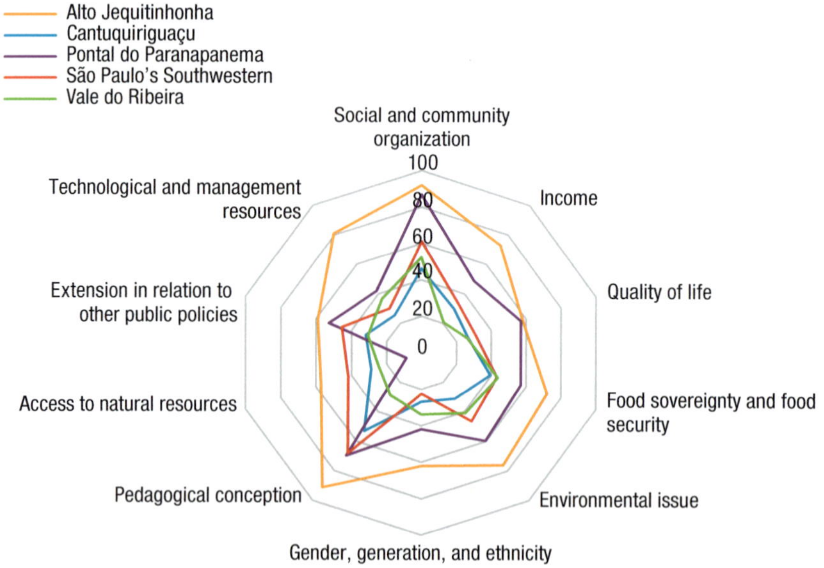

Source: Authors.

From the national-level results, we can see certain common patterns across the five territories. The main convergences among all territories observed relate to the best- and worst-evaluated indicators, namely the "food sovereignty and food security" indicator (the best) and the "gender, generation, and ethnicity" indicator (the worst). The indicator "food sovereignty and food security" reveals the importance of implementing extension and rural services programs that support policies for fighting hunger carried out by the federal government. Especially in rural areas, it shows that extension's actions to encourage diversified production for food security were perceived to be succeeding in eradicating the extreme poverty conditions that have historically affected these territories.

The performance of indicator "income" highlighted the dependence in all five territories on rural retirement, Bolsa Família welfare, and temporary jobs outside the family farms since these were determinant income sources for the interviewed families. The findings regarding the income indicator explain why in two territories the rating of the indicator "social and community organization" was low (Pontal do Paranapanema and Vale do Ribeira), while the score for income was superior. In these territories we interviewed the highest number of farmers who declared that at least one family member had paid employment outside the rural establishment or had a temporary job outside farming. In the other three territories, where most interviewed farmers declared that their income depended on on-farm activities, we saw the highest number of farmers who were members of cooperatives or associations. The phenomenon, apparently paradoxical, that ratings for the indicator "social and community organization" led to an inferior rating for the indicator "income" was because farmers who were working in partnership with associations and cooperatives reported payment delay from the federal government's public procurement programs. Also, logistical difficulties, lack of infrastructure, and other difficulties contributed to this situation.

The indicator "gender, generation, and ethnicity" (the worst evaluated in all territories) reveals the limitations that extension still faces. In the questionnaire we asked both open and closed questions related to specific activities developed by extensionists in communities with groups of women, young and older people, and ethnic groups, such as black rural and indigenous communities. Farmers responded with few or nonexistent activities aimed at these groups, while the extensionists reported that they develop none or insufficient activities in this field. The indicator "pedagogical conception" had the highest evaluation regarding extension practices in the five territories studied, and it was considered the most relevant guideline under extension for extensionists and farmers interactions. The extensionists reported their services based on participatory methodologies were effective in reaching out to the farmers, and the farmers also recognized such practices in the extensionist services as most valuable. The indicator "environmental issues" revealed the persistent limitations to promote agroecological principles. Although PNATER places highest importance on agroecological principles, reduced initiatives from extensionists, and limited appropriation from farmers, were observed in practice.

4.7 Conclusion and Recommendations

In Brazil the PNATER stands out as a political initiative most aligned with the contemporary global debates concerning the role of smallholder agriculture in ensuring food security, fostering rural development, and promoting natural resource preservation. Its elaboration and implementation process at the territory-level extension programs demonstrate the recognition of the importance of agricultural extension by the federal government and the recent resumption of investments in this area. This political decision has significant consequences, such as the creation of new institutional apparatuses (for example, the Technical Assistance and Rural Extension Department, Rural Extension Law, National Agency for Technical Assistance and Rural Extension, National Program for Technical Assistance and Rural Extension in Family Farming and Agrarian Reform), the establishment of a pluralistic system of agricultural extension, and the restructuring of the state-level public organizations providing extension and rural services in Brazil.

Recognition of family farming and traditional communities as a target audience for its actions is another essential part of PNATER contribution, which began to make visible these social categories that were historically marginalized, forgotten, or disregarded by Brazilian policymakers for a long time (Wanderley 1999). After the advent of PNATER, these marginalized groups became recognized as critical social agents demanding specific policies, including their legal institutionalization of their presence and their communities (Law 11,326/2006—Family Farming Law). The adoption of endogenous development and agroecological principles in PNATER has a significant implication for the development of rural extension and advisory services. As a result, since 2013 or so, several formal courses on these topics emerged throughout Brazil at different levels (undergraduate, postgraduate, technical education), and hundreds of research groups began focusing on the themes of family farming and agroecology (Balla, Massukado, and Pimentel 2014;

Cardoso et al. 2018; Petersen, Mussoi, and dal Soglio 2013; Petry 2017; Silva, Sousa, and Assis 2017).

PNATER was unable to succeed in making agricultural extension recognized outside the field of its stakeholders. It is evident in the budgetary constraints that the policy has faced since its implementation, in the significant variations in its budget, in the lack of investments to structure the MDA, and in the delay in establishing its institutional structures at various levels. This remains a challenge for the PNATER's full recognition as a strategic policy for national development. Brazilian researchers who studied PNATER commonly agree that the approaches to improve its implementation have resulted in significant distortions of its principles (Caporal 2011; Diesel and Miná Dias 2016; Diniz, Lima, and Almeida 2011; Silva and da Caporal 2016; Zarnott et al. 2017). For example, actions such as the adoption of Public Calls as a means of contracting service providers, the Rural Extension Law drafting, and the establishment of the National Agency for Technical Assistance and Rural Extension, among others, aimed to improve the efficiency of the Brazilian agricultural extension system. However, all of these actions present significant contradictions concerning PNATER's initial intentions. All interventions seem to lack social participation, since the decisions on implementation are made almost exclusively by a small group of policymakers.

Thus, in pursuit of efficiency in the execution of the policy, many of the innovations initially proposed (for example, endogenous development and agroecology) essentially disappeared in the text of the Rural Extension Law. Other PNATER principles have been neglected, such as universalization, social control, continuous service, and decentralized coordination. Therefore, the Brazilian agricultural extension system is once again acquiring a diffusionist approach that has already demonstrated its limits in the past. In this context the present case study sheds some new insights on the challenges faced by the current implementation of extension and rural services under PNATER. We summarize the key conclusions, recommendations, and lessons from this study. Our analysis indicates that the path to improving the efficiency of a pluralistic and innovative agricultural extension system, such as the one conceived under PNATER, should be based on the decentralization of its coordination and investments in implementation processes. Centralized management has demanded structures that MDA does not have and can be considered a main factor limiting system efficiency resulting in a high level of bureaucracy and slowness in releasing resources as well as in evaluation of field reports issued by the providers (Brazil, MDA 2004).

Between 2010 and 2014, services funded by PNATER reached approximately 550,000 farm families. Despite the concern that the Brazilian family farms are growing (4.3 million currently), the current reach of the PNATER shows that the agricultural extension services financed by the federal government directly serves a broad audience. This places Brazil among the countries with the most significant access to public services of agricultural extension around the globe. However, despite significant advances promoted by the policy, many of the guidelines and objectives envisaged initially are losing importance or being abandoned over time. Various factors such as inherent difficulties in implementing the policy's innovations, ideological differences between the different policymakers over time, pressures of the state's regulatory agencies, lack of priority for rural extension in the context of development policies, changes in the international and national political conjuncture (for example, the 2008 financial crisis, new political coalitions in Brazil), path dependence, pursuit of efficiency, and new actors interested in the policy (medium-sized farmers), among other factors, contributed to this situation. Thus we agree with Diesel and Miná Dias (2016), who suggest that the Brazilian state underestimated the challenges, especially within the field of political economy, in implementing a policy of high complexity such as PNATER.

It is essential to recognize that PNATER was part of a set of public policies aimed at family farming that directly impacted the lives of its beneficiaries, which, among other results, removed Brazil from the FAO hunger map (FAO and IFAD 2017). Furthermore, by recognizing family farming as the target audience of the public rural extension services, PNATER contributed to the empowerment of this social category, which has been perceived as a relevant social agency not only by the public authorities but also by the civil society. Another substantial impact was the expansion of dialogue between the state and civil society, which, despite the limitations and setbacks pointed out in this chapter, allowed remarkable social participation in the governmental spaces (for example, the National Council for Sustainable Rural Development and the National Conference for Technical Assistance and Rural Extension). As a result, preexisting councils with civil society participation, which were ineffective and only of a consultative nature, became responsible for the elaboration and evaluation of federal policies.

Moreover, considering that between 2010 and 2014 approximately 500,000 families received rural extension services funded by federal resources and that during the 1990s this situation was practically nonexistent, PNATER's reach to family farming is undeniable. However, there is little information on the direct impact of PNATER at the field level, either because of the difficulties of assessing the impacts of agricultural extension activities or because of the inability to establish a monitoring process for this purpose. PNATER is a public policy implemented with the support of federal resources, which are dependent on political decisions for their release. This can undoubtedly jeopardize the policy in the long term, as changes in government can lead to changes in priority. Disruptions in agricultural extension policies due to the change of government have occurred in the past and may happen in the future, even considering the existence of the Rural Extension Law.

The low recognition that PNATER has with the urban sectors of society further strengthens this possibility. The recognition of the importance of state-supported agricultural extension service is restricted to those sectors directly involved with or benefiting from the policy. The documents produced at the Technical Advisory and Rural Extension National Conferences and the Technical Advisory and Rural Extension National Assessment Seminar expressed concern about the political and financial sustainability of the policy.

It is clear that, despite the obstacles, PNATER has advanced in the construction of a national system of agricultural extension with participatory processes in which the development, operationalization, and evaluation of the policy are participative and democratic. This process has encouraged different social actors involved (farmers, advisers, universities, nongovernmental organizations, public service providers, social movements, etc.) to introspect the policy and commit to its continuity, which can be an essential guarantor of the PNATER's long-term sustainability. In other words, dismantling PNATER would not be an easy task, even if there was a political interest in doing so.

The existence within PNATER's structure of actors who self-finance a significant part of their activities should be highlighted, as in the case of public state-level agricultural extension organizations and private organizations that provide paid services. According to data from the Brazilian Association of State-Level Public Organizations for Technical Advisory and Rural Extension (ASBRAER 2014), in 2010 only 8 percent of the budget spent with agricultural extension in Brazil came from the federal government, with state governments accounting for 79 percent of these expenditures. From these data, we can infer that PNATER's primary function was not to finance agricultural extension in Brazil, but rather to organize this type of service demanded by Brazilian society, giving it the direction and political support it needed to reach the marginalized population.

The recent creation of the National Agency for Technical Assistance and Rural Extension in 2013 was a strategy of the federal government to guarantee greater sustainability of the policy in the long term. It was an attempt to offer a greater autonomy to its administrators to carry out its activities, but as different authors point out, this is a path with substantial built-in risks, which are mainly associated with the retrieval of the old rural extension model (for example, technology transfer processes, broader public-state service providers, limited social participation, centralized and homogenized processes) (Thomson, Bergamasco, and Borsatto 2017; Zarnott et al. 2017).

Since the mid-2000s, Brazil has developed and implemented a national policy for an agricultural extension with innovative characteristics. Yet, as highlighted by Diesel and Miná Dias (2016), the Brazilian case emphasizes the importance of the political economy dimension in the designing and implementation of innovations in agricultural extension systems. On the one hand, PNATER, by serving more than 500,000 family farmers, was part of a set of public policies that helped to eradicate hunger and reduce poverty in Brazil. On the other hand, its implementation continues to face great difficulties in fulfilling its ambitious original objectives. The study of the Brazilian case raises an important question: Is it possible to implement innovative agricultural extension systems for smallholder farmers that can simultaneously reduce poverty, ensure food security, foster rural development, and promote environmental preservation?

Unfortunately, the Brazilian case does not offer a final answer to this question. But it helps to understand the challenges of moving in this direction. Perhaps the greatest lesson we can draw concerns the role of the state in the system. We thus conclude with brief considerations regarding the state's role as financier, coordinator, and executor of agricultural extension and rural services. In the case of developing countries, to expect family farmers to self-finance their agricultural extension systems is at best utopian. The Brazilian evidence indicates that the state plays an essential role in promoting improvements in the productive systems and living situation of family farmers, which requires public investments. Evident in the case studied is the need for public finances to operationalize agricultural extension activities aimed at family farmers as well as to train technical staff to carry out this work. At the same time, the difficulties of making this agenda relevant to society in general and consequently in the political debates are apparent.

Regarding the coordination of the agricultural extension system, the Brazilian case highlights the importance of this process being decentralized and promoting greater social participation, which has the central role in breaking the path-dependence of the state and introducing innovations in agricultural extension systems. It is difficult for the state to coordinate a plural and complex system like the Brazilian one. Past research has shown that the higher the level of central coordination of the system, the more it moves away from the demands of the stakeholders. Moreover, when the coordination processes are shared, stakeholders tend to engage more with the policy, assuming responsibilities and commitments, and strengthening the policy. Also, the available evidence indicates that the state should act more as a promoter of a decentralized system (for example, structuring ways to understand and meet the demands of other stakeholders) rather than assume the role of its chief coordinator. The examples of the two Technical Advisory and Rural Extension National Conferences, in which tens of thousands of people participated in each, and the National Council for Rural Development, Agrarian Reform, and Family Farming demonstrate how much social participation contributes to the process of building and implementing innovations.

Regarding the execution of the extension and rural advisory services, the Brazilian case demonstrates that the state does not need to be, nor should it be, the leading service provider. Often, other types of organizations (such as NGOs, farmer-based organizations, and private organizations) can perform this service with higher quality and efficiency. However, this does not mean that public providers do not have an important role to play; their back-office structure allows for better integration between research and extension, preparation and publication of educational material, dissemination of activities, and support to other public policies, among other possibilities. The Brazilian case demonstrates that a pluralistic system of providers brings various advantages and should be a path to follow.

Policymakers should ensure that extension staff has the necessary resources for the job at hand. Most of all, there should be sufficient numbers of staff to carry out the policy mandate. In Brazil's case, the intention was to cover the entire country, which was not possible. Choices must be between quality and quantity; if too few agents are trying to cover too many clients, quality of their work will suffer. There is also a need to develop dialogue channels together with policies between extensionists and the local public authorities. Extensionists are in rural areas and understand well the problems of smallholders. Thus they can be essential interlocutors, supporting local farmers' organizations, and communicating rural communities' demands for services such as infrastructure related to transport, education, housing, and other needs.

It is essential that governments have coherent and complementary policies across ministries and departments. Policymakers need to ensure that extension staff are well-trained and have the necessary material resources and sufficient numbers for the job. More participatory and horizontal channels between extensionists and local public authorities are urgently needed so that extensionists can effectively act as intermediaries between family farmers and the government. Finally, the Brazilian experience shows that a change in government can suddenly affect agricultural development policies, revealing its high vulnerability. Finally, the Brazilian experience shows that even innovative policies may have complications in execution, possibly leading to watering down some of the original innovations. Many factors outside of the technical aspects of designing and implementing a policy play a role, and one must be aware of potential political economy elements that may step in to affect policy implementation, however well designed the program intervention is.

This chapter examined perceptions of family farmers and extensionists regarding extension services resulting from the agricultural extension policy in Brazil. The results show that participatory methodologies should guide service delivery, that a agroecological principles should be the technical guideline, and that family farming should be the primary beneficiary, receiving free and universal (and pluralistic) services.

References

Albaum, G. 1997. "The Likert Scale Revisited." *International Journal of Market Research* 39: 1–21. https://doi.org/10.1177/147078539703900202.

Allen, I. E., and C. A. Seaman. 2007. "Likert Scales and Data Analyses." *Quality Progress* 40: 64–65.

Alston, J. M., C. Chan-Kang, M. C. Marra, P. G. Pardey, and T. J. Wyatt. 2000. *A Meta-Analysis of Rates of Return to Agricultural R&D Ex Pede Herculem?* Washington, DC: International Food Policy Research Institute (IFPRI).

Anderson, J. R. 2007. "Agricultural Advisory Services." Background paper for *World Development Report 2008, Agriculture for Development.* Washington, DC: World Bank.

ASBRAER (Brazilian Association of State-Level Public Organizations for Technical Advisory and Rural Extension). 2014. *Assistência Técnica e Extensão Rural no Brasil: Um Debate Nacional sobre as Realidades e Novos Rumos para o Desenvolvimento do País.* Brasilia, Brazil: ASBRAER.

Balla, J. V. Q., L. M. Massukado, and V. C. Pimentel. 2014. "Panorama dos Cursos de Agroecologia no Brasil." *Revista Brasileira de Agroecologia* 9: 3–14.

Bellen, H. M. V. 2010. *Indicadores de Sustentabilidade: Uma Análise Comparativa*. 2nd edition. Rio de Janeiro: Editora FGV, 2010.

Benson, A., and T. Jafry. 2013. "The State of Agricultural Extension: An Overview and New Caveats for the Future." *Journal of Agricultural Education and Extension* 19 (4): 381–393. DOI: 10.1080/1389224X.2013.808502.

Bergamasco, S. M. P.P ., R. S. Borsatto, and C. R. Thomson. 2016. *Evaluation of Extension Reforms in Brazil*. Final report presented to IFPRI from the Research Study "Evaluation of Extension Reforms in Brazil." Washington, DC: IFPRI.

Bergamasco, S. M. P. P., C. R. Thomson, and R. S. Borsatto. 2017. "Da Extinção da Embrater à Criação da Anater: Os Desafios da Política de Assistência Técnica e Extensão Rural Brasileira." In *Agricultura Familiar Brasileira: Desafios e Perspectivas de Futuro*, edited by G. C. Delgado and S. M. P. P. Bergamasco, 312–340. Brasilia, Brazil: Ministério do Desenvolvimento Agrário.

Birkhaeuser, D., R. E. Evenson, and G. Feder. 1991. "The Economic Impact of Agricultural Extension: A Review." *Economic Development and Cultural Change* 39 (3): 607–640.

Birner, R., and J. R. Anderson. 2007. *How to Make Agricultural Extension Demand-Driven? The Case of India's Agricultural Extension Policy*. IFPRI Discussion Paper 729. Washington, DC: IFPRI.

Brazil, IBGE (Instituto Brasileiro de Geografia e Estatística). 2006. Censo Agropecuário 2006. "Resultados Preliminares." Accessed February 15, 2016. ftp://ftp.ibge.gov.br/Censos/Censo_Agropecuario_2006/>.

———. 2010. Censo 2010. www.censo2010.ibge.gov.br. Accessed August 12, 2010.

Brazil, MDA (Ministry of Agrarian Development). 2010. *Política Nacional de Assistência Técnica e Extensão Rural para a Agricultura Familiar e Reforma Agrária (PNATER) e o Programa Nacional de Assistência Técnica e Extensão Rural na Agricultura Familiar e na Reforma Agrária (PRONATER)*. Brasilia, Brazil: MDA.

Brazil, MDA and SAF (Secretaria de Agricultura Familiar). 2004. *Política Nacional de Assistência Técnica e Extensão Rural—PNATER*. MDA, SAF, Grupo de Trabalho Ater, 2004, 22p. Accessed March 18, 2015. https://www.ipea.gov.br/participacao/images/pdfs/conferencias/2CNDRSS/2cndrss%20poltica_nacional.pdf.

Caporal, F. R. 2011. "Lei de Ater: Exclusão Da Agroecologia e Outras Armadilhas." *Agroecologia e Desenvolvimento Rural Sustentável* 14 (1): 23–33.

Caporal, F. R., and J. A. Costabeber. 2002. "Construindo uma Nova Extensão Rural no Rio Grande do Sul." *Agroecologia e Desenvolvimento Rural Sustentável* 3: 10–15.

Cardoso, I. M., N. A. de Sousa, M. V. de Aguiar, L. Damigo, and C. Amâncio. 2018. "Núcleos de Agroecologia: Tecendo redes de solidariedade, diversidade e resistência." *Revista Brasileira de Agroecologia* 13: 3–7.

Chianca, T. 2008. "The OECD/DAC Criteria for International Development Evaluations: An Assessment and Ideas for Improvement." *Journal of Multidisciplinary Evaluation* 5: 11.

Chowdury, A. H., H. H. Odame, and C. Leeuwis. 2014. "Transforming the Roles of a Public Extension Agency to Strengthen Innovation: Lessons from the National Agricultural Extension Project in Bangladesh." *Journal of Agricultural Education and Extension* 20 (1): 7–25. DOI: 10.1080/1389224X.2013.803990.

Christoplos, I., P. Sandison, and S. Chipeta. 2012. *Guide to Evaluating Agricultural Extension*. Lindau, Switzerland: Global Forum for Rural Advisory Services (GFRAS).

Compagnone, C., and B. Simon. 2018. "Cooperation and Competition among Agricultural Advisory Service Providers: The Case of Pesticides Use." *Journal of Rural Studies* 59: 10–20. https://doi.org/10.1016/j.jrurstud.2018.01.006.

Corrêa da Silva, H. B. 2011. "Innovations in Extension and Advisory Services for Alleviating Poverty and Hunger: Lessons from Brazil." Paper presented at the International Conference Innovations in Extension and Advisory Services, Nairobi, Kenya, November 15–17.

Davies, H. T. O., S. M. Nutley, and P. C. Smith, eds. 1997. *What Works? Evidence-Based Policy and Practice in Public Services*. Bristol, UK: Policy Press.

Dewey, J. 1927. *The Public and Its Problems*. Athens, OH, US: Swallow Press.

Diesel, V., and M. Miná Dias. 2016. "The Brazilian Experience with Agroecological Extension: A Critical Analysis of Reform in a Pluralistic Extension System." *Journal of Agricultural Education and Extension* 22: 415–433. https://doi.org/10.1080/1389224X.2016.1227058.

Diesel, V., M. Miná Dias, and P. S. Neumann. 2015. "PNATER (2004–2014): Da concepção à materialização." In *Políticas Públicas de Desenvolvimento Rural no Brasil*, edited by C. Grisa and S. Schneider, 107–128. Porto Alegre, Brazil: Editora da UFRGS.

Diniz, P. C. O., J. R. T. Lima, and A. Almeida. 2011. "Chamadas públicas de ATER: Primeiras reflexões." In *Congresso Internacional da Associação Latinoamericana de Sociologia*, 28. Recife. Anais eletrônicos. Recife, Brazil: Universidade Federal de Pernambuco.

FAO (Food and Agriculture Organization of the United Nations) and IFAD. 2017. *Overcoming Hunger and Rural Poverty: Brazilian Experiences*. Brasilia, Brazil: IFAD; Rome: FAO.

Faure, G., Y. Desjeux, and P. Gasselin. 2012. "New Challenges in Agricultural Advisory Services from a Research Perspective: A Literature Review, Synthesis and Research Agenda." *Journal of Agricultural Education and Extension* 18 (5): 461–492. https://doi.org/10.1080/13892 24X.2012.707063.

Feder, G., A. Willett, and W. Zijp. 2001. "Agricultural Extension: Generic Challenges and the Ingredients for Solutions." In *Knowledge Generation and Technical Change: Institutional Innovation in Agriculture*, edited by S. Wolf and D. Zilberman, 313–356. Boston, MA, US: Kluwer.

Howlett, M., and S. Giest. 2015. "Policy Cycle." In *International Encyclopedia of the Social and Behavioral Sciences*, 288–292. New York: Elsevier. https://doi.org/10.1016/B978-0-08-097086-8.75031-8.

Howlett, M., M. Ramesh, and L. Pearl. 2009. *Studying Public Policy: Policy Cycles and Policy Subsystems*.3rd edition. New York: Oxford University Press.

Jannuzzi, P. M. 2009. *Indicadores Sociais do Brasil*. 4th edition. Campinas, Brazil: Editora Alínea.

Klerkx, L., F. Landini, and H. Santoyo-Cortés. 2016. "Agricultural Extension in Latin America: Current Dynamics of Pluralistic Advisory Systems in Heterogeneous Contexts." *Journal of Agricultural Education and Extension* 22 (5): 389–397. DOI: 10.1080/1389224X.2016.1227044.

Klerkx, L., and A. Proctor. 2013. "Beyond Fragmentation and Disconnect: Networks for Knowledge Exchange in the English Land Management Advisory System." *Land Use Policy* 30: 13–24. DOI:10.1016/j.landusepol.2012.02.003.

Knierim, A., P. Labarthe, C. Laurent, K. Prager, J. Kania, L. Madureira, and T. H. Ndah. 2017. "Pluralism of Agricultural Advisory Service Providers—Facts and Insights from Europe." *Journal of Rural Studies* 55: 45–58. http://dx.doi.org/10.1016/j.jrurstud.2017.07.018.

Labarthe, P., and C. Laurent. 2013. "Privatization of Agricultural Extension Services in the EU: Towards a Lack of Adequate Knowledge for Small-Scale Farms?" *Food Policy* 38: 240–252. http://dx.doi.org/10.1016/j.foodpol.2012.10.005.

Liverman, D. M., and S. Vilas. 2006. "Neoliberalism and the Environment in Latin America." *Annual Review of Environmental Resources* 31: 327–363. https://doi.org/10.1146/annurev.energy.29.102403.140729.

Miná Dias, M. 2008. "Políticas Públicas de Extensão Rural e Inovações Conceituais: Limites e Potencialidades." *Perspectivas Em Politicas Públicas* 1:101–14.

Nettle, R., A. Crawford, and P. Brightling. 2018. "How Private-Sector Farm Advisors Change Their Practices: An Australian Case Study." *Journal of Rural Studies* 58: 20–27. https://doi.org/10.1016/j.jrurstud.2017.12.027.

OECD (Organisation for Economic Co-Operation and Development). 1991. "Principles for Evaluation of Development Assistance." DAC Principles for Evaluation of Development Assistance. OECD, Paris.

Oladele, I. O. 2011. *Agricultural Extension Policy: The Missing Link in Innovations in Extension and Advisory Services.* Paper presented at the International Conference "Innovations in Extension and Advisory Services," Nairobi, Kenya, November 15–17.

Otero, J., and D. Selis. 2016. "La Revista 'Extensión En Las Américas' Influencia de Los EUUU Em Los Servicios de Extensión Rural Latinoamericanos." *Extensão Rural* 23 (1): 42–57.

Ozor, N., A. E. Agwu, N. A. Chukwuone, M. C. Madukwe, and C. J. Garforth. 2008. "Cost-Sharing of Agricultural Technology Transfer in Nigeria: Perceptions of Farmers and Extension Professionals." *Journal of Agricultural Education and Extension* 13 (1): 23–37. DOI: 10.1080/13892240601162072.

Peixoto, M. 2008. *Extensão Rural no Brasil— Uma Abordagem Histórica da Legislação, Núcleo de Estudos e Pesquisas do Senado.* Brasilia, Brazil: Consultoria Legislativa do Senado Federal.

Petersen, P., E. M. Mussoi, and F. dal Soglio. 2013. "Institutionalization of the Agroecological Approach in Brazil: Advances and Challenges." *Agroecology, Sustainability, and Food Systems* 37: 103–114. https://doi.org/10.1080/10440046.2012.735632.

Petry, C. 2017. "As Urban Citizens, How to Achieve Autonomy and Food Security?" *Acta Horticulturae* 423–426. https://doi.org/10.17660/ActaHortic.2017.1189.82.

Pettan, K. B. 2010. *A Política Nacional de Assistência Técnica e Extensão Rural (PNATER): Percepções e Tendências.* 393 f. Tese (Doutorado em Engenharia Agrícola na Área de Concentração em Planejamento e Desenvolvimento Rural Sustentável)–Universidade Estadual de Campinas, São Paulo.

Portes, A., and K. Hoffman. 2003. "Latin American Class Structures: Their Composition and Change during the Neoliberal Era." *Latin American Research Review* 38: 41–82. https://doi.org/10.1353/lar.2003.0011.

Prager, K., P. Labarthe, M. Caggiano, and A. Lorenzo-Arribas. 2016. "How Does Commercialisation Impact on the Provision of Farm Advisory Services? Evidence from Belgium, Italy, Ireland and the UK." *Land Use Policy* 52: 329–344. https://doi.org/10.1016/j.landusepol.2015.12.024.

Purcell, D. L., and J. R. Anderson. 1997. *Agricultural Extension and Research: Achievements and Problems in National Systems.* Washington, DC: World Bank.

Qamar, K. 2005. *Modernizing National Agricultural Extension Systems: A Practical Guide for Policy-Makers of Developing Countries.* Rome: FAO.

Rivera, W., and G. Alex, eds. 2005. *National Strategy and Reform Process: Case Studies of International Initiatives.* World Bank Extension Reform for Rural Development Series, Volume 5. Washington, DC: World Bank.

Silva, L. M. S., R. P. Sousa, and W. S. Assis. 2017. "A Educação Superior e a Perspectiva Agroecológica: Avanços e Limites Dos Núcleos de Agroecologia Das IES No Brasil." *Redes* 22 (2): 250–274.

Silva, T. C., and F. R. da Caporal. 2016. "Lei de Assistência Técnica e Extensão Rural em Alagoas-Brasil: Contribuições ou descontinuidade?" *Extensão Rural* 23: 7. https://doi.org/10.5902/2318179621288.

Soares, S., L. Souza, W. de Silva, F. Silveira, and A. Campos. 2016. *Perfil da pobreza: Norte e Nordeste rurais*. Brasilia, Brazil: Centro Internacional de Políticas para o Crescimento Inclusivo (IPC-IG)/PNUD.

Sulaiman, R. V., and A. Hall. 2005. "Extension Policy at the National Level in Asia." *Plant Production Science* 8 (3): 308–319.

Terrapon-Pfaff, J., C. Dienst, J. König, and W. Ortiz. 2014. "A Cross-Sectional Review: Impacts and Sustainability of Small-Scale Renewable Energy Projects in Developing Countries." *Renewable and Sustainable Energy Reviews* 40: 1–10. https://doi.org/10.1016/j.rser.2014.07.161.

Thomson, C. R., S. M. P. P. Bergamasco, and R. S. Borsatto. 2017. "O histórico de criação da Agência Nacional de Assistência Técnica e Extensão Rural (ANATER) e Os Desafios Impostos a Sua Consolidação na Conjuntura Política de 2017." *Retratos de Assentamentos* 20: 70–92.

Wanderley, M. de N. B. 1999. "Raízes Históricas do Dampesinato Brasileiro." In *Agricultura Familiar: Realidades e Perspectivas*, edited by J. Carlos Tedesco. 3rd edition. Passo Fundo, Brazil: Universidade de Passo Fundo.

World Bank. 2010. *Gender and Governance in Rural Services: Insights from India, Ghana, and Ethiopia*. Washington, DC: World Bank and IFPRI.

Zarnott, A. V., V. P. Dalbianco, P. S. Neumann, and M. A. V. Fialho. 2017. "Avanços e Retrocessos na Política de Extensão Rural Brasileira: Análise Crítica Sobre a ANATER." *Revista De La Facultad De Agronomía* 116: 107–119.

Chapter 5

UGANDA

Ephraim Nkonya, Nana Afranaa Kwapong, Edward Kato, Patience Rwamigisa, Bernard Bashaasha, and Margaret Mangheni

5.1 Introduction

Recent efforts to address poverty and food security in Uganda have been directed to increasing agricultural productivity through enhancing agricultural research and extension, both of which saw declining investment in the 1980s and 1990s (Beintema and Stads 2006, 2011). A number of countries took bold steps to modernize agriculture through enhancing agricultural extension, a sector that is key to increasing farmer knowledge on new agricultural technologies and practices, among other objectives (see the broader definition in Chapter 1). Uganda is one of the countries in Africa south of the Sahara that invested significantly in improving provision of agricultural extension and other agricultural development programs.

Uganda was among the pioneers of the pluralistic and demand-driven agricultural extension services when it launched the National Agricultural Advisory Services (NAADS) in 2001. Initial impact of NAADS showed the new approach had a strong and robust impact (Benin, Nkonya, Okecho, Randriamamonjy, Kato, Lubade, and Kyotalimye 2012; Benin, Nkonya, Okecho, Randriamamonjy, Kato, Lubadde, Kyotalimye, and Byekwaso 2011). Despite these significant achievements, the NAADS program experienced significant changes starting in 2013 when Operation Wealth Creation was created—and implemented—by the Uganda military.[1] Operation Wealth Creation's mandate was to monitor and provide logistical support in distribution of agricultural farm inputs among beneficiary farmers. The NAADS program has also been used as an instrument for the Ugandan government's agenda of socioeconomic transformation over the years. The program was characterized by significant political interventions (Kwapong and Nkonya 2015). The model of public funding and privately delivered agricultural

1 Operation Wealth Creation was created in 2013 and the Cabinet reformed the NAADS program in 2014, which resulted in establishment of the single spine extension system in 2015.

extension services as promoted by the NAADS program was abandoned. Provision of advisory services by private and nongovernmental providers, however, continued but with limited government support to specialized areas that the government was not addressing.

The objective of this chapter is to contribute to the policy debate on the changing landscape of agricultural extension and advisory services in Uganda. Particularly, we investigate the effectiveness of different modes of advisory services implemented in Uganda. We compare the effectiveness of pluralistic and demand-driven advisory services with the traditional supply-driven advisory services, which operated along the NAADS approach from 2001 to 2014. Government-affiliated advisory services continued to offer supply-driven advisory services in subcounties where the NAADS program was not operating. We explore the effectiveness of the current agricultural advisory services approach used—with emphasis on gender—to reflect the key role women play in both agricultural production and extension services. We discuss the enabling environment and policies, in which we investigate the historical context of extension services in Uganda.

The contribution of this study to the literature is its use of data collected from extension and advisory services to analyze the effectiveness of traditional and new advisory services. Most studies in the past evaluating effectiveness of the traditional and new agricultural extension services use household surveys (for example, Benin et al. 2010; Davis et al. 2012). About 60 percent of the extension providers interviewed in this study worked under the traditional system before the NAADS program. This allowed them to provide an informed perception of both systems. The results will inform policymakers on more effective approaches for providing agricultural extension services.

The rest of the chapter is structured as follows. Section 5.2 discusses the enabling environment and policies and describes the characteristics of the Uganda extension system, including its past and current policies as well as the institutional landscape. Section 5.3 presents the organization and management of extension and advisory methods, and Section 5.4 the advisory methods used. Section 5.5 summarizes the data sources and analytical methods used to investigate the effectiveness of different modes of advisory services implemented in Uganda. Section 5.6 presents the evidence of access to extension services by service provider types. Section 5.7 summarizes the evidence on effectiveness and impact of extension services, and Section 5.8 analyzes the drivers of targeting of extension services. Section 5.9 provides some concluding remarks and policy implications.

5.2 Enabling Environment and Policies

Since the late 1990s, Uganda's agricultural extension services have experienced dramatic changes that have served as a natural experiment (Chapman and Tripp 2003) of agricultural extension systems in Africa south of the Sahara. Uganda's agricultural services have undergone nine phases over the past century, during which remarkable successes and challenges were observed. Our review of these historical changes draws lessons for Uganda and other countries. We talk about the overall policy environment before addressing organization as well as management and advisory methods.

Semana (1998) identifies seven phases (or "epochs") of agricultural extension and research development in Uganda: 1898–1907, 1920–1956, 1956–1963, 1964–1971, 1971–1980, 1981–1991, and 1992–1998. This chapter includes two more phases: 1999–2009 and 2010–2019. Although the discussion by Semana involved mainly changes on extension service delivery approaches, this chapter analyzes the changes in agricultural policies over time. Changes in agricultural development strategies in Uganda can be linked to the evolution in the rest of Africa south of the Sahara. Since 1998, more than ever before, global forces are influencing agricultural policy development processes in the developing countries of Africa south of the Sahara. These global forces include privatization, liberalization, decentralization, and democratization. Uganda is one of the countries that adopted far-reaching policy and institutional reforms to conform to these global policies.

Early colonial period (1898–1907). Uganda was first colonized by the British around 1860. It became a British protectorate in 1894, and the signing of Buganda Agreement in 1900 effectively institutionalized British rule. One of the economic priorities of the British colonial government was the production of export agricultural crops including coffee, cotton, tea, and tobacco. Accordingly, research and extension focused on developing and producing the export crops by mainly using imported technologies with little effort to develop local technologies (Delgado 1995). The belief was that existing Western technology could substantially increase agriculture production. Teaching farmers how to use modern tools and technologies was the main component of this strategy.

Enforcement of colonial policy (1920–1956). Following the introduction of export crops, the mode of agricultural development in Uganda was mainly regulatory. Agricultural regulations and bylaws were established and enforced by local chiefs who wielded powers. The emphasis was on distribution of planting materials of the major crops and simple messages related to crop production technologies. The bylaws required farmers to produce certain crops in accordance with specified agricultural practices, soil conservation, and storage of famine food reserves. The colonial policy therefore relied heavily on regulations and included use of corporal punishment to attain compliance, which had long-lasting psychological effects on the farmers. Farmers were directed on *what* to do and *how* to do it, with no concrete explanation on *why* they had to do it. For example, coffee—regarded as the major export crop promoted by the colonial rule—was nicknamed *kiboko* (whip or cane), since its production practices were forcibly enforced on farmers. Likewise, farmers growing crops on steep slopes were forced to construct soil and water conservation structures. One of the independence celebrations was destruction of the soil and water conservation structures (NEMA 2007). The use of heavy-handedness may have affected the mind-set of farmers who held negative attitudes toward extension service providers.

Use of progressive farmers (1956–1963). In an effort to promote technology development and transfer, the colonial administration adopted the policy of using progressive farmers to enhance production. Extension was concentrated on progressive farmers with the hope that their improved performance would serve as motivation for the rest of the community to adopt the same techniques. The emphasis was to concentrate technical advice and support in the form of inputs and credit on selected progressive farmers. The expectation was that the improved performance of these farmers would have a demonstration and multiplier effect for increased production and productivity. Although this strategy was desirable given the limited number of extension staff, it failed to achieve the expected results because the community perceived the progressive farmers as a privileged class whom they could not emulate. In some instances, progressive farmers were not willing to share the information and resources with the rest of the community. During this period the colonial policy emphasized an advisory approach to agricultural extension service delivery but also relied heavily on regulatory methods. The agricultural extension staff worked closely with chiefs to enforce bylaws to ensure that the advice given was put into practice. This period saw a surge in agricultural production and productivity, particularly of export crops, although the overall agricultural growth slowed down (World Bank 1993).

Postindependence period (1964–1971). After gaining independence in 1962, the postcolonial government embarked on a radical departure from the colonial policy. The colonial regulation and *kiboko* approach left a negative legacy, which the newly independent government had to address. The new government encouraged an educational and advisory approach to agricultural extension and technology development. Many agricultural colleges were established and/or upgraded to produce skilled personnel, and courses for practical and communication skills were introduced in the curriculum of agricultural colleges. This was further complemented by in-service trainings within the public service. To ensure technological advancement, existing research stations/institutes were strengthened, and new ones were established. District Farm Institutes were established all over the country to train farmers, and the government encouraged and supported the research–extension–farmers link to attain technological advancement. A number of pilot projects and programs were initiated to establish the most applicable educational methods.

The new government expanded provision of agricultural advisory services to nonexport crop farmers, and smallholder farmers became the target for extension services (Ellis and Biggs 2001). Extension agents were posted to administrative centers to offer advisory services on all major crops, livestock, and forestry. This was due to a conceptualized approach of various line ministries and departments. However, there was weak harmonization of extension messages by the different line ministries. This period registered an annual agricultural growth rate of 10 percent (PMA 2000).

The nondirectional period (1971–1980). This period is also referred to as the dormant phase, characterized by misrule and breakdown of law and order following military takeover of the government in January 1971. The agricultural growth that started after independence was brought to a halt due to civil strife, economic mismanagement, disintegration of public infrastructure and services, lack of private sector investment, scarcity of foreign exchange for agricultural inputs, and the collapse of the emerging commercial agricultural sector. In addition, the research facilities and service delivery systems fell dormant, which weakened the research–extension–farmers linkage. During this period there was no clear agricultural policy to guide agricultural development (Semana 1998), causing the nondirectional period. Consequently, output growth rates fell from an average of 10 percent to –2 percent annually (PMA 2000).

Recovery period (1981–1991). The Idi Amin regime was overthrown in 1979, after which three civilian presidents briefly led Uganda between April 1979 and July 1985.[2] Despite the short-lived presidential regimes, the government launched an economic recovery program, including an agricultural policy with a strong focus on training and better links with research,

2 These were Yusuf Lule (April–June 1979); Godfrey Binaisa (June 1979–May 1980); and Milton Obote (July 1980–December 1985) (Uganda, State House 2019).

farmers, and other related institutions. Most projects/programs designed at the time emphasized smallholder farmers with the main objective of providing agricultural inputs, supporting adaptive research and extension, and strengthening selected institutions. Consequently, from 1981 to 1983 the country experienced a 17 percent growth rate, most of which occurred in the agricultural sector. Renewed political crisis, however, led to negative growth rates of approximately −4 percent in 1984 and −2 percent in 1985 and 1986 (World Bank 1993).

Until 1991, there were parallel extension services in different line ministries or departments and nongovernmental organizations that had mushroomed to fill the gaps left by the public sector during the nondirectional period. This resulted in duplication, conflict, and confusion in the service delivery system. To address this, the government introduced a new policy on agricultural extension services. The Ministries of Agriculture and Animal Industry were merged to form the Ministry of Agriculture, Animal Industry and Fisheries (MAAIF)—a ministry that has retained its composition and structure to date. The idea behind the merger was to rationalize, integrate, and harmonize the use of scarce resources as well as to professionalize extension education through learning and teaching (Semana 2002). The new approach emphasized partnerships and dialogue to promote participation. It marked the beginning of the use of participatory methods and techniques in agricultural research and extension. During this period farmers' indigenous knowledge was tapped through participatory methods. There was a revival in agricultural growth with an annual growth rate of approximately 5 percent from 1986 to 1991, although the export crop gross domestic product declined to −0.8 percent annually, reflecting poor incentives for coffee production (World Bank 1993). The World Bank–sponsored structural adjustments in the 1980s, which focused on macroeconomic balance through aggregate demand management with no strong agricultural component and reduced budgetary allocation to agricultural extension, further weakened an already fragile rural service.

Introduction of institutional reforms (1992–1998). This was a period of consolidating participatory methods and techniques in technology development and dissemination. Extension remained largely educational in its approach under its unified extension service delivery with support from the World Bank and other donors. The approach provided a single line of command from frontline extension staff to the district extension coordinator and then to the ministry headquarters. The approach provided strong linkages between extension and research, with increased sharing of information and joint planning with the farmers as well as preseason workshops. The training-and-visit system was adopted for effective service delivery, and bimonthly training workshops were organized.

However, institutional reforms—including liberalization, decentralization, and privatization and the postconstitutional restructuring of the MAAIF—severely affected these initiatives. The field staff were significantly downsized to the extent that by 1998 the ratio of extension staff to households had declined to only about 1:1,800 (Uganda, MAAIF 1998). Furthermore, the system was complicated by the adoption of the decentralization policy in 1995, where MAAIF extension responsibilities were delegated to the district local governments. The districts lacked capacity to implement MAAIF roles, which severely affected farmers' access to services. The policy reforms introduced during this period had adverse effects on service delivery, which prompted the government of Uganda and its development partners to adopt institutional and administrative changes that culminated in the development of the Plan for Modernization of Agriculture in 2000. This plan was accompanied by radical institutional and administrative reforms within the agricultural sector. In the process, the World Bank reduced support to extension and research in favor of the plan (Semana 2002).

Implementation of institutional reforms (1999–2009). The weakening agricultural extension services gave nongovernmental organizations a bigger role, leading to more diversity of approaches (Friis-Hansen et al. 2003). The concept of popular participation in extension, and in development more broadly, was widely accepted in the 1990s and was incorporated into traditional agricultural extension services approaches (Chipika and Friis-Hansen 2004). It is from this mix of participatory approaches that the idea of demand-driven extension approach became popular. Semana (2002) describes this period as one of uncertainty with a mixture of conflicting views and ideas. In line with the government policies of liberalization, decentralization, and privatization, this period witnessed increased private-sector involvement in the agricultural sector from production through processing to marketing chains. The thrust of reforming agricultural extension during these years can be traced from the development of the Poverty Eradication Action Plan. The Plan for Modernization of Agriculture was used as a platform for implementing the Poverty Eradication Action Plan in the agricultural sector. One of the Poverty Eradication Action Plan's pillars was "enhancing production, competitiveness and incomes," which focused on the transformation of agriculture from subsistence to commercial. To achieve this, the Plan for Modernization of Agriculture prioritized seven policy areas for intervention: research and technology development; national agricultural advisory services; agricultural education; improving access to rural finance; agro-processing and marketing; sustainable natural resource utilization and management; and physical infrastructure.

The Plan for Modernization of Agriculture assigned priority to boosting agricultural extension services through NAADS, which started in 2001. Emphasis was placed on improving extension and research efforts with increased responsiveness to farmers' needs (Uganda, MAAIF and MFPED 2000). The NAADS approach marked a transition from a public-sector supply-driven approach to a private-sector demand-driven and public-funded and privately delivered extension service delivery system. In addition, NAADS advisory services covered postharvest and marketing topics, which were ignored by the traditional agricultural extension services. NAADS also developed and enhanced farmer groups, thus increasing the capacity of farmer-controlled institutional structures and processes for managing NAADS at all levels. Particularly, the mandate of planning, implementation, monitoring, and evaluation of advisory services was vested to farmers, with the local governments playing a predominantly facilitatory role. The new system was designed to ensure farmer empowerment, poverty targeting, gender mainstreaming, deepening decentralization, commercialization, fostering participation, managing natural resource productivity, increasing institutional efficiency, privatization, and market access (Uganda, MAAIF and MFPED 2000). NAADS received worldwide attention from both scholars and practitioners in agricultural extension (Chapman and Tripp 2003). The program received significant financial and technical support from development partners and the government of Uganda. In other words, NAADS became one of the most far-reaching and promising agricultural extension reform initiatives in Uganda and elsewhere in Africa south of the Sahara.

An impact assessment of NAADS's first phase, which ran from 2001 to 2009, showed that direct participation in the NAADS program increased agricultural income by 37 percent to 95 percent, while indirect participation—in which farmers who did not belong to NAADS farmer groups received advisory services from NAADS-affiliated advisory services—led to an increase of agricultural income by 27 percent to 55 percent. The rate of return from NAADS investment was 8 percent to 49 percent (Benin, Nkonya, Okecho, Randriamamonjy, Kato, Lubade, and Kyotalimye 2012). About 77 percent of NAADS farmer groups also reported to be empowered to make their own decisions on the provider of advisory services and to demand specific agricultural technologies. Participation in NAADS increased the propensity to demand improved crop varieties, crop management practices, soil conservation, livestock breeds, postharvest practices, and marketing information (Benin et al. 2010).

However, NAADS's relationship with the Plan for Modernization of Agriculture and other upper institutional structures were not well formulated, which posed challenges. NAADS operated outside the framework of the plan and no efforts were made to rationalize, integrate, and harmonize these projects into the new agricultural framework. NAADS was under the MAAIF, while the Plan for Modernization of Agriculture was under the Ministry of Finance, Planning and Economic Development. Overall, MAAIF as an institution was detached from the plan's establishment, which impacted negatively on the overall implementation of the Plan for Modernization of Agriculture and outcomes of the NAADS program. On the research side, the National Agricultural Research System policy was established in 2003. Like NAADS, the NARS policy advocated for a market-responsive, client-oriented, and demand-driven national agricultural research system consisting of public and private institutions working in tandem for sustainable agricultural growth of Uganda (NARS 2003).

The emergence of the pluralistic agricultural extension service delivery system (2010–2019). Toward the end of NAADS phase I, there was negative press coverage of its performance related to mismanagement and corruption; limited access to extension services; declining agricultural productivity; and increased costs of service delivery. Consequently, the president suspended NAADS twice—in September 2007 and July 2010. Rwamigisa (2014) revealed several factors that were both of design and implementation in nature that affected the NAADS program. These included:

1. The program was designed and implemented on the assumption that the reform would be implemented holistically as envisaged under the Plan for Modernization of Agriculture. As it turned out, five of the seven pillars of the plan were never implemented.

2. During implementation, it was expected that all institutions would play their complementary role, but there were no financial provisions made to ensure institutional compliance.

3. The program design assumed that the public extension staff in local governments would be laid off and retooled to form a critical mass of service providers in the private sector. This was never realized as other broad agricultural functions had to be attended to in local governments.

4. It was also assumed that the local governments had sufficient technical capacity to implement the program. Decentralization, however, had not taken root and local governments were grappling with capacity challenges.

5. The reform assumed that the smallholder farmer had reached a level of development to demand, own, and control the service, which was not the case in practice.

6. Expectations of stakeholders beyond the mandate of the NAADS program were partly due to conflicting messages about the program and external influence. This led to the tendency by NAADS to play multiple roles beyond what would ordinarily be expected of an extension program. For example, the procurement and "distribution" of agricultural inputs was introduced under the Prosperity for All program.

7. Failure of the NAADS program to integrate fully with the existing decentralized extension services created a semblance of two parallel extension service delivery systems that were in conflict.

In view of these factors, political interventions in agricultural advisory services intensified, and the government introduced significant changes to the NAADS program in 2014 and 2015. NAADS's new mandate was "to provide support for the management of agricultural input distribution chains, promotion of strategic commodity interventions, agricultural value chain development, and farmer access to agricultural financing" (NAADS 2018).

The strategic objectives of the new NAADS are:

1. To increase access to agricultural inputs to smallholder farmers. Under this objective, vulnerable groups—namely women, youth, older persons, and people with disability—will be targeted;

2. To support development of agriculture value chains by enhancing agribusiness, value addition, and marketing linkage services;

3. To increase access to agricultural financial services by facilitating development of farmer groups/cooperatives; and

4. To strengthen institutional and collaboration frameworks to enhance operational effectiveness and efficiency.

Achieving the first objective involved input distribution—an activity that is implemented through Operation Wealth Creation. The Uganda military has been given the mandate for Operation Wealth Creation. However, Operation Wealth Creation's mandate is to monitor distribution of agricultural farm inputs among the beneficiary farmers, while provision of technical advisory services is done by public agricultural extension and advisory services by district local governments and coordinated by MAAIF. For objective (2), 12 strategic commodities were selected: tea, cocoa, fruits, maize, beans, rice, cassava, banana, dairy, meat, poultry, and fisheries (NAADS 2018). This is about half of more than 40 enterprises promoted under NAADS phase I (Benin et al. 2012).

The agricultural advisory services have gone beyond NAADS reforms. New policies and strategies for providing technical agricultural services were formulated. The government adopted a new model in 2015 under the name "the Single Spine Agricultural Extension System" (SSES) to provide technical advisory services. MAAIF was restructured and the Directorate of Agricultural Extension Services (DAES) was reestablished. The extension function was transferred from the NAADS secretariat back to DAES. The NAADS secretariat roles were redefined to include input chain management, strategic interventions, and promotion of value-addition technologies. In local governments the extension function was integrated into the production and marketing departments. SSES is a public sector–led extension service delivery system with built-in strong collaborative mechanisms with private-sector involvement in service delivery (Uganda, MAAIF 2016). SSES's objective is commercialization of smallholder farmers and other value-chain actors. Like NAADS, SSES uses farmers' groups and other actors in agriculture value chains across the country. SSES has two main dimensions:

1. Integrated, coordinated, and harmonized pluralistic extension system with multiple providers addressing diverse needs.

2. Extension meets needs along the entire value chain and integrates with other agricultural support services for optimum return on investment.

From the policy perspective, the desired end state is that the public extension system will naturally evolve from direct service delivery to service regulation as commercialization takes root. To provide a clear policy direction, MAAIF spearheaded development of the National Agricultural Extension Policy 2016 and the National Agricultural Extension Strategy 2016/2017–2020/2021. The policy process was highly consultative to ensure domestic ownership of the reform program. The reform builds on lessons learned from previous extension approaches including the NAADS program and incorporates many of the fundamental ideas and practices required by the extension services to deliver quality and timely information, share knowledge, and build linkages across key commodity value chains. Several policy instruments were developed to guide implementation of the new reform model. These included the Ethical Code of Conduct for agricultural extension and advisory service providers, a process of registration and accreditation of agricultural extension and advisory service providers, guidelines and standards for agricultural extension services, procedures for developing agricultural extension materials, and a detailed manual for farmer institutional development among others.

It is too early to make assessment of the performance of SSES in terms of number and type of farmers reached, changes in productivity and income, and other outcomes. However, involvement of the army to distribute free or highly subsidized inputs poses challenges to development of more efficient private agricultural inputs dealers as well as creating dependence by farmers on government support. Now that we have set the scene, the analysis below will dwell with the different extension systems that existed before SSES.

5.3 Organization and Management of Extension

The National Agricultural Extension Policy 2016 advocated for a pluralistic agricultural extension service delivery system with multiple actors both in public and private sectors playing complementary roles. A large number of agricultural extension providers exist in Uganda. In addition to the public system (where different MAAIF departments and local governments provide services), there are farmer organizations, universities, private agrodealers, farmer leaders, and a host of nongovernmental providers and development partners. Inadequate coordination leads to duplication of services and resource waste as well as conflicting messages (Rwamigisa, Kakuba, and Massawe 2019). One of the main objectives of the policy is to establish a well-coordinated, harmonized pluralistic agricultural extension delivery system for increased efficiency and effectiveness.

With multiple challenges confronting agriculture (including climate change, new pests, and diseases), success in the sector requires up-to-date information, knowledge, and skills that farmers can access through an effective and well-organized agricultural extension delivery system. Farmers can only benefit from productivity-enhancing inputs such as improved seeds, fertilizer, and pesticides when they have the knowledge and skills to apply them.

Therefore, to bridge this gap and contribute to the agricultural sector and socioeconomic transformation, an effective institutional framework and organization and management of the services are needed to implement the policies mentioned previously in this chapter.

A number of good agricultural extension practices are promoted under the new policy that influence the organizational and institutional arrangements. Some of the practices include:

1. Professional ethical values in conduct of delivery of extension services.
2. Organizational learning in service delivery.
3. Satisfaction by all parties in the course of service provision.
4. Participation and involvement of all actors in agricultural extension processes including public-private partnerships.
5. Whole family concept is core in agricultural extension service delivery.
6. Observance of cultural differences shall be key in design and implementation of agricultural extension programs.
7. Service delivery is anchored on interest and need of all parties.
8. Cooperation with clientele.
9. Agricultural extension is predominantly a public good.
10. Public and private sector involvement, separately or in partnership.
11. Market-oriented interventions (market is the primary driver).
12. Value-chain approach.
13. Quality assurance in extension services is mandatory.

Based on these underlying good practices, the Single Spine Extension System is structured and implemented at national, zonal, district, and subcounty levels. At the national level, MAAIF is the main actor responsible for developing policy, regulation, and strategy. The Directorate of Agricultural Extension Services is responsible for the management and coordination of SSES at the national and local government levels. Other MAAIF directorates and agencies support the technical functions of SSES. Key actors include line ministries; nonstate actors (including NGOs and civil society organizations, agro-input dealers, traders, and processors); universities; training institutions; and development partners.

At the zonal level, Zonal Agricultural Research and Development Institutes (ZARDIs) coordinate regional/zonal stakeholder platforms for issues that cut across districts. They coordinate regular zonal meetings with district production and marketing departments and nonstate actors. They coordinate multistakeholder innovations platforms, conduct adaptive trials on new technologies in a collaborative manner, and facilitate flow of technologies from research stations to farmers for adoption.

At district level, the district production and marketing departments are responsible for providing agricultural extension services. They plan and conduct training of farmers and other value-chain actors in application of agricultural technologies, awareness campaigns on technologies, markets and marketing information as well as support planning processes at all levels of the value chains. Other technical departments and district level nonstate actors provide technical support. Political and technical administrators oversee and monitor implementation. At the subcounty level, there are at least three technical extension staff (for crop, livestock, and, where necessary, fisheries) that provide extension services directly to the beneficiaries. Other responsibilities include knowledge sharing, supervising, and quality assurance of services provided by nonstate actors. The policy has enhanced the coordination role of frontline extension staff who have to work collaboratively with nonstate actors for harmonized delivery of agricultural extension services.

At the farmer level, farmers are organized into groups or associations according to agricultural enterprises of common interest to facilitate delivery of agricultural extension services. These groups aggregate into high-level farmer organizations at the subcounty, district, and national levels. An unpublished assessment conducted by MAAIF and the Uganda Cooperative Alliance in 2017 on capacity needs by farmer organizations showed that farmer institutions are weak on management, financial systems, and business management. They are engaged only in primary production and weak in value addition and bulk marketing. The SSES therefore has prioritized farmer institutional development in the transformation agenda. At each level, multi-actor platforms shall be formed to, among other things, improve interactions of the various actors and joint planning, monitoring, evaluation, learning, and review by fostering partnerships and linkages, building networks, assessing challenges and opportunities, and undertaking lobbying and advocacy. Figure 5.1 summarizes the SSES institutional arrangements and linkages.

The DAES is structured into two main departments, the Department of Agricultural Extension and Skills Management and the Department of

FIGURE 5.1 Institutional arrangements and linkages for the Single Spine Agricultural Extension System

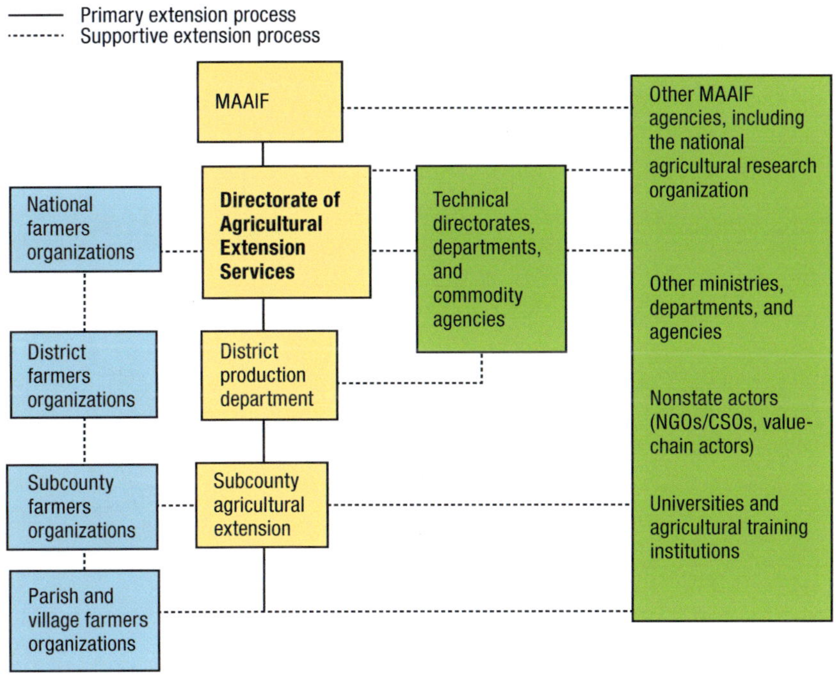

Source: Adapted from Uganda, MAAIF (2018).

Note: CSO = Civil Society Organization; MAAIF = Uganda's Ministry of Agriculture, Animal Industry and Fisheries.

Agricultural Investments and Enterprise Development (Figure 5.2). Figure 5.3 shows the district production and marketing department structure.

The pluralistic nature of the SSES gives the DAES a complex coordinating role as depicted in Figure 5.4, which illustrates that the public agricultural extension system under SSES is not the traditional public extension service as commonly known, but rather a newly oriented system designed to function in a pluralistic extension service delivery system.

Funding for advisory services is provided mainly by the government of Uganda, although there is some support from bilateral donors, the International Fund for Agricultural Development, and the World Bank. Although budget allocation for agricultural extension and advisory services has been increasing—from about US$10 million in 2015/2016 to about US$36 million in 2018/2019 (MAAIF Agricultural Planning and

FIGURE 5.2 Directorate of Agricultural Extension Services

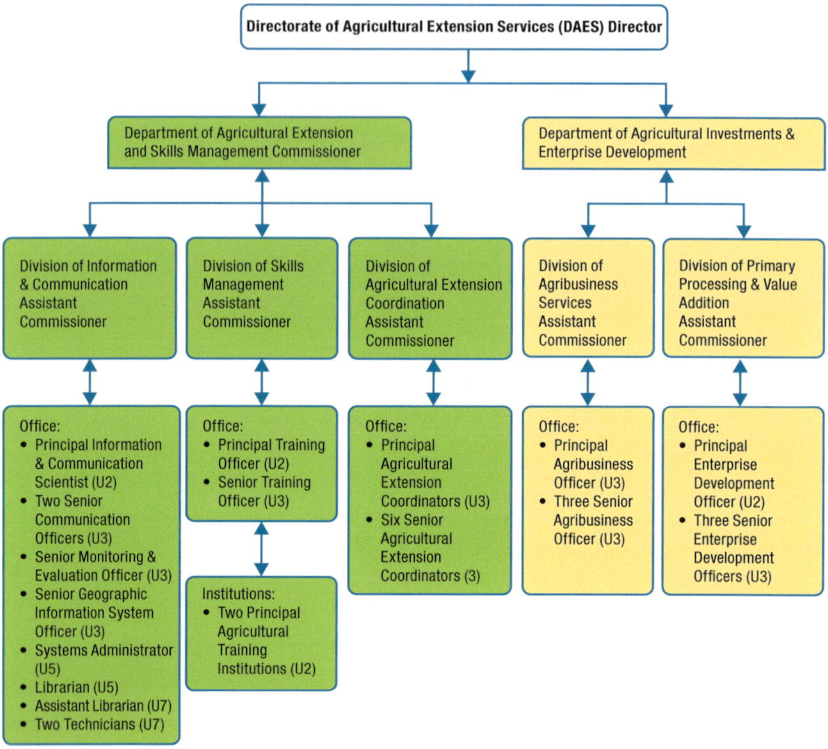

Source: Uganda, MAAIF (2016).

Development Department in Rwamigisa, Kakuba, and Massawe 2019)—public funding is still inadequate to meet the goals of the national extension policy. For instance, the MAAIF budget for 2019/2020 is about US$67 million, but the approved budget is only about US$36 million (Rwamigisa, Kakuba, and Massawe 2019). Thus the funding gap is about 54 percent of MAAIF budget estimates. In addition to being inadequate, funding is often late (Rwamigisa, Kakuba, and Massawe 2019).

Due to this inadequate budget, staff recruitment for extension personnel is not complete. DAES is operating at 38 percent staffing level and local governments at 78 percent. Because the 2019/2020 budget is unchanged from the previous year, there are limited funds to recruit new staff (Rwamigisa, Kakuba, and Massawe 2019). However, government has been slowly recruiting new staff and by December 2018 had recruited 3,872 (or 78 percent) of targeted 5,000 extension workers. This has reduced the ratio of extension worker

FIGURE 5.3 Structure of the District Production and Marketing Department

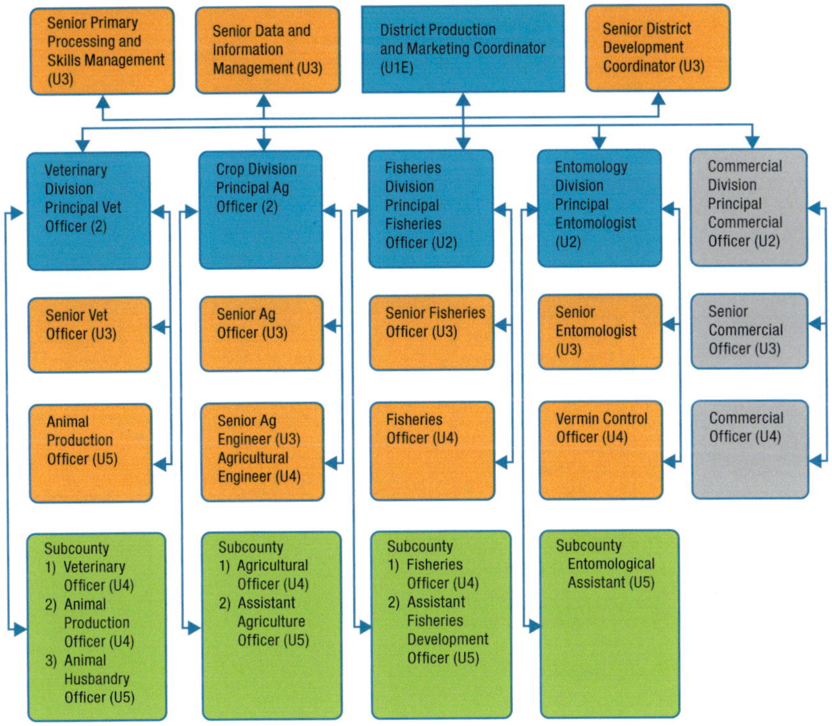

Source: Uganda, MAAIF (2016).

to households from 1:5,000 in 2014 to 1:1,800 (Uganda, MAAIF 2018 in Rwamigisa, Kakuba, and Massawe 2019). The recommended level is 1:500.

Also hindering extension's effectiveness is lack of transport. Although the government provided 137 motor vehicles and 1,061 motorcycles to local governments in March 2019 with support from the International Fund for Agricultural Development, this covered only 20 percent of what was needed (Rwamigisa, Kakuba, and Massawe 2019).

5.4 Advisory Methods

As most countries today, Uganda has a pluralistic set of extension methods to meet diverse goals and serve various types of clientele. Mass media methods include radio, television, leaflets, fliers, and short message services. In addition, demonstrations and shows are also used to reach large audiences.

FIGURE 5.4 Coordination of the system by the Directorate of Agricultural Extension Services (DAES)

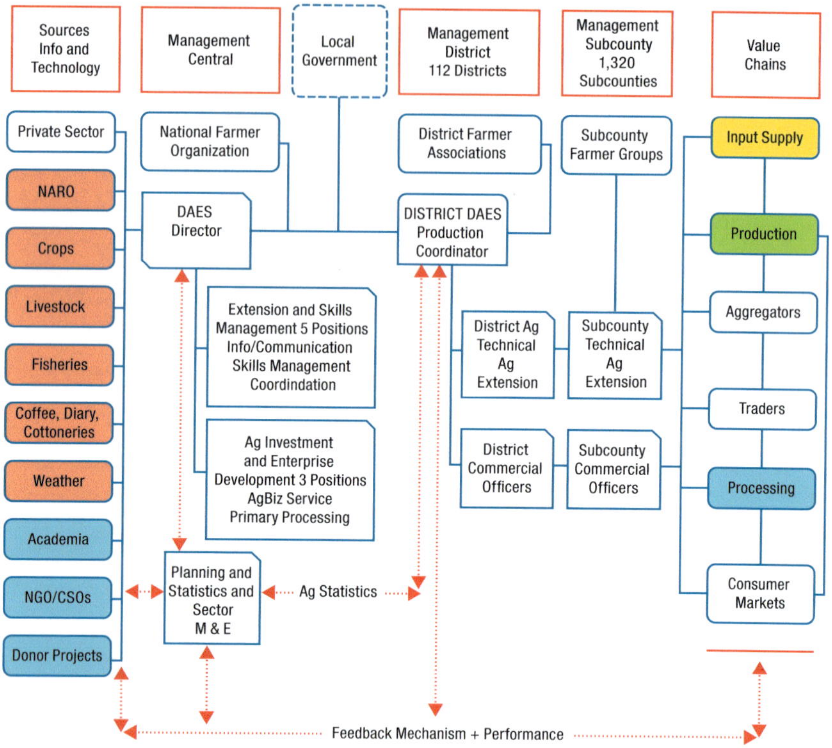

Source: Uganda, MAAIF (2016), working papers.

Note: CSO = Civil Society Organization; M&E = monitoring and evaluation; NARO = National Agricultural Research Organisation. Dotted lines represent feedback; solid lines represent coordination.

Information and communications technologies are useful in that they can be reused and can reach large audiences at little cost. More interactive approaches using information and communications technologies include interactive voice response and participatory radio and video. In addition, more sophisticated methods such as the use of drones are also being piloted to support crop research data acquisition and interpretation services (UFAAS 2019). Group approaches are very common. In particular, farmer field schools are popular as a group-based, adult education approach that is participatory and empowering. The field schools are used with young people and pastoralists as well as with more traditional crop farmers.

As a means of reducing costs and extending reach, the use of farmers and local community members as extension providers is becoming predominant in Uganda. The "village agent model" is being quickly scaled up across the country. The model was started by Sasakawa Global 2000 and modified through support from the United States Agency for International Development (USAID). MAAIF is currently in the process of training some 32,000 village agents over the next few years. Apart from delivering advisory services, village agents are to be at the center of an elaborate system whereby farmers make seasonal production contracts with agents and their traders to be verified and enforced by the government agricultural extension service at the district local government level.

Now that we have covered the history of Uganda's extension policies and enabling environment, the organization and management of extension, and methods used, we look at the performance of extension using several analytical methods. Recent research has shed light on the effect of some of these methods. Van Campenhout, Spielman, and Lecoutere (2018) compared a control group, a group that was shown a video about maize practices, a group that was shown a video and also given access to an Interactive Voice Response (IVR) service, and a group that receives all the former in addition to an SMS text campaign that reminds the farmer about the IVR service. As a result of being shown the video, farmers increased knowledge. However, researchers did not find incremental effects for the addition of IVR and SMS reminders. The researchers also examined gender effects and found that (1) there was evidence that showing video only to the female co-head reduced household-level knowledge as opposed to when the information is given to the man only, (2) there was evidence that targeting the couple instead of only the man leads to an increased likelihood of adopting practices; and (3) there was evidence that changing the messenger attribute (for example, videos where males, female, or couples were acting) did not have any effect on household-level outcomes (Van Campenhout, Spielman, and Lecoutere 2018).

5.5 Data Sources and Methods

Two datasets were used in this study as described in Table 5.1—one from farmers and the second from providers of agricultural advisory services.

We use the 2015–2016 Living Standard Measurement Survey (LSMS) data to determine the gender-disaggregated access to agricultural extension services from providers with different institutional affiliation. We also use the same data to determine the type of extension messages given to farmers from

TABLE 5.1 Datasets used in this chapter

Dataset	Description	Year collected	Analytical method
Agricultural extension agents' survey	A survey of agricultural extension agents was done by IFPRI and Makerere University in 2007/2008. The survey covered agents with four different institutional affiliations—including (with percentage of respective institution in brackets) government (63%); National Agricultural Advisory Services (NAADS) (22%); nongovernmental organizations (10%); and others (5%). The others include cooperatives and private providers (for example, input traders, etc.)	2007/2008	• Censored Least Absolute Deviations estimator (CLAD)—to determine share of all farmers served in the area of jurisdiction and female farmers served • Probit model—to determine type of technologies promoted
Living Standard Measurement Survey	Nationally representative data collected by Uganda Bureau of Statistics. The survey devotes a section on agricultural extension services which collects data on agricultural extension agents and their institutional affiliation and the type of extension messages they provide. The survey also collects information on performance of the extension agents on providing supply and demand-driven advisory services	2016/2017	Descriptive analysis—to determine: 1. the gender-disaggregated access to agricultural extension services from providers with different institutional affiliation and 2. the type of extension messages given to farmers from agents with different institutional affiliation

Source: Authors.

extension agents with different institutional affiliations. We use descriptive analysis for the LSMS data. Using data collected from agricultural advisory service providers in 2008, we analyze the effectiveness of agricultural extension approach using four outcomes:

1. Farmers served by an agricultural extension agent as share of farmers in area of jurisdiction;
2. Share of female farmers served by an agricultural extension agent;
3. Type of technologies promoted; and
4. Type of farmers targeted by an agricultural extension agent.

Since the share of all farmers served in the area of jurisdiction and female farmers served are double censored (0 and 1) dependent variables, we use a two-limit tobit model. However, the two-limit tobit model assumes homoscedasticity (which occurs when the error does not change across observations of the independent variables) and normality (Long 1997). To address

TABLE 5.2 Type of econometric models used

Outcome	Type of dependent variable	Model
Farmers served as share of all farmers in area of jurisdiction	Double censored (0–1)	Tobit and CLAD
Female farmers served as share of total female farmers in area of jurisdiction	Double censored (0–1)	Tobit and CLAD
Type of technologies promoted	Dichotomous (0,1)	Probit or logit
Targeting poor or female farmers	Dichotomous (0,1)	Probit or logit

Source: Authors.
Note: CLAD = Censored Least Absolute Deviations.

potential violation of these assumptions, we used the Powell (1984) Censored Least Absolute Deviations (CLAD) estimator, which takes into account the censored nature of the data and addresses violations of the normality and homoscedasticity assumptions (Vijverberg 1987). The Censored Least Absolute Deviations estimator bootstraps the standard error to achieve robust estimations (Vijverberg 1987). The other models are estimated using probit or logit specifications since their dependent variables are dichotomous. Table 5.2 summarizes the type of models used to analyze the drivers of each outcome or rural services.

We also examined the interaction terms of affiliation with sex of agricultural extension agents (AEA) in order to assess the effectiveness of female extension agents under different affiliations (Table 5.3). The drivers of the share of farmers served and share of female farmers served are estimated using the following model:

$$Y = b_0 + \mathbf{b}_i\mathbf{X}_j + e_i$$

Where Y = the farmers served by an agricultural extension agent as share of total number of farmers in the agent's area of jurisdiction. $0 \leq Y \geq 1$; X is a vector of covariates affecting share of farmers served. The vector of the covariates affecting proportion of farmers served by extension, their expected sign, and justification of the expected sign are summarized in Table 5.3.

$$e_j \text{ error term, } e_j \sim N(0,1).$$

The same covariates are used for analyzing the type of technologies promoted and type of farmers targeted. Since there are many types of technologies promoted and several types of farmers targeted, the justification of their expected signs is not given for brevity reasons. Agricultural extension agents serving in areas with better rural services or in high agricultural potential and

TABLE 5.3 Covariates and their expected signs for share of farmers and female farmers served

Covariate	Sign	Justification
Female agent (cf male)	– (prop of farmers) + (prop of female)	Limited resources to serve but greater likelihood to serve female farmers (Lahai, Goldey, and Jones 2000; FAO 2011).
Age of agent	–/+	Older extension agents may have higher administrative position, which reduces the number of farmers served. But they can also have more resources to serve more.
Ethnicity	+	Extension agents with same ethnicity as area of jurisdiction will serve more farmers due to knowledge of language and familiarity with culture.
Level of education	–	Extension agents with higher education tend to serve at district offices and therefore serve fewer farmers.
Affiliation with government	–	Extension agents affiliated with government historically are less effective (Uganda, MAAIF and MFPED 2000).
High agricultural potential	+	Greater density of farmers, better market access.
Better rural services	+	Better services enhance effectiveness of extension agents.
Female agent x NAADS, NGO	+	Female agricultural extension agents will serve better under NAADS and NGOs due to better facilitation.

Source: Authors.

Note: NAADS = National Agricultural Advisory Services of Uganda.

older agents are also expected to have better access to Internet, motorbikes, and training opportunities.

5.6 Access to Extension Services by Type of Service Providers

LSMS 2015–2016 data has one section focused on agricultural extension services. The questions covered are on whether a household received advisory services from agricultural extension agents affiliated with NAADS, input suppliers, NGOs, cooperatives or farmer associations, large-scale farmers, and other providers. Figure 5.5 shows the institutional affiliation of service providers.

To determine the gender-disaggregated access to agricultural extension services from providers with different institutional affiliation, we use the same data to determine the type of extension messages given to farmers from

FIGURE 5.5 **Institutional affiliation of providers of agricultural advisory services (% of service providers)**

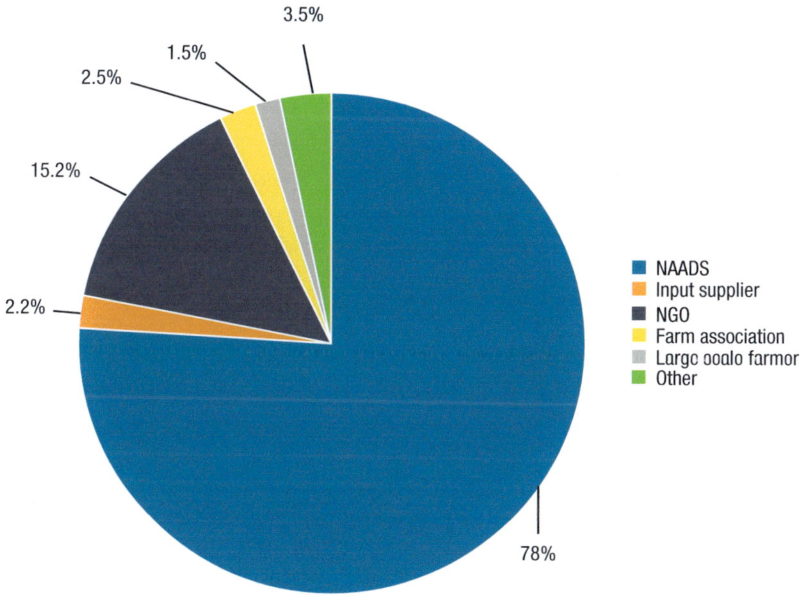

Source: World Bank (n.d.).

Note: NAADS = National Agricultural Advisory Services of Uganda. Because of rounding, numbers might not add up precisely to 100.

agricultural extension agents with different institutional affiliation. To determine the drivers of effectiveness of institutional affiliation in coverage, type of advisory services given, and targeting women and poor farmers, we use data drawn from agricultural extension agents. A total of 208 agricultural extension agents were selected from 16 districts. Since availability of rural services plays a key role in effectiveness of providing rural services, we developed an index of rural services and used the index as a covariate of the outcomes of interest. The rural service index was used to select case study districts.

We used seven rural services and examined their change over time. The seven rural services included in the index were distance to all-weather roads, bank, secondary schools, primary schools, health centers, agricultural input shops, and agricultural extension services. These data were obtained from a household survey conducted by Uganda Bureau of Statistics. We used the varimax rotation methods (Kaiser 1958) to identify major factors that represent the seven rural services included in this analysis. Varimax rotation maximizes the sum of the variances of the squared correlation (loadings) so that each individual rural service can be linearly represented using a single index. Factors with eigenvalues of greater than 1 were retained (Stata 2007). Only four factors were identified, which explained 77 percent of the variance. We then used the principal component factors generated from factor analysis to identify the districts with comparable level of rural services. An average hierarchical method was used to define the number of clusters (Stata 2007).

Table 5.4 shows the three indexes of rural services: (1) poor rural services; (2) medium rural services; and (3) best rural services. Rural services are best in the central region around Lake Victoria and are generally poor to medium in the northern districts. Western and eastern districts generally have medium to high level rural services. Figure 5.6 shows the results of quality service on a map. The figure shows very poor services in the northeast area of the Karamoja, which was not included in our study due to insecurity concerns during the study period. We then purposively selected districts from three of four rural service indexes. However, we did not sample agricultural extension agents from north and east districts. All districts in the north and east districts have very poor rural services. Purposive selection of districts was made to ensure that districts from each of the four administrative regions (Central, East, North, and West) of Uganda were represented. In addition, agricultural potential of the districts was considered in selecting the districts. Table 5.4 reports the selected districts and the index of rural services under each of the four geopolitical zones.

In each of the districts selected, a sampling frame was developed by listing all agricultural extension agents operating in a given district. A few agents were affiliated with the crop development authority or other institutions but their numbers were too few to form an independent group (Table 5.5). Likewise, there were only a few private agricultural extension agents working independently of NAADS and therefore not included in the analysis. But a number of independent agents worked under NAADS and were thus treated as NAADS-affiliated. NGO-affiliated agents were put into two groups. The first group included NGOs affiliated with NAADS and the other, those unaffiliated with NAADS. The table reports the percentage of agents and their affiliation across the four geopolitical regions.

About 63 percent of the 208 agricultural extension agents interviewed were affiliated with the government, while 22 percent and 10 percent, respectively, were associated with NAADS and NGOs (Table 5.5). The government agricultural extension agent affiliation was larger than expected largely due to the survey being conducted in 2008, during which the government was reemploying extension to prepare NAADS phase II. However, during the

TABLE 5.4 Selected districts and performance of rural services development under the geopolitical zones

Central		East		North		West	
District	Rural service performance	District	Rural service performance	District	Rural service performance	District	Rural service performance
Masaka	Best	Kumi	Medium	Lira	Medium	Bushenyi	Medium
Luweero	Best	Soroti	Poor	Gulu	Medium	Isingiro	Best
Mubende	Best	Mbale	Poor	Yumbe	Poor	Kabale	Medium
Nakasongola	Best	Katakwi	Medium	Pader	Unclassified	Kamwenge	Poor

Source: Authors.
Note: Sampling was done when Uganda had 56 districts. Uganda now has 127 districts.

TABLE 5.5 Type of extension service providers across regions

Affiliation	Percentage of total interviewed (n = 208)
Government	63
NAADS	22
Nongovernmental organization	10
Others	5

Source: Agricultural extension agent survey.
Note: NAADS = National Agricultural Advisory Services of Uganda.

time of the survey, government-affiliated agricultural extension agents were not yet working with NAADS. Other types of agents (crop authorities, independent and farmer organizations) only account for 5 percent of the total agent sample. Female agents accounted for only 11 percent of the agricultural extension agents in the country, and their number was significantly smaller than the equivalent of male agents in all types of affiliations (Table 5.6). This is comparable with the national average, which is 12 percent. Consistent with Swanson (2008), who noted that NGOs providing agricultural advisory services are focused on supporting women and other forms of social capital formation, the NGOs reported the largest share of female agents (26 percent), while the government reported the lowest share (10 percent).

The majority of the agricultural extension agents have a diploma (Table 5.7), which is a certificate obtained after high school or after ordinary-level secondary school followed by a college training that gives a certificate. The level of education is not significantly different across types of affiliation, but the share of government agricultural extension agents with first degree

FIGURE 5.6 Index of performance of rural services at district level

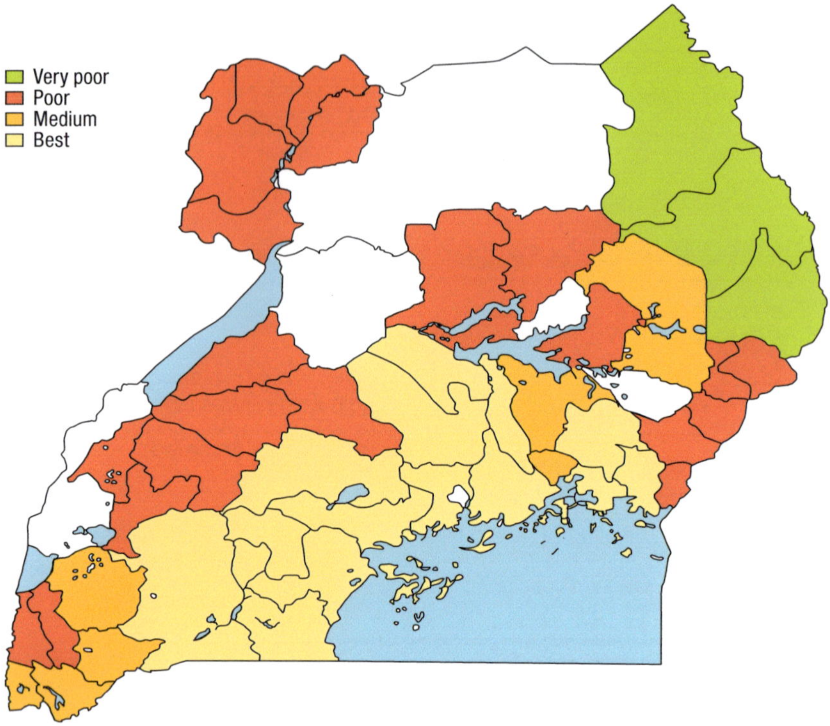

Source: Authors collected primary data and created the map.

Note: White space represents missing data: no information was collected from these districts.

TABLE 5.6 Gender of agricultural extension agents

Affiliation	Male (%)	Female (%)	Paired test (P-value)
Government (n = 137)	89.7	10.3	0.000***
NAADS (n = 48)	83.3	16.7	0.000***
Nongovernmental (n = 23)	73.1	26.9	0.015**
All (n = 208)	89.0	11.0	0.000***

Source: Agricultural extension agents survey.

Note: NAADS = National Agricultural Advisory Services of Uganda. Asterisks denote significance levels: *, ** & *** shows significant at P = 0.10, 0.05 & 0.01 respectively.

TABLE 5.7 **Age and education level of providers**

		Education level (% holding)				
Affiliation	Age (years)	Below college	College diploma	Bachelor's degree	Master's degree	Other
Government	44.0	5.1	37.2	48.7	7.7	1.3
NAADS	42.8	0.0	50.0	42.6	7.4	0.0
Nongovernmental	32.2	7.7	53.9	38.5	0.0	0.0
All	42.0	4.2	42.0	46.2	6.8	0.9

Source: Agricultural extension agents survey.
Note: NAADS = National Agricultural Advisory Services of Uganda.

or postgraduate education is highest. Consistent with Swanson (2008), nongovernmental agricultural extension agents reported the youngest age, suggesting that NGOs tend to hire younger agents. Government-affiliated agents reported the oldest age but highest share of agricultural extension agents with bachelor's degrees (Table 5.7). This reflects the government's long-term investments in educating agricultural extension agents. Agents affiliated with NGOs reported the lowest share of agents with college degrees. The young age of the agricultural extension providers could be the reason for such lower level of education.

Table 5.8 shows that the government agricultural extension agents serve the largest number of farmers but the smallest share of female-headed households (7 percent), while NGOs serve the smallest number of farmers. The share of female and male subsistence farmers met is comparable across all three types of affiliations. The major difference across types of affiliation is the share of farmers met. Agents affiliated with NGOs met about 43 percent of the farmers in their area of jurisdiction, while NAADS and government-affiliated agents met only about 20 percent of the farmers in their area of jurisdiction. The results underscore the smaller coverage of NGOs and their apparent targeting of women (Swanson 2008).

As is the case in Nigeria (Nkonya et al. 2010) and other countries, the major topic promoted across all types of affiliation remains improved varieties, which at least 62 percent of the agricultural extension agents promoted (Table 5.9). Promotion of agrochemicals is the second most important topic. Of interest is that 44 percent of the NAADS-affiliated agents promoted use of herbicide, while only 28 percent of government-affiliated agents and 19 percent of NGO-affiliated agents promoted herbicide use.

TABLE 5.8 Farmers in the area of jurisdiction of agricultural extension agents

Affiliation	Number of farmers served	Percentage of farmers met	Percentage of female-headed households met	Percentage of female subsistence farmers met	Percentage of male subsistence farmers met
Government	29,705	22.2	7.2	74.9	54.9
NAADS	13,529	19.2	9.9	76.7	53.1
Nongovernmental	5,005	43.2	16.1	71.5	59.1

Source: Agricultural extension agents survey.
Note: NAADS = National Agricultural Advisory Services of Uganda.

TABLE 5.9 Major topics promoted

	Percentage reporting		
Topics promoted	Government (n = 137)	NAADS (n = 48)	Nongovernment (n = 23)
Improved seed varieties	85.3	81.5	61.5
Agrochemicals	65.4	62.9	61.5
Herbicides	28.2	44.4	19.2
Plant protection techniques	20.5	16.7	19.2
Organic fertilizer	15.4	20.4	11.5
Chemical fertilizers	10.0	11.1	7.7
Agroforestry	8.3	11.1	0.0
Soil conservation technologies	0.6	1.9	7.7

Source: Agricultural extension agents survey.
Note: NAADS = National Agricultural Advisory Services of Uganda.

Promotion of agroforestry was done by only 8 percent of the government-affiliated agents and 11 percent of NAADS-affiliated agents. Similarly, promotion of soil erosion control was quite low (fewer than 8 percent). This is a cause for concern since advisory services seem focused on improved varieties, plant protection, and, to a limited extent, fertilizer. Promotion of organic soil fertility management practices is limited. The econometric analysis takes a more rigorous analysis of the descriptive statistics discussed previously.

5.7 Evidence of Effectiveness and Impact

About 12 percent of households received agricultural advisory services in 2015–2016 (Table 5.10). NAADS remains the largest provider of agricultural advisory services in Uganda. The program reached 78 percent of farmers who

TABLE 5.10 Institutional affiliation of providers of extension services

Institutional affiliation of provider	Percentage of households reporting: Central	Eastern	Northern	Western	National	Female-headed house-hold	Male-headed household	P-value female-headed households = male-headed households
Any provider	7.9	15.5	15.6	7.6	12.0			
NAADS	67.4	88.3	71.4	81.6	78.0	70.2	80.8	0.044**
Input supplier	0.0	0.0	3.3	6.1	2.2	0.0	3.0	0.107*
Nongovernmental	9.3	8.7	23.5	14.2	15.2	20.2	13.4	0.071*
Farm association	9.3	0.0	2.5	2.0	2.5	1.1	3.0	0.358
Large-scale farmer	4.6	0.9	1.6	0.0	1.5	1.1	1.7	0.732
Other	9.3	3.8	2.5	0.0	3.5	7.1	2.1	0.0341**

Source: Computed from LSMS 2015–2016.

Note: NAADS = National Agricultural Advisory Services of Uganda. Responses for access to advisory services for each institution excludes households who didn't have access to advisory from any provider. Asterisks denote significance levels: *, ** & *** shows significant at P = 0.10, 0.05 & 0.01 respectively.

received advisory services in 2015–2016. At the national level, male-headed households had significantly (at p = 0.05) greater access to NAADS than female-headed households—suggesting that NAADS's targeting of women in implementing its strategic objectives has not yet shown favorable results. However, NGOs—which are the second major providers of advisory service—reached significantly higher (at p = 0.10) share of female-headed households than male-headed households—underscoring the nongovernmental organization comparative advantage in targeting women. The eastern region reported the highest access to NAADS advisory services, while the central region reported the lowest access. This is surprising given that the central region has the highest market access.

Given that only NAADS and NGOs are the largest providers of advisory services—and thus with enough observations to conduct statistical tests—the rest of the discussion ignores the other providers. It is striking to observe that participation in NAADS training and farmer groups is below 10 percent at the national level (Table 5.11). In 2007 about 716,000 households—or 20 percent of national farming households—participated in NAADS (Benin, Nkonya, Okecho, Randriamamonjy, Kato, Lubadde, Kyotalimye, and Byekwaso 2011). This means participation in the program has decreased by about 50 percent since 2007. This could reflect reduced budget or declining trust, perception, and effectiveness of NAADS, as recently reported (for example, Turyahikayo and Kamagara 2016; Barungi, Guloba, and Adong

TABLE 5.11 Participation in NAADS training and farmer groups (%)

Region	Participated in NAADS training			Belong to NAADS farmer group		
	Female-headed households	Male-headed households	P-value	Female-headed households	Male-headed households	P-value
Central	4.2	3.8	0.768	4.8	2.4	0.053**
East	8.4	10.7	0.334	4.4	7.1	0.159
North	4.0	10.0	0.003***	4.7	7.5	0.128
West	4.3	10.3	0.009***	2.4	5.8	0.049**
National	5.1	8.7	0.000***	4.2	5.7	0.075*

Source: Computed from LSMS 2015–2016.
Note: NAADS = National Agricultural Advisory Services of Uganda. Asterisks denote significance levels: *, ** & *** shows significance at P = 0.10, 0.05 & 0.01, respectively.

2016). The introduction of the Single Spine Agricultural Extension System (SSES) could have contributed to NAADS's reduced coverage of farmers.[3]

Table 5.11 shows that about 9 percent of male-headed households participated in NAADS training—a level that is significantly higher (at p = 0.01) than female-headed households' participation. Likewise, the share of male-headed households who participated in NAADS training in the north and west regions was significantly higher (at p = 0.01) than the case for female-headed households. There is gender parity in NAADS participation in the central and eastern regions (at p = 10). Compared to male-headed households, a significantly higher (at p = 0.05) share of female-headed households in the central region were members of NAADS groups. This demonstrates that it is possible to achieve gender parity in areas with high market access.

Likewise, even though 61 percent of households reported to be aware of NAADS training, only 16 percent participated in training (Figure 5.7). Only 37 percent of households were aware of NAADS groups and only 18 percent of those aware joined the farmer groups. This is consistent with the results reported in Figure 5.7, which shows limited participation in NAADS training. Awareness of the 12 strategic enterprises is even lower, as only 17 percent of households reported being aware and only 22 percent of those aware participated in their production. Despite the broad consultation done to select the

3 Although farmer group training under NAADS has decreased, giving way to SSES, farmers seem not to have realized the institutional changes and apparently think SSES is NAADS. This could have contributed to the high percent of farmers reporting that they received advisory services from NAADS in Table 5.10, but the very low participation in NAADS farmer groups in Table 5.11, as NAADS farmer groups no longer provide training in its former structure.

FIGURE 5.7 Awareness of NAADS training, groups, and priority enterprises

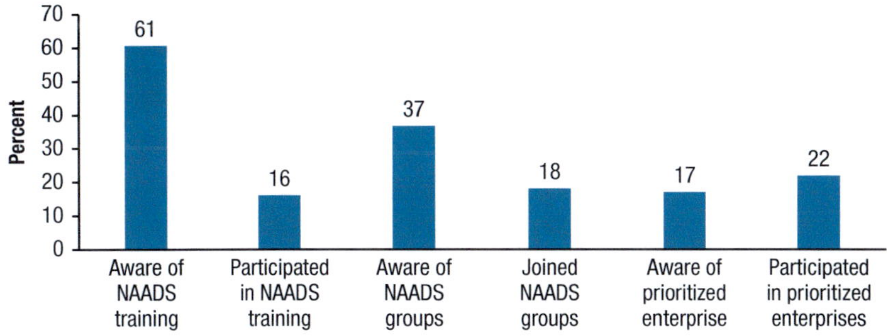

Source: Computed from World Bank (n.d.).
Note: NAADS = National Agricultural Advisory Services of Uganda.

TABLE 5.12 Demand-driven and supply-driven provision of advisory services

	Total number of visits per year	
Service provider	Solicited	Unsolicited
NAADS	99	81
Input supplier	2	0
Nongovernmental organization	57	55
Cooperative/farmer group	8	3
Large-scale Farm	2	6
Other extension	8	14

Source: Computed from LSMS 2015–2016.
Note: NAADS = National Agricultural Advisory Services of Uganda.

strategic enterprises, the low share of farmers producing them calls for strategies to determine relevance and enhance greater uptake.

The key approach of NAADS phase I was demand-driven agricultural extension services. Accordingly, training of farmers was done to increase their capacity to demand advisory services (Benin, Nkonya, Okecho, Randriamamonjy, Kato, Lubadde, Kyotalimye, and Byekwaso 2011). Respondents reported higher frequency of solicited visits by NAADS and NGOs than unsolicited visits—underscoring higher capacity to demand and seek advisory services (Table 5.12).

Of interest is the type of extension messages given. Consistent with its strategic objective of developing agriculture value chains and facilitating market linkages, NAADS delivered advisory services covering the entire value chain

TABLE 5.13 Type of extension messages given by providers, by their affiliation (%)

Extension topics	NAADS (n = 254)	Input supplier (n = 7)	NGOs (n = 49)	Cooperative/ farmers association (n = 8)	Large-scale farmer (n = 5)	Others (n = 11)
Agricultural production	95.6	71.4	91.8	100	80.0	63.6
Agricultural prices	41.7	28.5	57.1	75.0	80.0	45.4
Agro-processing	32.6	28.5	51.0	37.5	40.0	54.5
Crop marketing	37.4	28.5	65.3	62.5	20.0	72.7
Livestock marketing	31.8	28.5	38.7	25.0	0.0	54.5
Fishing production	10.2	0.0	18.3	12.5	0.0	9.0
Meat production	36.2	0.0	26.5	12.5	0.0	18.1
Milk production	23.6	14.2	26.5	25.0	0.0	36.3
Livestock diseases	47.2	57.1	42.8	25.0	20.0	63.6

Source: Computed from LSMS 2015–2016.

Note: NAADS = National Agricultural Advisory Services of Uganda. Analysis restricted to those who received extension services from a given provider.

and all 12 strategic enterprises (Table 5.13).[4] Agricultural production was the most popular topic promoted by all providers. The share of NAADS message recipients was lowest for fish advisory services. This is expected given the low importance of fishing enterprise compared to crops and livestock. It is interesting to note that nongovernmental organizations provided higher advisory services on agricultural prices and processing and crop marketing than NAADS. This further illustrates the comparative advantage of NGOs providing specialized advisory services.

In summary, the household-level results show the impact and effectiveness of NAADS in providing advisory service reforms in Uganda has been reduced. The LSMS 2015–2016 data show the negative short-term effect of the reforms on NAADS effectiveness. Particularly, the coverage of farmers by NAADS has been reduced by about 50 percent, while the role of private extension agents has been reduced. Similarly, farmer groups formed under NAADS phase I have been reduced significantly. These results cast concerns on the ongoing extension reforms and direction of the NAADS program. This chapter takes a deeper analysis of data drawn from agricultural extension agents. We examine the drivers of share of farmers served, targeting of women and poor households, and type of technologies provided. The analysis is done across institutional affiliation of agricultural extension agents.

4 However, the table aggregates all crops into one group.

Drivers of Share of Farmers and Female Farmers Served by Agricultural Extension Agents

For brevity, we focus the discussion on policy-relevant covariates, which include gender, level of education, institutional affiliation of agricultural extension agents, and their access to rural services. Given the key role women play in agricultural production and provision of advisory services, the discussion emphasizes the gender role and its impact. Consistent with other studies (for example, Lahai, Goldey, and Jones 2000; World Bank 2010), female agents were more likely to provide advisory services to female farmers than male agents (Table 5.14). Consistent with a priori expectation, agricultural extension agents with bachelor or postgraduate degrees provided less advisory services than those with a diploma or lower education. This is due to placement, where agents with a degree or a master's degree (46 percent of the total agents interviewed) are given supervisory roles at the district or subcounty level. The agricultural extension agents with master's level education accounted for only 7 percent and therefore are given strategic supervisory roles. The diploma and certificate holders have less specialized professions and are meant to do most of the legwork to advise farmers.

NAADS- and NGO-affiliated agricultural extension agents are more likely to provide advisory services to women than government-affiliated agents. This is consistent with Swanson (2008), who noted the NGO bias toward women and other vulnerable groups. This result is consistent with what farmers reported (Table 5.10). Benin, Nkonya, Okecho, Randriamamonjy, Kato, Lubadde, Kyotalimye, and Byekwaso (2011) observed that female-headed households benefited more indirectly from NAADS than male-headed households. The results underscore the weak capacity of government-affiliated agricultural extension agents to provide advisory services to women farmers and highlight the need to increase their capacity given that the changes in the advisory services are reinstating provision of advisory services to government-affiliated agents. NGO-affiliated agents were more likely to offer advisory services to a larger share of farmers than was the case for the government-affiliated agents. This could be due to better incentives provided by NGOs, small number of farmers served, better working facilities (for example, all of them reported having a motorbike and cell phone), and their young age, all of which lead to more efficient delivery of rural services.

Surprisingly, agricultural extension agents in areas with poor rural services were more likely to provide advisory services to higher proportions of farmers than those in moderate or best rural services. These unexpected results are

TABLE 5.14 Determinants of proportion of farmers served by extension providers in the community

	Proportion of farmers served		Proportion of female farmers served	
Variable	**Two-limit tobit**	**CLAD**	**Two-limit tobit**	**CLAD**
Female agent	7.958	–4.781	1.768	5.498**
Age of agent	–0.233	–0.066	0.054	0.047
Ethnic group (cf. Luganda)				
Nyakitara	3.267	0.056	–0.165	–0.781
Northern	–13.706	–8.147***	–0.963	–3.120*
West Nile	–2.483	–7.159***	–2.96	–3.753*
Eastern bantu	–5.015	3.714*	–0.557	–0.451
Eastern non-bantu	23.754***	13.998***	9.151***	5.230***
Level of education (cf. certificate)				
Diploma	–11.187	–4.776	1.838	–0.208
Bachelor degree	–15.332	–8.115***	1.259	–1.017
Postgraduate degree	–24.988*	–9.751**	2.978	1.98
Other education	–45.702	–0.413	–4.605	–6.459*
Institutional affiliation (cf. Government)				
NAADS	–0.264	1.737	6.404***	1.341
NGO	7.082	4.328*	6.228*	2.919
High agricultural potential	2.086	2.197	–3.054	–0.785
Performance of rural services (cf. Poor)				
Medium performance	–5.515	–4.931**	2.272	–0.875
Best performance	–10.154	–5.898**	0.389	–2.096
NAADS x female agent	–14.114	–2.631	–8.017	–6.287*
NGO x female agent	1.408	21.977***	–1.772	7.633*
Constant		17.391***		1.627

Source: Agricultural extension agents survey.
Note: cf = control or reference group. NAADS = National Agricultural Advisory Services of Uganda. CLAD = Censored Least Absolute Deviations estimator. Asterisks denote significance levels: *, ** & *** shows significant at P = 0.10, 0.05 & 0.01 respectively.

supported by descriptive statistics, due to the large number of farmers in areas with good rural services (Table 5.15). It is likely that agricultural extension agents in areas with medium or best market access are engaged in other activities and do not devote enough time to provide services. However, there is need to examine these puzzling results.

TABLE 5.15 Farmers across levels of rural service development groups

Rural services	Number of farmers in area of jurisdiction	Share served (%)
Poor	38,445	20.7
Moderate	14,143	18.1
Best	13,452	14.5

Source: Agricultural extension agents survey.

Female agricultural extension agents working under nongovernmental organizations provide advisory services to a larger share of all farmers and to women farmers than is the case of female agents affiliated with government. This is additional evidence of the orientation of NGOs to provide advisory services to women farmers.

5.8 Drivers of Targeting of Advisory Services to Different Groups of Farmers

Disregarding institutional affiliation, female agricultural extension agents were less likely to target any group—including women. However, when affiliated with NGOs or NAADS, female agents were more likely to target poor farmers, large-scale farmers, women, export farmers, and young farmers (Table 5.16). Consistent with its objectives of targeting poor farmers, NAADS-affiliated agricultural extension agents were more likely to target poor farmers than government-affiliated agents. NGO-affiliated agents were more likely to target export crop farmers but less likely to target large farmers, female farmers, and livestock farmers. The results conflict with those observed previously. Probable reason for these results could be the inclusion of the interaction terms discussed above. Accordingly, the NGO times female agent and NAADS times female agent interaction terms show positive impact on targeting poor farmers, female farmers, and young farmers.

Drivers of the Type of Technology Provided by Agricultural Extension Agents

While female agricultural extension agents generally were less likely to provide advisory services on fertilizer, agroforestry, and soil erosion control,

TABLE 5.16 Determinants of type of farmer targeted by provider (marginal effects)

Variable	Poor farmer	Large-scale farmer	Export farmer	Female farmer	Young farmer	Livestock farmer
Female agent	−0.250***	−0.095***	−0.053***	−0.164***	−0.133***	−0.035
Age of agent	0.002	−0.002	−0.001	−0.002	0.003	−0.001
Ethnicity of agent (cf. Ganda)						
Nyakitara	−0.013	−0.018	−0.051***	−0.106***	−0.062	−0.125**
Northern	−0.108*	−0.081***	−0.096***	−0.042	−0.049	0.011
West Nile	−0.038	−0.002	−0.050***	0.06	0.046	−0.05
Eastern bantu	−0.138***	−0.03	−0.074***	−0.045	−0.052	−0.002
Eastern non-bantu	−0.065	−0.014	0.011	0.053	0.055	−0.003
Highest level of education (cf. Certificate)						
Diploma	0.014	0.305***	−0.015***	0.028	0.458***	−0.142*
Bachelor degree	0.056	0.346***	0.025	0.016	0.459***	−0.089
Postgraduate degree	−0.002	0.538***	−0.022***	0.048	0.746***	0.045
Other education	−0.184***	−0.026***	0.012***	−0.100***	0.087***	−0.160***
Affiliation of AEA (cf. Government)						
NAADS	0.174**	0.078	0.037	0.066	0.074	0.129
NGO	0.065	−0.055***	0.020***	−0.134***	−0.002	−0.177***
High agricultural potential	0.03	0.006	−0.024**	0.015	−0.011	0.035
Performance of rural services (cf. Poor)						
Medium performance	−0.057	−0.060**	−0.013***	0.037	−0.015	0.101
Best performance	−0.148**	−0.059*	−0.052***	−0.033	−0.004	0.009
NAADS x female agent	0.715***	0.530***	0.068***	0.710***	0.731***	−0.083
NGO x female agent	0.730***	0.534***	0.280***	0.873***	0.066***	−0.033***
Number of observations	**202**	**202**	**202**	**202**	**202**	**202**

Source: Agricultural extension agents survey.

Note: cf. = control or reference group. NAADS = National Agricultural Advisory Services of Uganda. AEA = agricultural extension agents. Asterisks denote significance levels: *, ** & *** shows significant at P = 0.10, 0.05 & 0.01 respectively.

they were more likely to provide advisory services on agroforestry, soil erosion control, and fertilizer if they work under NAADS or nongovernmental organizations (Table 5.17). This gives further evidence of the likelihood of NGOs to provide advisory services on sustainable land management technologies observed by Nkonya et al. (2005). Contrary to Nkonya et al. (2004), however, NGO-affiliated agricultural extension agents were less likely to provide advisory services on agroforestry. Results show that better educated agents were more likely to provide advisory services on

TABLE 5.17 Determinants of technology provided by extension provider (marginal effects)

Variable	Improved seeds	Agrochemical	Plant protection	Herbicide	Fertilizer	Organic fertilizer	Agroforestry	Soil erosion control
Female AEA	−0.034	0.026	−0.039	−0.036	−0.154***	0.135	−0.082***	−0.031***
Age of AEA	−0.002	0.001	−0.003	0.003	0.006***	0.002	0.003*	−0.004***
Ethnicity of AEA (cf. Luganda)								
Nyakitara	−0.138	−0.061	0.113	0.195*	−0.027	−0.067	0.112	0.007
Northern	0.138*	−0.291**	−0.032	−0.07	−0.043	−0.147***	−0.044	−0.142***
West Nile	0.049	0.1[illegible]1	0.155	−0.032	−0.024	−0.001	−0.081***	−0.099***
Eastern bantu	0.059	−0.121	−0.055	0.07	−0.072**	−0.093**	−0.014	−0.063***
Eastern non-bantu	0.098*	0.059	−0.069	0.019	0.096	−0.121***	0.022	−0.038***
Level of education (cf. Certificate)								
Diploma	0.257***	0.011	0.065	0.097	0.478***	0.113	0.013	−0.092***
Bachelor's degree	0.183***	0.053	0.014	0.185	0.474***	0.115	−0.007	−0.148***
Postgraduate degree	0.065	−0.106	−0.013	−0.081	0.755***	0.292	−0.028	−0.033***
Other education	0.190***	−0.23	0.11	0.373	0.245***	−0.145***	0.067***	−0.025***
Affiliation of AEA (cf. Government)								
NAADS	−0.042	0.017	−0.103	0.164*	−0.017	0.094	0.012	0.018
NGO	−0.339***	0.084	−0.037	0.059	−0.024	0.089	−0.077***	−0.016
High agricultural. potential	−0.082	−0.078	−0.018	−0.043	−0.088***	−0.042	−0.01	0.012

(continued)

TABLE 5.17 Continued

Variable	Improved seeds	Agrochemical	Plant protection	Herbicide	Fertilizer	Organic fertilizer	Agroforestry	Soil erosion control
Rural service performance (cf. Poor)								
Medium	0.156***	0.04	0.009	−0.012	0.03	−0.072	−0.059*	−0.129***
Best	0.267***	0.162*	−0.003	−0.089	−0.014	−0.137***	0.001	−0.182***
NAADS x female AEA	0.082	−0.209	0.039	−0.07	0.760***	−0.093	0.016***	0.019***
NGO x female AEA	0.034	−0.174	−0.096	−0.12	0.788***	−0.092	0.677***	−0.025***
Number of observations	**202**	**202**	**202**	**202**	**202**	**202**	**202**	**202**

Source: Agricultural extension agents survey.

Note: cf. = control or reference group; AEA = agricultural extension agents; NAADS = National Agricultural Advisory Services of Uganda. Asterisks denote significance levels: *, ** & *** shows significance at P = 0.10, 0.05 & 0.01 respectively.

improved crop seeds and chemical fertilizer but less likely to provide advisory services on soil erosion control.

This suggests that better educated agents seem more oriented toward providing advisory services on short-term benefits, which could lead to land degradation. Since the agricultural extension agents with higher education were the supervisors of the agents with a certificate in agriculture, their low propensity to provide advisory services on technologies that lead to long-term benefits—namely agroforestry and soil and water conservation practices—is a concern and calls for reevaluation of their capacity to provide such technologies. One of the solutions to addressing the low propensity to provide sustainable land management practices is to provide on-the-job training.

NAADS was more likely to give advisory services on herbicide than government-affiliated agricultural extension agents (Table 5.17). This reflects NAADS's commercial orientation. Best rural services were positively associated with provision of improved seeds but negatively associated with propensity to provide soil erosion control. The high propensity to provide improved seeds in areas with best rural services is likely due to the higher access of improved seeds in areas with high market access. This further raises concerns on the capacity of advisory services to provide sustainable land management advisory services.

5.9 Conclusions and Policy Implications

This study comes at a time when Uganda is implementing reforms in provision of its agricultural extension services. Even though the data used are dated, the results show the negative short-term effect of the reforms on NAADS effectiveness. The results are corroborated by most recent studies that reach the same conclusion. Particularly, the coverage of farmers by NAADS was reduced by about 50 percent, while the role of private extension agents also reduced. Similarly, farmer groups formed under NAADS phase I reduced significantly. This calls for the need to improve the institutional landscape of agricultural advisory services in Uganda—a country that trailblazed the advisory service approach. On the positive side, the higher share of farmers who received advisory services from NAADS phase II on topics other than production has increased significantly.

Results drawn from the agricultural extension agents survey as well as the household survey reaffirm the important role played by pluralistic extension services, since each type of institutional affiliation seemed to have a comparative advantage. This requires reconsideration of some of the new strategies of giving a greater role to advisory services by NGOs and private agricultural extension agents. As seen in the results, for example, female agents affiliated with NGOs and NAADS (under phase I) had greater propensity to provide advisory services to the poor and to female farmers and to offer advisory services on sustainable land management than males. Female extension agents hold the key to provision of agricultural advisory services to female farmers and the poor and to providing more sustainable land management extension messages.

Female agents serving under NAADS or NGOs were more likely to target their advisory services to women and poor farmers than their male counterparts. These favorable outcomes suggest the need to increase the number of female agents to exploit their great potential to reach women and the poor. Currently, only about 11 percent of the agricultural extension agents in Uganda are women. Most of them are located in the southern and western regions, underscoring the need to create incentives to reach the poorest region in the north. Female agents working under NAADS and nongovernmental organizations were more likely to provide advisory services on soil erosion control and agroforestry than their male counterparts. Our study shows that the share of agricultural extension agents providing messages on control of soil erosion and agroforestry was among the lowest but among those most demanded by farmers. This shows a big gap of provision of advisory services on organic soil fertility management practices. This is a common problem in other countries in Africa south of the Sahara. For example, Nkonya and Koo (2017) and Nkonya, Koo, and Kato (2017) showed that such practices—used in combination with chemical fertilizer—are more profitable and more sustainable than use of fertilizer alone, but their adoption rate in Kenya, Niger, Nigeria, and Uganda was lower than adoption of fertilizer only. Hence recruitment of more female agents will address this gap and contribute to achieving the Development Strategy Investment Program objective of ensuring sustainable land management.

Training of agricultural extension agents serving under the Single Spine Agricultural Extension System is required to increase their knowledge on sustainable land management and marketing and postharvest knowledge. Our results showed that agricultural extension agents with higher levels of education and those affiliated with government had a lower propensity to provide advisory services on organic soil fertility management practices than those with a certificate or those affiliated with NAADS or NGOs. In addition, the focus of agricultural extension agents affiliated with government is on production technologies. This suggests their weak capacity to provide advisory services on marketing and postharvest technologies. Such advisory services are key to achieving objectives of several policies and strategies seeking to transform subsistence farming to commercial farming (for example, the National Agricultural Extension Policy 2016 and the National Agricultural Extension Strategy 2016/2017–2020/2021). Another new aspect that calls for retraining is the provision of demand-driven advisory services and provision of gender-sensitive advisory services, both of which show weaknesses.

References

Barungi, M., M. Guloba, and A. Adong. 2016. "Uganda's Agricultural Extension Systems: How Appropriate Is the Single Spine Structure?" *Economic Policy Research Centre (EPRC) Research Report* 16. Economic Policy Research Centre, Kampala, Uganda.

Beintema, N., and G. Stads. 2006. *Agricultural R&D in Sub-Saharan Africa: An Era of Stagnation*. Agricultural Science and Technology Indicators (ASTI) background report. Washington, DC: International Food Policy Research Institute (IFPRI).

———. 2011. *African Agricultural R&D in the New Millennium: Progress for Some, Challenges for Many*. Food Policy Report, Agricultural Science and Technology Indicators (ASTI). Washington, DC: IFPRI.

Benin S., E. Nkonya, G. Okecho, J. Randriamamonjy, E. Kato, G. Lubade, and M. Kyotalimye. 2010. "Returns to Spending on Agricultural Extension: The Case of the National Agricultural Advisory Services (NAADS) Program of Uganda." *Agricultural Economics* 42 (1): 249–267

———. 2012. "Impact of the National Agricultural Advisory Services (NAADS) Program of Uganda: Considering Different Levels of Likely Contamination with the Treatment." *American Journal of Agricultural Economics* 94 (2): 386–392.

Benin, S., E. Nkonya, G. Okecho, J. Randriamamonjy, E. Kato, G. Lubadde, M. Kyotalimye, and F. Byekwaso. 2011. *Impact of Uganda's National Agricultural Advisory Services Program*. IFPRI Research Monograph. Washington, DC: IFPRI.

Boteler, F. E. 2007. "Building Disaster-Resilient Families, Communities, and Businesses." *Journal of Extension* [Online] 45 (6). Article 6FEA1.

Chapman, R., and R. Tripp. 2003. *Case Studies of Agricultural Extension Programs Using Privatized Service Provision*. London: Agren.

Chipika, S., and E. Friis-Hansen. 2004. *Transformation of Agricultural Extension under Participatory District Planning: Comparative Experience in Shurugwi and Gwanda Districts.* Volume 3. Demand-Driven Approaches to Agricultural Extension: Case Studies of International Initiatives. Agricultural and Rural Development Discussion Paper 10. Extension Reform for Rural Development. Washington, DC: World Bank.

Davis, K., E. Nkonya, E. Kato, D. A. Mekonnen, M. Odendo, R. Miiro, and J. Nkuba. 2012. "Impact of Farmer Field Schools on Agricultural Productivity and Poverty in East Africa." *World Development* 40 (2): 402–413.

Delgado, C. L. 1995. *Africa's Changing Agricultural Development Strategies: Past and Present Paradigms as a Guide to the Future.* IFPRI 2020 Discussion Paper 3. Washington, DC: IFPRI.

Ellis, F., and S. Biggs. 2001. "Evolving Themes in Rural Development 1950s–2000s." *Development Policy Review* 19 (4): 437–448.

FAO (Food and Agriculture Organization of the United Nations). 2011. *State of Food and Agriculture 2010: Women in Agriculture, Closing the Gender Gap for Development.* Rome: FAO.

Friis-Hansen, E., C. Aben, M. Ameu, and J. R. Okoth. 2003. "Evolution of Extension Approaches in Soroti District, Uganda." Proceedings from workshop entitled "Extension and Rural Development: A Convergence of Views on International Approaches?" November 12–15, 2002, Washington, DC, World Bank.

Hambleton, R. 1992. "Decentralization and Democracy in UK Local Government." *Public Money and Management* (July–September): 9–20.

Kaiser, H. 1958. "The Varimax Criterion for Analytic Rotation in Factor Analysis." *Psychometrika* 23 (3): 187–200.

Kwapong, N. A., and E. Nkonya. 2015. "Agricultural Extension Reforms and Development in Uganda." *Journal of Agricultural Extension and Rural Development* 7 (4): 122–134.

Lahai, B., P. Goldey, and G. E. Jones. 2000. "The Gender of the Extension Agent and Farmers' Access to and Participation in Agricultural Extension in Nigeria." *Journal of Agricultural Education and Extension* 6 (4): 223–233.

Long, S. 1997. *Regression Models for Categorical and Limited Dependent Variables.* Advanced quantitative techniques in the social sciences. Thousand Oaks, CA, US: Sage Publications.

NAADS (National Agricultural Advisory Services). 2000. *Master Document of the NAADS Task Force and Joint Donor Group. Final Report 2000.* Entebbe, Uganda: MAAIF.

———. 2018. *NAADS Strategic Plan 2016–2019/20: Sustainable Household Food Security and Incomes.* Entebbe, Uganda: MAAIF.

NARS (National Agricultural Research System). 2002. *Reform Program: Master Document of the NARS Review Task Force. Draft Report 2002.* Entebbe, Uganda: MAAIF.

———. 2003. National Agricultural Research System Policy. Entebbe, Uganda: MAAIF.

NEMA (National Environment Management Authority). 2007. *State of Environment Report for Uganda.* Kampala, Uganda: NEMA.

Nkonya, E., and J. Koo. 2017. "The Unholy Cross: Profitability and Adoption of Climate-Smart Agriculture Practices in Africa South of the Sahara." In *A Thriving Agricultural Sector in a Changing Climate: Meeting Malabo Declaration Goals through Climate-Smart Agriculture,* edited by A. De Pinto and J. M. Ulimwengu, 103–113. ReSAKSS Annual Trends and Outlook Report 2016. Washington, DC: IFPRI.

Nkonya, E., J. Koo, and E. Kato. 2017. "Climate Risk Management through SLWM in Sub-Saharan Africa." 2017. In *Climate Smart Agriculture—Building Resilience to Climate Change*, edited by D. Zilberman, L. Lipper, N. McCarthy, S. Asfaw, and G. Branca, 445–476. New York: Elsevier.

Nkonya E., J. Pender, P. Jagger, D. Sserunkuuma, C. Kaizzi, and H. Ssali. 2004. *Strategies for Sustainable Land Management and Poverty Reduction in Uganda.* IFPRI Research Report 133. Washington, DC: IFPRI.

Nkonya, E., J. Pender, C. Kaizzi, K. Edward, and S. Mugarura. 2005. *Policy Options for Increasing Crop Productivity and Reducing Soil Nutrient Depletion and Poverty in Uganda.* Environmental and Production Technology Division Discussion Paper 138. Washington, DC: IFPRI.

Nkonya, E., D. Phillip, T. Mogues, J. Pender, and E. Kato. 2010. *From the Ground Up: Impact of a Pro-Poor Community Driven Development Project in Nigeria.* IFPRI Research Monograph. Washington, DC: IFPRI.

PMA (Plan for the Modernisation of Agriculture). 2000. *Eradicating Poverty in Uganda: Government Strategy and Operational Framework.* August. Kampala, Uganda: MAAIF and Finance Planning and Economic Development.

Powell, J. L. 1984. "Least Absolute Deviations Estimation for the Censored Regression Model." *Journal of Econometrics* 25: 303–325.

Rwamigisa, B. P. 2000. "A Critical Analysis of the Plan for Modernization of Agriculture." Unpublished paper. Kampala, Uganda.

———. 2014. "The Political and Institutional Dimensions of Agricultural Extension Reform Process in Uganda: Experiences of the National Agricultural Advisory Services." PhD thesis, Makerere University, Kampala, Uganda.

Rwamigisa, B. P., T. Kakuba, and S. Massawe. 2019. "Uganda's Single Spine Agricultural Extension System: What It Is, Progress and Early Lessons on Its Implementation." Unpublished policy note, IFPRI, Washington, DC.

Semana, A. R. 1998. "Experiences of Agricultural Extension in the 100 Years of Agricultural Research in Uganda." Unpublished paper. Department of Agricultural Extension/Education, Makerere University, Kampala.

———. 2002. "Agricultural Extension Services at Crossroads: Present Dilemma and Possible Solutions for Future in Uganda." Paper presented at the CODESTRIA-IFS Sustainable Agriculture Initiative Workshop. Kampala, Uganda, December 15–16.

Stata. 2007. Statistical Data Analysis. College Station, TX, US.

Swanson, B. 2008. *Global Review of Good Agricultural Extension and Advisory Service Practices.* Rome: Food and Agriculture Organization of the United Nations.

Thompson, T. S. 2002. "A History of Extension at USAID: Case Study." Workshop presented at "Extension and Rural Development: A Convergence of Views on International Approaches?" Washington, DC, November 12–15.

Turyahikayo, W., and E. Kamagara. 2016. "Trust, Perception and Effectiveness of Extension Services in Uganda: A Case of National Agricultural Advisory Services (NAADS)." *Journal of Agricultural Extension and Rural Development* 8 (11): 224–231.

UFAAS (Uganda Forum for Agricultural Advisory Services). 2019. "Drone Based Crop Research Data Acquisition and Interpretation Services Now Available in Benin, DRC, Ghana, Tanzania and Uganda." Blog. Accessed July 19, 2019. www.ufaas-ugandacf.org/updates/drone-based-crop-research-data-acquisition-and-interpretation-services-now-available-benin.

Uganda, MAAIF (Ministry of Agriculture Animal Industry and Fisheries). 1998. *Draft Project Completion Implementation Report of the Agricultural Extension Project (CR 2424-UG).* Entebbe, Uganda: MAAIF.

———. 2003. *Ministry of Agriculture Animal Industry and Fisheries Policy Statement for Financial Year 2003/2004.* Entebbe, Uganda: MAAIF.

———. 2010. *Agriculture for Food and Income Security. Agricultural Sector Development Strategy Investment Plan: 2010/11—2014/15.* Entebbe, Uganda: MAAIF.

———. 2016. *National Agriculture Extension Strategy 2016–2021.* Entebbe, Uganda: MAAIF.

———. 2018. *Agriculture Extension Users Manual.* Entebbe, Uganda: MAAIF.

Uganda, MAAIF and MFPED (Ministry of Finance, Planning and Economic Development). 2000. *Plan for Modernization of Agriculture: Eradicating Poverty in Uganda.* Entebbe, Uganda: MAAIF; Kampala, Uganda: MFPED.

Uganda, State House. 2019. "Past Presidents of Uganda." Accessed November 22, 2019. www.statehouse.go.ug/past-presidents.

Uphoff, N. 1993. "Grassroots Organizations and NGOs in Rural Development: Opportunities with Diminishing States and Expanding Markets." *World Development* 21 (4): 607–622.

Van Campenhout, B., D. J. Spielman, and E. Lecoutere. 2018. *Information and Communication Technologies (ICTs) to Provide Agricultural Advice to Smallholder Farmers: Experimental Evidence from Uganda*. IFPRI Discussion Paper 1778. Washington, DC: IFPRI. http://ebrary.ifpri.org/cdm/singleitem/collection/p15738coll2/id/133022.

Vijverberg, W. 1987. "Non-Normality as Distributional Misspecification in Single-Equation Limited Dependent Variable Models." *Oxford Bulletin of Economics and Statistics* 49 (4): 417–430.

World Bank. 1993. *Uganda; Agriculture, a World Bank Country Study*. Washington, DC: World Bank.

———. 2010. *Gender and Governance in Rural Services: Insights from India, Ghana, and Ethiopia*. Washington, DC: World Bank. DOI: 10.1596/978-0-8213-7658-4.

———. n.d. Uganda LSMS data. Washington, DC.

Chapter 6

ETHIOPIA

Guush Berhane, Catherine Ragasa, Gashaw Tadesse Abate, and Thomas Woldu Assefa

6.1 Introduction

Ethiopia has made enormous progress in terms of economic growth, agricultural development, and poverty reduction since the famines of the 1970s and 1980s (Hill et al. 2016; Bachewe et al. 2018; Dorosh and Rashid 2012). Sustained increases in agricultural productivity contributed to overall economic growth and poverty reduction in recent years (Bachewe et al. 2018; Hill et al. 2016). Ethiopia has witnessed rapid and sustained economic growth, averaging 11 percent between 2004 and 2014, pulling millions out of poverty and slashing headcount poverty from 44 percent in 2000 to 30 percent in 2011 (World Bank 2015). Land and labor use have expanded significantly, and total factor productivity grew by about 2.3 percent per year in the past decade, with modern input use more than doubling. The expansion in modern input use appears to have been driven by several factors, including high public expenditures in the agriculture sector, especially on agricultural extension and advisory services, an improved road network, higher levels of rural education, and favorable domestic and international price incentives (Bachewe et al. 2018).

Despite the rapid economic and agricultural growth of the past two decades, however, Ethiopia remains one of the poorest countries in Africa, with a significant proportion of its population still relying on traditional farming practices, which prevents rapid agricultural transformation and structural changes from taking place. Many constraints and bottlenecks remain, and the challenges that the country currently faces are how to build on the earlier successes and how to accelerate development to maintain rapid overall economic growth and enter into a sustainable structural transformation of the economy. Experts have highlighted the crucial role that extension services will continue to play in this rural transformation (Bachewe et al. 2018).

This chapter takes stock of the provision of extension services in line with the country's development and suggests ways such services might contribute to accelerating agricultural growth and poverty reduction in the years ahead. Ethiopia has invested significantly in its extension system, in line with national policies that place a high priority on boosting agricultural production and productivity. Ethiopia is considered a leading country globally in the provision of extension services to farmers and has the highest extension agent-to-farmer ratio of any country. Ethiopia presents an interesting case, as it is one of few African countries to have placed agriculture at the forefront of its economic development policies, having invested heavily in the sector over the past two decades. Unlike many countries where private-sector extension services expanded following drastic public funding cuts to their extension systems in the 1980s and 1990s (Zhou and Babu 2015), Ethiopia's extension system remains predominantly public.

A unique feature of Ethiopia's agriculture sector is that it has received unprecedented high-level political commitment since the current government assumed power in 1991. The Agriculture Development–Led Industrialization (ADLI) strategy was developed in the mid-1990s to serve as a roadmap to transform smallholder agriculture. Rural education, health, infrastructure, agricultural research, and extension services were among its top priorities. Ethiopia is one of only four African countries to have implemented the Comprehensive Africa Agriculture Development Programme (CAADP) agreement of a 10 percent target of annual government expenditures earmarked for agriculture over the 2003–2013 period (Benin 2014). More recently, Ethiopia's transformation agenda elaborates its ambitious five-year Growth and Transformation Plan (GTP), placing significant importance on the agriculture sector in general and on the extension system in particular.

Although substantial progress has been made since the public official document that outlined the current extension services model was issued in 2002 (Ethiopia 2002), Ethiopia's extension system faces several challenges, mainly relating to the way it is organized. The fact that it covers vast and scattered geographic areas implies that it is poorly resourced and has weak links with knowledge centers, hampering the quality of the extension services delivered. Evidence suggests a substantial increase in the number of farm households reached with extension services (Bachewe et al. 2018); however, it is less clear whether productivity gains have been achieved and poverty has been reduced in consequence (Dercon et al. 2009; Davis, Swanson, and Amudavi 2009; Nisrane et al. 2011; Spielman, Mekonnen, and Alemu 2012; Krishnan and Patnam 2014; Abay et al. 2017; Berhane et al. 2018).

This chapter reviews and provides evidence on the performance of the extension system in Ethiopia and the impact it has on productivity and other development outcomes. Data and information from various sources were analyzed to describe the enabling policies, landscape, organizational structure and management systems, capacities and incentives of different actors, approaches and tools to identify bottlenecks in the extension system, and areas for improvement. It presents the results of this stock taking using elements of the best-fit framework (Birner et al. 2009) presented in Chapter 2.

6.2 Data Sources

Four datasets—two large household-level and two development agent–level datasets—are used in this study (described in Table 6.1).

The first dataset is extracted from a large longitudinal dataset collected to evaluate Ethiopia's Agricultural Growth Program (AGP) implemented between 2010 and 2015. The data were collected collaboratively between the Central Statistical Agency (CSA) of Ethiopia and the International Food Policy Research Institute (IFPRI) and cover 7,500 farm households that were visited twice (in 2011 and again in 2013). For the survey, 61 districts were randomly selected from among the 83 AGP districts in the four main regions of the country: Amhara; Oromia; Southern Nations, Nationalities and Peoples (SNNP); and Tigray. In parallel, 32 districts were randomly selected from among non-AGP districts in the same regions. Three enumeration areas were randomly selected in each district, from which 26 households were similarly selected from each. The selection of survey households was carried out randomly using a fresh list of households within each selected enumeration area. This dataset included a wealth of household- and plot-level information on use and application of extension and advisory services as well as household-level perceptions on how well these services were received.

The second household-level dataset comes from a large nationally representative agricultural sample survey run each year by the CSA. This dataset provides the bigger picture about the extent to which these services were provided and the growth over time. These two household-level datasets enabled us to dissect into the demand side of the story and were used to triangulate the evidence from the development agent–level (supply-side) datasets discussed below. Specifically, the longitudinal household dataset is deployed in the econometric analysis run by Berhane et al. (2018) reported in the results section. The third dataset originated from a unique survey of 237 development agents (DAs) selected from eight districts in seven regions: Oromia, Amhara, SNNP, Tigray, Gambella, Afar, and Benishangul-Gumuz. Data were collected in 2009 by IFPRI and the Ethiopian Economics Association. These data are uniquely designed to fathom the profiles and roles played by DAs in Ethiopia.

TABLE 6.1 Datasets used in this chapter

Dataset	Description	Year collected	Extension focus	Analytical method
AGP survey data	Large-scale longitudinal dataset collected to evaluate Ethiopia's Agricultural Growth Program (AGP) implemented between 2010 and 2015, covering 7,500 households	2011, 2013	Survey contains two pages of questions on agricultural extension, and linked to technology adoption and input use, production, and other outcome indicators	Descriptive analysis; econometric analysis conducted by authors, both presented in this chapter for the first time and citing authors' published papers using same dataset
Ethiopia CSA survey data	Annual large-scale data collection by the Central Statistical Agency (CSA) on agriculture in Ethiopia	Those used in the paper: 2004–2014	Includes questions on access to extension services and input usages	Descriptive analysis; citing papers using more in-depth analysis using this dataset
IFPRI/EEA DA survey data	Data covering profile, roles, capacity, and motivation of 237 development agents	2009	Focus on extension	Descriptive analysis; citing papers using more in-depth analysis using this dataset
Digital Green DA survey data	Baseline data as part of the impact evaluation of a Digital Green pilot project in Ethiopia, covering 896 development agents	2017	Focus on extension	Descriptive analysis; citing papers using more in-depth analysis using this dataset

Source: Authors' compilation from various datasets.

The fourth dataset was obtained from a survey conducted for an impact evaluation of a Digital Green pilot project in Ethiopia. A quantitative survey was administered to 896 development agents involved in the project. Digital Green is working with different partners in Ethiopia to introduce a community-centric participatory video approach to the provision of extension services. The pilot project aims to improve the efficacy of the country's public extension system by broadening its reach through cost-effective information and communications technologies (ICT). Digital Green's survey of development agents provides information on characteristics of the human capital of DAs, their activities, their job satisfaction, and their overall motivation. These two DA-level datasets offer a wealth of information on how well the system is equipped and enabled us to gauge the quality and quantity of services provided by the system. The results from these two datasets are presented in the organization and effectiveness sections of this chapter.

6.3 Governance Structures and Policies

Context

The delivery of extension services in Ethiopia takes place within a complex public bureaucratic structure that involves diverse interactions that, together with the distinct features of the agriculture sector, tend to be a challenge in terms of achieving effective extension services provision. First, a substantial share of the Ethiopian population depends on agriculture for its livelihood, and this population consists mainly of smallholder farmers. Thus extension services for the entire farming community would require extensive coverage across the country, placing considerable demand on public, human, and financial resources.

Second, given the wide geographic dispersion of activities, with farmers widely scattered, reaching them with extension services through conventional means is time consuming and would be a complex undertaking for DAs. For example, rainfall varies significantly between mountains and valleys, even across short distances. Large variations exist in altitude, rainfall, connectivity, and cropping patterns, and there is no single dominant crop for Ethiopia's food consumption (as is, for example, rice in Asia, maize in Latin America, and wheat in cooler climates). Thus five cereals are cultivated on a wide scale in Ethiopia: teff (an indigenous crop widely grown only in Ethiopia and Eritrea), wheat, maize, sorghum, and barley. Coffee, cultivated in the rainfall-sufficient southern highlands, is Ethiopia's major export crop. Livestock—mainly cattle, sheep, and goats—are the major sources of meat and livelihood for the pastoralist and agro-pastoralist populations.

Third, although major strides have been made in Ethiopia in expanding road infrastructure throughout the country, transport infrastructure within rural areas is still such that traversing space to contact a large number of farmers within a given area continues to be difficult for DAs. Throughout the country the population is more concentrated along major road networks, although much of Ethiopia's population remains remote: 45 percent of the population lives more than five hours from a city of 50,000 (Schmidt and Kedir 2009). Furthermore, the rugged terrain in much of the highlands makes transport and communication difficult.

The above three features—large number of potential extension services users, geographic dispersion, and infrastructure constraints—place a strain on public resources if extension services are to reach a large segment of the agricultural population. It is also a challenge to monitor the quality and quantity of extension services provided. One approach would be to manage the resource and quality control challenges by significantly standardizing the modality of delivery and content of extension services to facilitate not only the provision of DA training at scale but also the identification of objectives and standards against which agent performance can be measured.

A fourth essential aspect is the extent to which Ethiopian agriculture is suitable to such approaches, given its richly diverse agroecological conditions, including soil, climate, and altitude. As such, Ethiopia is a particularly unique case, exemplifying much of Africa in terms of ecological processes applied to agricultural production systems. The result is that Ethiopia produces even more diverse crop products than does Asia.[1] As such, delivery modalities of extension services, as well as the content of advisory services, are much harder to standardize in Ethiopia's diverse conditions than in countries with more homogenous agricultural settings. This diversity calls for frontline extension services providers to have a relatively large degree of discretion in tailoring agricultural advice to the agro-physical and socioeconomic peculiarities of the areas they serve, especially in view of their better knowledge compared with those at higher levels of the agricultural bureaucracy.

However, giving greater discretion to DAs complicates the effective monitoring of their performance and may widen the gap between the skills they need and the skills they acquire in agricultural education and from standard training packages and guidelines. Moreover, greater discretion would require them to provide varied services, which requires a variety of knowledge centers to backstop extension services. Therefore, the choice between alternative designs of extension systems—in particular, extensive standardization versus delegation of content and modality to the frontline provider—constitutes a range of trade-offs. Much of the literature on the design of extension systems, which heralds the benefits of giving frontline DAs a larger degree of flexibility in conducting their work, does not adequately discuss the operational challenges in conducting quality control. Similarly, the de facto substantial standardization of delivery mechanisms and content in many country policies does not necessarily account for the issues in matching advice to farmers' local conditions. Described below is Ethiopia's standardized delivery system, its evolution over the past two decades, and its challenges, despite the country's diverse agroecology.

---

1 Over the past several decades, this has meant that a set of agricultural technologies, such as improved varieties, developed to be suitable for one area in Asia has been applicable to much wider areas within that region, whereas technologies have had geographically much more circumscribed applicability in Africa (World Bank 2007: 55), including Ethiopia.

Ethiopia's investment in agriculture mainly focuses on the provision of "advisory and training services" through a public extension services structure that spans from the federal ministry to the regions and trickles down to *kebeles* through frontline development agents.[2] Over the past two decades, Ethiopia's extension system underwent a number of experiments in terms of delivery arrangements within the structure as well as in the nature of extension services provided (Ethiopia 2002; Gebremedhin, Hoekstra, and Tegegne 2006). With regard to delivery, the system has evolved from one that was largely top-down and managed by the federal ministry to one owned and operated by the regions and *woredas* (districts), within the general wave of decentralized public services.[3]

While Ethiopia's extension system remains predominantly public and unimodal in terms of services provided, the government nevertheless has given special attention to the agriculture sector. With the earlier launching of the ADLI strategy, the government of Ethiopia placed agriculture and rural development at the core of its development policy agenda and committed a substantial share of the national budget to the sector.[4] This emphasis on smallholder agriculture, based on the ADLI approach, continued with the addition of the Sustainable Poverty Reduction Strategy of 2002 (Ethiopia, MoFED 2002) and the Plan for Accelerated and Sustained Development to End Poverty in 2006 (Ethiopia, MoFED 2006).

Past, Present, and Future Policies

The history of extension services in Ethiopia goes back to the introduction of the land grant extension system by the Imperial Ethiopian College of Agriculture and Mechanical Arts (IECAMA) in 1953 (Belay 2003).[5] It was followed by the Comprehensive Package Program (1967), the Minimum Package Project I (1971–1979), the Minimum Package Project II (1980–1985), and the Peasant Agricultural Development Program (PADEP) (1985–1995). Despite some variations, these early efforts had limited focus on the most

2 A *kebele* is the lowest administration unit in Ethiopia. There are around 15,000 *kebeles* in Ethiopia.

3 For example, starting from 2003, each *woreda* was required to identify its priority market-oriented commodity production, along with a detailed projection of input requirements to be delivered through the extension system (Gebremedhin, Hoekstra, and Tegegne 2006).

4 Agricultural development, which set the pace not only for the industrial sector but also for other sectors, provided the blueprint for national development. An emphasis was placed on the provision of extension, rural education, and the strengthening of public agricultural research.

5 The history of extension in Ethiopia before 1991 is well documented (Gebremedhin, Hoekstra, and Tegegne 2006; Davis et al. 2010; Spielman, Kelemwork, and Alemu 2011).

TABLE 6.2 Historical evolution of agricultural extension services in Ethiopia

Period	Program/event	Objectives/highlights	Remarks
1953–1963	Establishment of IECAMA (Imperial Ethiopian College of Agriculture and Mechanical Arts)	The establishment of the IECAMA is said to be the start of extension services in Ethiopia and its effort was around the areas where it had agricultural experimental stations.	Major constraints were limited manpower and outreach and lack of complementary services such as inputs and credit.
1963	Transfer of extension services mandate to the Ministry of Agriculture	The mandate to provide extension services was moved to the then Ministry of Agriculture (Ethiopia, MoA), structured as a department at the national level and extension personnel assigned at the provincial level.	This new structure did not become active until 1968.
1967–1975	Comprehensive Integrated Package Projects	Several pilot comprehensive package extension programs were implemented.	Since all of these programs and projects were operational in only small areas, the vast majority of the country was out of their reach.
1971–1979	Minimum Package Project I (MPP-I)	The MPP-I established minimum package areas within a 10-kilometer radius of all-weather roads and within a 50 kilometer to 75 kilometer distance designed to serve about 10,000 households each. Each minimum package area used 5 extension agents, about 5 input supply workers, and 1 extension supervisor.	The major drawbacks of the MPP-I included minimal attention given to the livestock sector, not benefiting smallholders, and not being able to reach the vast majority of farmers.
1980–1985	Minimum Package Project II (MPP-II)	In MPP-II, extension services responsibility was given to the commodity-based specialized departments in the ministry. Regions also adopted a similar structure and *woredas* became the lowest structures where extension personnel were located.	The MPP-II failed to achieve its objectives due to shortage of extension personnel, and burdening extension agents with activities such as tax collection and organization of cooperatives.
1985–1994	PADEP (Peasant Agricultural Development Program) using Training and Visit system	PADEP aimed to provide inputs, credit, and extension services to smallholders organized into approximately 2,900 farmer service cooperatives using a Training and Visit extension approach.	By the end of the Derg regime (1971–1991), extension services had been reduced to instruments of political control over the peasantry (Wubneh 2007).
1993–1994	Sasakawa-Global 2000 pilot program (SG-2000)	The major objective of SG-2000 was to increase agricultural food production at the level of small-scale farmers and stimulate the linkage between research and extension services so that agricultural technologies within the country could be made available to the extension system.	Major productivity increases were achieved, which convinced the government of Ethiopia to adopt and expand it as a National Agricultural Extension Intervention Program (NAEIP) through the PADETES in 1995.

Period	Program/event	Objectives/highlights	Remarks
1995–2008	Participatory Demonstration and Training Extension System (PADETS or PADETES)	PADETS or PADETES is promoted as the national extension system and builds on the success of the SG-2000. It falls under the National Agricultural Extension Intervention Program (NAEIP).	Limited evaluation and impact assessment (available studies show mixed results[a]).
2004	Agricultural Technical and Vocational Education and Training colleges (ATVETs) and farmer training centers in each *kebele*	In line with decentralization processes, ATVETs were established to train a new cadre of extension services workers and farmer training centers (FTCs) were established to become the focal point of extension services support in every *kebele* in the country.	Limited evaluation and impact assessment. Available studies show mixed results. But a recent study by Bachewe et al. (2018) attributes agricultural productivity growth to investments in extension services programs as a main driver.
2006–2012	Rural Capacity Building Project (RCBP)	The RCBP aimed to increase agricultural productivity, focusing on investments in building technical, physical, and managerial capacity of the extension system and its linkages with the research system.	Unclear impact on agricultural productivity and quality of extension services based on project documents. Recent study shows some positive impacts. Women farmers' access and satisfaction with extension services increased significantly immediately after the start of the project, but that effect did not last into the medium term (Buehren et al. 2019).
2009/2010	Participatory Extension System (PES)	Trainings and demonstration of improved technologies and best practices were carried out on model farmers' fields to advance their skills.	No evaluation yet.
2010; 2016	Growth and Transformation Plan (GTP); GTP II	GTP I aimed for extension services reach to 14.6 million beneficiaries by 2014/2015; GTP II aims to increase beneficiaries to 18.237 million by 2019/2020.	GTP reached 95% of target.

Source: Authors' compilation from various sources,

Note: [a] Dercon et al. (2009); Davis, Swanson, and Amudavi (2009); Nisrane et al. (2011); Elias et al. (2013); Krishnan and Patnam (2014); Abay et al. (2017); and Berhane et al. (2018).

accessible and high-potential areas of the country.[6] The evolution of Ethiopia's thinking on the provision of extension services to the wider rural communities grew in the next two decades (1967–1985), albeit at a slower pace, and remained largely foreign funded (Gebremedhin, Hoekstra, and Tegegne 2006; Davis et al. 2010). Table 6.2 provides the key historical milestones in the evolution of the extension system over the past six decades.

6 For example, when it started, the college had only two development agents, increasing to 132 working in 77 extension locations (Gebremedhin, Hoekstra, and Tegegne 2006).

From 1993 through 2006

In 1993 a pilot extension services program was initiated by Sasakawa Africa Association and Global 2000 of the Carter Center (SG-2000), a nongovernmental organization. The pilot program operated in collaboration with extension services staff from the Ministry of Agriculture for two years and was tested in four regions: Amhara, Oromia, SNNP, and Tigray. The program promoted the use of productivity-enhancing technologies by providing inputs and credit as well as training using demonstration plots (0.25–0.50 hectares), supervised by researchers and DAs. SG-2000 demonstrated that with sufficient inputs, supervision, and management, farmers could double or triple their yields of maize and wheat (Davis et al. 2010; Gebremedhin, Hoekstra, and Tegegne 2006). At program inception, there were 160 farmers, increasing to 1,600 in 1994.

In 1995 the transitional government of Ethiopia adopted the SG-2000 pilot program as its national extension system, referred to as the Participatory Demonstration and Training Extension System (PADETES), later relabeled as the National Agricultural Extension Intervention Program (NAEIP). PADETES/NAEIP was the first national extension services program fully funded by the Ethiopian government. The goal of PADETES/NAEIP is to improve incomes by increasing productivity, ensuring self-sufficiency in food production, establishing farmers' organizations, increasing production of export crops, conserving natural resources, and increasing women's participation in development. The program was initially implemented in seven regions with technology packages for wheat, maize, sorghum, and teff in high rainfall areas. The program later expanded its coverage and number of technology packages. Technology packages related to crop production for moisture-stressed areas, livestock, high-value crops, postharvest technology, and agroforestry, among others (Gebremedhin, Hoekstra, and Tegegne 2006) (Box 6.1). The aim of the program was to reach about nine million farmers, using the adapted training-and-visit model, initially promoted by the World Bank in various countries.

PADETES/NAEIP, which remains the core program implemented today, follows a technology or extension services package approach for agricultural development that incorporates information on agricultural technology, provision of inputs and credit, and communication methods, all facilitated and supervised by the DAs. Farmers who agree to participate in the program make a 25–50 percent down payment on the inputs used, with the remainder to be paid following harvesting. The rates for advance payment vary according to

BOX 6.1 Technology packages under the Participatory Demonstration and Training Extension System (PADETES)/National Agricultural Extension Intervention Program (NAEIP)

The dryland, cereal-based extension package covers Ethiopia's major staple food crops, such as maize, wheat, teff, and sorghum. The contents of the package are improved seeds, chemical fertilizers (500 kilograms of DAP and 50 kilograms of urea per hectare), herbicides, and tied ridging for moisture conservation. Local seeds are also considered one of the options, together with improved technologies in areas where farmers believe that their own local varieties are superior to improved ones (Abesha, Waktola, and Aune 2000). The high-value crop extension package includes pulses, vegetables, and oil crops. The livestock development extension package includes items for dairying for milk production, fattening for meat production, poultry for egg production, and apiculture (beekeeping) for honey production. Three types of packages are disseminated: integrated household, regular, and minimum.

Integrated household package extension programs. These are based on the selection of a package of technologies from a menu of package choices provided to farmers. Ideally, the needs of households to increase overall farm productivity and household income are assessed and a set of complementary extension packages are identified and planned to be made available to the household. In practice, this proved difficult to implement and was not implemented as planned. The aim is to arrive at certain income levels per year and to help eventually transform subsistence farmers to some form of specialization for market-oriented agricultural development. The overall result was mixed.

The major components of the extension packages are (1) crop production packages (of cereals, pulses, fruits, and vegetables) including crop protection and irrigation; (2) livestock production technologies (feeds, dairy, fattening, animal health, fishery, poultry, apiculture); and (3) natural resources management (agroforestry, soil conservation, and water harvesting).

Regular package extension programs. These are aimed at enabling farmers to adopt improved seeds with commercial fertilizer, improved management practices, and soil moisture conservation practices.

Minimum package extension programs. These stipulate that farmers adopt improved seeds with traditional soil fertility management (for example, application of compost and manure) and soil moisture conservation practices. The minimum package implements the recommended standard packages of a commodity without necessarily integrating different options or elements of other packages that a given household could profitably use. Most farmers involved in the extension package program ended up implementing this package.

Source: Authors' compilation from various sources.

the types of technologies to be used and the resource level of the farming communities. Loans for crop-based packages are repaid immediately after the harvest, whereas for other extension services packages, various repayment options can be made depending on the types of development program undertaken.

To deliver knowledge, extension services make use of the following: individual visits by DAs to farmers' homes or farms; group-based approaches by working with communities, farmers' organizations, and other groups; and mass-media approaches including radio and some print media. In some *woredas,* extension services messages are transmitted at church/mosque gatherings during religious holidays or other occasional social gatherings, indicating the need to ensure the effectiveness of such fora in reaching the intended recipients.

Since the development agent was at the center of all these efforts, the number of DAs was increased from 2,500 in 1995 to 15,000 in 2002 and, as a result, households covered by extension services increased from 32,000 to 4.2 million in the same period. But the yield levels realized by farmers through the public extension services program were not as high as the original demonstration plots of the SG-2000 pilot program, mainly due to the lack of sufficient knowledge and supervision by extension services staff as done in the pilot (Gebremedhin, Hoekstra, and Tegegne 2006; Davis et al. 2010).

From 2006 On

Based on PADETES/NAEIP assessments, the government realized the need for additional extension services staff in the system. This was in line with the government's broader decentralization program. Therefore, Agricultural Technical and Vocational Education and Training colleges (ATVETs) were founded to increase the number of development agents assigned at the decentralized level (Davis et al. 2010). As shown in Figure 6.1, between 2003/2004 and 2014/2015, about 80,000 DAs were trained in different fields of studies in these ATVETs; 60 percent of these DAs graduated over a four-year period from 2003/2004 to 2006/2007. The number of graduates significantly dropped in 2008/2009 and has continued to be low since. The decline suggests that the training of additional development agents and retraining of existing DAs did not continue as planned. That said, DA exits following a summer and continuing education program offered by higher education institutions that allowed development agents to study other fields (outside of their initial trainings) could be an additional explanation for the declining number of DA graduates in ATVETs in recent years. Anecdotal evidence suggests that

FIGURE 6.1 Development agents graduated from agricultural technical and vocational education and training colleges, by field of study (2003/2004–2014/2015)

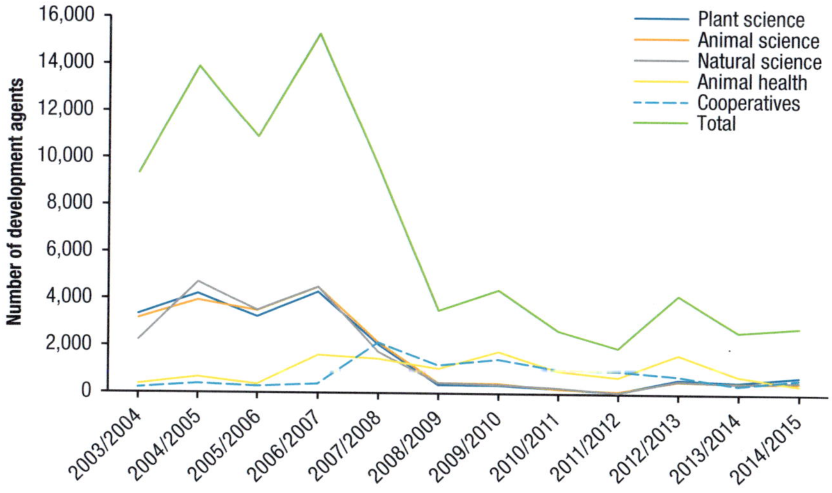

Source: Authors' compilation based on secondary data obtained from the Ethiopian Ministry of Agriculture.

the expanded opportunities development agents received from this program contributed to a high turnover of DAs who often wanted to change fields as they moved higher in their studies.

Estimates over the 2010–2016 period indicate that Ethiopia has one of the most extensive extension systems in the world in terms of its extension agent-to-farmer ratio. As shown in Figure 6.2, by 2010, Ethiopia's DA–farmer ratio was estimated at one development agent per 476 farmers—that is, 21 development agents per 10,000 farmers. In comparison, figures for Tanzania stood at 1 DA per 2,500 farmers—that is, 4 development agents per 10,000 farmers (Figure 6.2). According to recent data from Ethiopia's Ministry of Agriculture and Natural Resources (MoANR), the number of DAs has increased to about 72,000 reportedly serving about 16.7 million smallholder farmers in 2016/2017—that is, 1 development agent per 230 farmers or 43 DAs per 10,000 farmers. This coverage is consistent with the Bachewe et al. (2018) study showing that about 80 percent of farm households have been reached by extension services. Nevertheless, the quality of services provided remained largely the same.

In addition to the human capital, the government of Ethiopia established more than 15,000 farmer training centers (FTCs) throughout the country

FIGURE 6.2 Number of agricultural extension agents per 10,000 farmers in selected countries

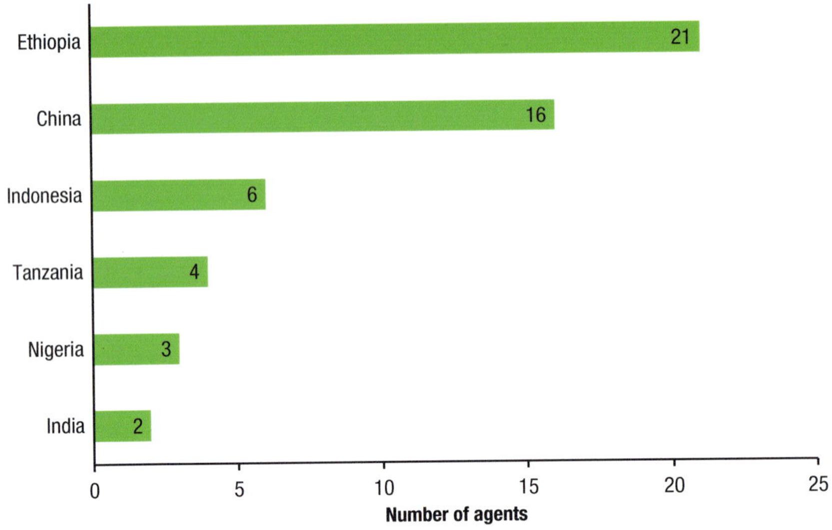

Source: Davis et al. (2010).
Note: For Ethiopia, figures in 2016/2017 show a higher ratio of 43 development agents per 10,000 farmers.

(about one FTC at each *kebele*) during this period. These FTCs were designed as local-level focal points for farmers to receive information, training, demonstrations, and advice, and included classrooms and demonstration fields. How fully resourced and functional these farmer training centers are has always been an issue and needs a stand-alone detailed study. Davis et al. (2010) estimated that only about 30 percent of these FTCs are functional. Anecdotal recent evidence and additional field visits suggest that many are still poorly resourced or even completely abandoned.

In addition to the DAs stationed in the *kebeles*, roughly 7,000 subject matter specialists (SMSs) and 4,000 supervisors are employed in the *woreda* and regional extension services offices. On average, about 30 agricultural officers work in 9 divisions or units within each *woreda* agriculture office, including 10 or more SMSs, who are expected to provide technical support and training to the DA staff in the *kebeles*. In practice, recent focus group discussions suggest that given the limited resources available and the scale and technical complexities involved, implementing these technical support activities in full scale has proved difficult.

In 2011 a new extension services delivery approach was designed to ensure more efficient delivery of extension services to farmers. In addition to the existing system of training through farmer training centers and development agent visits to farmers, farmer development groups were created. These groups consist of 20 to 30 farmers. Each development group has subdevelopment groups organized with five members, led by a model farmer (otherwise known as *the one-to-five network or syndicate*). In addition to FTC training, demonstration of improved technologies and best practices were planned to be carried out on model farmers' fields to advance their skills. This is expected to help create a better learning process among farmers through group settings and facilitates a favorable environment for scaling up of best practices.

The subdevelopment groups led by model farmers are expected to meet periodically to discuss key extension services messages. Rather than one-to-one meetings with each farmer, the DA is subsequently expected to meet and convey extension services messages through the group leaders. Given the amount of work development agents are expected to accomplish and the considerable transportation and communication challenges they face in their farm visits, it is believed that this new structure will eventually improve the efficacy of the extension system, although to what extent is yet unknown.

Under the assumption of scaling up of this recent group approach, the government of Ethiopia later launched another ambitious national five-year Growth and Transformation Plan (GTP) (2010–2015), aiming to further expand the extension services. By 2014/2015 it was estimated that extension services would reach 14.6 million beneficiaries (Ethiopia, MoFED 2010). Accordingly, 13.95 million farmers were reached by the end of the plan period—95 percent of the target set for 2014/2015 (NPC 2016). Again, the quality of the service provided by the extension services remained unimproved and official estimates later indicated results were below expectations. The second Growth and Transformation Plan (GTP II), another ambitious national plan launched in May 2016, set targets even higher—the total extension services beneficiaries were to increase from 13.95 million in 2014/2015 to 18.237 million by 2019/2020 (NPC 2016). Official midterm assessments of GTP II indicated results were again below expectations. Over the years the human resources devoted to extension services grew substantially, from approximately 200 development agents in the 1950s to almost 72,000 in 2016/2017, reaching about 13 million beneficiaries with extension services technology packages in 2014/2015 (Table 6.3).

TABLE 6.3 Milestones in Ethiopia's agricultural extension services since 1950

Year	Extension services posts/FTCs	Extension agents or development agents (number)	Beneficiaries	Sources
1953	77	132	—	IPMS
1971	346	330	—	IPMS
1993/1994	—		1,600	SG-2000
1995	—	2,500	32,000	IPMS
2002	—	15,000	4.2 million	IPMS, Davis et al. (2010)
2008	6,486	47,522	9 million	Spielman, Mekonnen, and Alemu (2012)
2014/2015	11,000	—	13 million	GTP I
2016/2017	>15,000	72,402	16.7 million	Ethiopian Ministry of Agriculture
Plan for 2019/2020	18,000	—	18.2 million	GTP II

Source: Various project documents and reports.
Note: — = data not available. Extension services = agricultural extension service; FTC = farmer training center; IPMS = Improving Productivity and Market Success; SG = Sasakawa Global; GTP = Growth and Transformation Project.

By way of reviewing GTP II, the government has recently drafted a new extension services strategy in collaboration with the Ethiopian Agricultural Transformation Agency (ATA). The strategy is built around the following nine pillars: (1) strengthening and building the capacity of FTCs; (2) enhancing agricultural knowledge and information systems; (3) enhancing client-oriented and multiactors' extension services; (4) facilitating market linkages; (5) mainstreaming gender, youth, and nutrition; (6) enhancing environmental management and sustainability; (7) enhancing institutional coordination and linkages among partners; (8) enhancing the development and utilization of human resources; and (9) establishing strong and dynamic result-based monitoring, evaluation, and learning systems. The new strategy is yet to be implemented.

Ethiopia's extension system is also systematically testing the role of ICT to improve the quality and coverage of extension services. The recently scaled-up Agricultural Information Hotline—Interactive Voice Response and a Short Message System—is one example; it provides real-time and immediate access to vital agronomic information. Farmers can call the hotline for free and receive information on a wide range of topics on all major crops grown in the country. It also sends customized content (in case of drought, pest, and disease) to callers based on crop and geography data captured upon registration.

A community-centric video approach to extension services provision is another ICT-based method the government is piloting with Digital Green. On the front end, the approach produces localized videos (that is, videos that feature local farmers and are tailored to local needs) on agricultural technologies and practices. Extension agents facilitate the video dissemination or screening. On the back end, extensive monitoring data are collected and analyzed for course corrections and tracking performance.

6.4 Organizational Capacity and Management

Overall Structure

Ethiopia is a federal republic with five administrative tiers: federal level, regional, zone, *woredas*, and *kebeles*. In 1992 the government executed its first decentralization exercise, making the regions responsible for policy implementation and providing them with broad discretionary authority. In 2001 and 2002 the ambitious second wave took place, further devolving responsibility for many public goods and services, including extension services, to district governments in the four regions (Amhara, Oromia, SNNP, and Tigray comprising 86 percent of Ethiopia's total population). The government directed the districts in these four regions to expand their extension services to every *kebele* with at least three development agents specializing in crops, livestock, and natural resource management. In some cases, additional DAs specializing in beekeeping, veterinary health, cooperatives, or other areas are added and usually serve multiple *kebeles*.

Woreda offices are staffed with subject matter specialists (SMSs) in the areas of extension services communication, crop production and protection, livestock development and animal husbandry, natural resources, and irrigation agronomy. Each SMS supervises at least three FTCs. *Kebeles* are divided into sub-*kebeles*. Each DA is assigned to a sub-*kebele*. Development agents are responsible for extension services activities in their sub-*kebele*. DAs are also members of a *kebele* task force. The task force, comprising 10 members including the DAs, is fully accountable to the administration of the *kebele*.

Capacity of Development Agents

Based on Digital Green's 2017 survey of 896 DAs in the four largest regions, all development agents have postsecondary education, and a random selection of them reveals a good mix of expertise on crops and livestock production and natural resource management. DAs are trained for 12 years, the last two years being provided at ATVETs. However, some areas of DA training, experience, or skills can be further improved in the following ways:

- Although most DAs (92 percent) have received extension services–related training, 8 percent have not. Recent training has focused heavily on crop production; only 11 percent mentioned training on water and soil conservation, and 6 percent mentioned animal production.
- Of the DAs surveyed, 10 percent do not have any farming experience. Having farm demonstrations and/or extensive training for these DAs may be important.
- Fewer than half of DAs (46 percent) are computer literate. Training in computer literacy can help as can equipping FTCs with computers and Internet access.
- Fewer than half of DAs (42 percent) own a smartphone. Providing smartphones to DAs could assist them to be more efficient and to reach out to more farmers in remote areas.

Approaches Used by Agricultural Extension Service Providers

Several extension services methods are often applied and combined, including door-to-door extension services, farm-to-farm extension services, development group meetings, community meetings, field demonstrations, and FTC training. The results from the 2009 DA survey show that community and development group meetings were more prominent than other approaches. Also, in earlier surveys, contact with farmers via social events (for example, weddings, funerals, and other religious activities) was emphasized by 15 percent of DAs as essential in reaching farmers.

Almost all DAs in the 2008 and 2016 surveys stated they used model farmers. In the 2016 survey, almost all DAs reported collaborating with development groups. However, the medium of delivery is dominated by word (speech, verbal), and only half use pictures or illustrations. Only 14 percent use an audio/sound system (including a radio), and only 7 percent use a video (sound and picture). A recent study by Bernard et al. (2016) shows significant potential for video in terms of technology adoption and productivity. The expansion and promotion of the use of low-cost ICT, including smartphones, instant messaging, radio, interactive voice response, and video, is an area that the extension system can improve on. A simple starting point would be to reimburse DAs' expenses for their use of mobile calls and text messages

TABLE 6.4 Incentives, work resources, and environment for development agents

Questions asked to Development Agents	Tigray	Amhara	Oromia	SNNP	Total
Fixed salary, monthly average in ETB	1,988	2,712	2,151	2,191	2,280
Housing or housing allowance, % yes	81.5	56.4	29.5	18.7	40.9
Transport allowance, % yes	5.0	2.2	6.6	3.2	4.6
Health allowance, % yes	4.2	0.4	1.6	1.1	1.6
Annual leave taken in 2016, days on average	2.8	3.3	2.3	1.5	2.5
Is there a performance-related award or prize for DAs?, % yes	41.2	42.3	40.2	45.6	41.9
Received award or prize, % yes	57.1	37.5	39.0	32.9	39.6
Type of the main award received, % yes					
Financial (money)	39.3	50.0	3.5	17.9	24.1
Certificate (recognition)	28.6	13.9	57.9	35.7	37.6
Education opportunity	28.6	27.8	26.3	28.6	27.5
Promotion	3.5	0.0	3.5	7.1	3.4
Transfer to preferred location	0.0	8.3	8.8	10.7	7.4
Satisfied with the existing DA incentive structure, % yes	20.2	29.1	25.6	31.6	27.0
Access to bicycles, % yes	3.4	18.9	22.9	2.1	14.9
Access to motorbikes, % yes	9.2	3.5	4.1	1.1	4.0
Access to extension services materials					
Leaflets, % yes	57.9	66.9	49.0	53.5	55.7
Slides, % yes	42.9	14.5	14.3	19.3	19.2
Package booklets, % yes	86.6	88.6	64.7	63.6	73.4
Annual reports, % yes	82.4	74.9	67.2	88.8	75.7
Have enough resource to fully carry out extension services related work, % yes	10.1	8.4	3.3	9.1	6.7

Source: Digital Green DA Survey, conducted in 2017.
Note: SNNP = Southern Nations, Nationalities and People' Region.

to farmers, an issue DAs routinely complain about when it comes to financial incentives.

Target Groups

In terms of target groups, 72 percent of development agents reported targeting the head and spouse within the same household, while 26 percent reported targeting only the head. This is worse in Oromia, in which only 66 percent of DAs interviewed targeted both senior adults in the household. A full 51 percent and 60 percent of DAs reported organizing training specifically for women and youth, respectively. These efforts provide for greater gender inclusion, although attention to head and spouse together should be further improved, especially in Oromia.

Incentives for Development Agents

In terms of incentives, only 27 percent of DAs interviewed said they are satisfied with the incentive structure within the public extension system (Table 6.4). Rewards and prizes are given for good performance, such as financial rewards, educational opportunities, certificates, and promotions. In fact, 40 percent of DAs said they received some reward for good performance in the past three years. However, promotion, benefits, and allowances appear not to have been the same for all; most development agents receive a housing allowance (or rural housing facility) in Tigray, but only 19 percent do so in SNNP (Table 6.4). Across all regions, only 5 percent receive a travel allowance and only 2 percent a health allowance.

Sanctions for underperformance were reported by 62 percent of DAs. Sanctions are mostly warnings or decreased salary (2008 DA survey). The main issue may be the lack of resources to carry out extension services work. Only 7 percent reported that they have enough resources to fully carry out their work. Only 15 percent have access to a bicycle and 4 percent have access to a motorcycle, while the majority deliver extension services on foot (Table 6.4). This situation has not changed much since the 2008 DA survey. About 73 percent have access to at least extension services materials/booklets, although this is lower in Oromia and SNNP regions (64–65 percent)—an improvement over the 2008 DA survey, in which only half of the DAs interviewed reported some access to extension services materials.

When these data are combined with the information on the use of pictures or illustrations (only half of DAs interviewed use them), it appears that most extension services materials are mainly text without illustrations. This implies that significant potential remains to improve methods of agricultural extension delivery and ensure message retention through more creative visual applications.

Workload

Despite recent efforts to clarify DAs' specific roles and tasks (Ethiopia, MoA 2011), the amount and type of work they are expected to perform is daunting and increases with each intervention, leaving development agents overburdened and poorly prepared to do regular extension services activities. Although extension services have been decentralized to the administrative control of regional governments and *woreda* administrations, the continued imposition of targets at the national level and the lack of local capacity have prevented, so far, the emergence of a dynamic, demand-driven system (Spielman, Mekonnen, and Alemu 2012). DAs are positioned at the frontline and, as such, are at the very end of the government structure, where they are at the end of the receiving line in terms of workflow around activities that do not necessarily relate to their specific tasks. As can be seen in Figure 6.3, every intervention flowing to the *kebeles*, for example, programmatic interventions such as the Productive Safety Net Program (PSNP), AGP, Feed the Future, and other meso- and micro-interventions, involves engagement by the DA, including those from nongovernment actors. This leaves development agents overburdened and underresourced to achieve their goals.

In terms of overall workload, only 62 percent of DAs reported that it is manageable, while the rest reported that their workload is too much. During planting season, development agents work 10 to 11 hours per weekday and 5 to 6 hours on weekends on average. A DA works with 19 development groups and with about 500 households on average. They are involved in many activities, which include visiting farmers, training farmers, receiving training and office work, administering farm taxes and credit repayments, collecting data, supplying inputs, and mobilizing community work. They are also involved in nonagriculture-related activities in the community. Moreover, development agents spend almost 8 hours per week (10 percent of their work time) in the office preparing and writing reports. On average, a DA prepares 50 reports in a year, in addition to the hours spent on collecting data and monitoring farms according to indicators (approximately 10 to 20 percent of their work time). Therefore, the time devoted to working with and advising farmers was only approximately 35 percent in 2016 and 46 percent in 2009, if one adds up the time spent by DAs on field work, visiting farmers, and training farmers.

Figure 6.3 shows the heavy work burden that DAs carry and the need to balance the expectations placed on them with the support they are given. In particular, ways to ease transportation time through the provision of motorcycles or vehicles and expanding the technologies and tools they use through expanding the use of radio, video, and smartphones would make extension services delivery more efficient and useful. Given the amount of data that DAs collect, and time spent on collecting data, monitoring, and report writing, these data can be used more systematically to improve or refine local extension services strategies and approaches. Digital Green's "Connect Online—Connect Offline" (COCO) platform can be further tested as an alternative to the current data collection system (Bernard et al. 2016). COCO data are

FIGURE 6.3 A typical work week for a development agent in Ethiopia

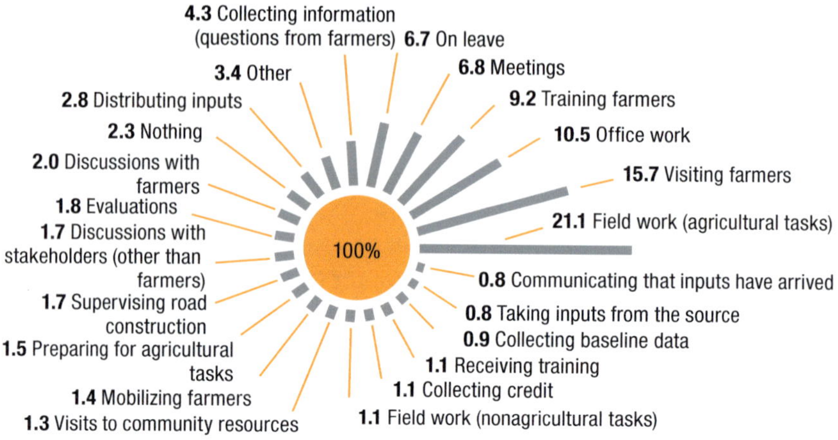

Source: Authors' calculations from IFPRI/EEA DA survey data 2009.

typically collected by DAs on paper and later digitalized at the *woreda* level. Data are automatically synchronized with Digital Green's main database via Internet connection. All data are integrated under the COCO platform. At the *woreda* level, the COCO platform automatically computes relevant statistics on technology dissemination and farmers' adoption and presents them on a dedicated user-friendly dashboard.

Linkages with National Agricultural Research Systems

An extension system is not only the most important vehicle to convey knowledge generated at research centers to farmers' fields but can also bridge the reverse information flow gap from farmers to researchers (Birkhaeuser, Evenson, and Feder 1991). As such, it is important that extension services are carefully synced with research centers. However, studies indicate that research–extension links are generally weak in many developing countries, partly due to poorly aligned incentive structures between researchers and extension managers, as each of them fall under separate management structures (Anderson and Feder 2004). In short, in such contexts, the priorities of national agricultural research centers are ill-aligned with those providing extension services or the farmers themselves, leading to undesirable outcomes in many instances (Purcell and Anderson 1997). The Ethiopian context is no different. Table 6.5 provides a summary of the literature on the evolution of research–extension–farmers linkages in the recent history of Ethiopia. A key takeaway from this summary is that the link between research and extension (and its feedback loop to farmers) has been inadequate and ad hoc despite efforts made at several points in time to address this challenge. The main observations are fourfold:

- Although the various efforts summarized in Table 6.5 aimed at interlinking research and extension, most of them continued to view research and extension as distinct processes. In some instances this was even manifested by organizing research and extension in different organizations or institutional structures.
- Most of these efforts to integrate research and extension lacked proper legal status to align incentive structures between the two entities and facilitate decision-making processes. The bodies that facilitated the interlinkage were organized in the form of platforms or ad hoc committees that lacked proper institutionalization to convey research outcomes to the extension system or the required accountability to enforce the research–extension linkage. Instead, such committees focused on rather broader issues of planning and execution of activities outlined by extension services providers. Relatedly, such initiatives heavily relied on project funds and often ceased to exist when projects phased out.
- More important, the end users of the technologies and extension services were not properly represented in most of such research–extension linkage platforms.
- Such efforts were also hampered by the limited capacities of national research centers to generate critical knowledge that addresses local conditions. Although rigorous studies evaluating the relevance and impacts of agricultural research investments in Ethiopia is yet to come, suggestive evidence indicates that research centers in Ethiopia, despite improvements, are highly underresourced both financially and skillwise, and lack proper coordination among centers themselves (Ethiopia, MoA 2014; Abate 2006; Belay and Degnet 2004). This is further discussed below.

The Ethiopian agricultural research system has a decentralized structure that includes three types of institutes: (1) National Agricultural Research Centers; (2) Regional Agricultural Research Centers; and (3) Agricultural Research Centers in Higher Education Institutions (Kassa and Alemu 2016).

TABLE 6.5 Historical evolution of research–extension–farmers links in Ethiopia

Period	Program/event	Objectives/highlights	Remarks
1952–1965	Integrating education, research, and extension	Formal research and extension services were started in 1952 when the Agricultural and Technical School at Jimma and the College of Agriculture and Mechanical Arts were established. Later, extension mandate was transferred to the Ethiopia, Ministry of Agriculture.	The system was modeled after the US land grant university system.
1966	Establishment of the Institute of Agricultural Research (IAR)	Following establishment of the IAR in 1966, research was divorced from education and extension without setting a mechanism for coordination of research and extension.	By compartmentalizing research and extension activities, a linear research–extension–farmers model was adopted.
1974–1977	Institute of Agricultural Research Extension Project Implementation Department of the Ministry of Agriculture (IAR/EPID)	The joint IAR/EPID program was mainly initiated for agricultural technology package testing and formulation of research recommendations.	The program was discontinued in 1977 due to budget problems and reinitiated in 1980/1981 as IAR/ADD (Agricultural Development Department), albeit it was not successful for various reasons.
1980s	Farming System Research (FSR) research–extension linkage	Conducted multidisciplinary surveys and focused on providing feedback to researchers on the characteristics of technologies, conveying farmers' problems, formulating recommendations appropriate to smallholder farmers, and generating useful recommendations for policy-makers.	Followed the FSR model, but the program was found to be expensive and time-consuming and was phased out as project funds ran out.

Ethiopia has 62 federal and regional agricultural research centers, excluding university research institutes, well spread across the various agroecologies of the country. This geographic dispersion makes it possible to link these centers with extension services located in each of these localities. However, recent studies indicate the absence of clear functional delineation and lack of formal or institutionalized mechanisms of collaboration among these three types of agricultural research centers. Cognizant of this, the Ethiopian Agricultural Research Council—an umbrella organization to coordinate national agricultural research—was established in 2014 (Deneke and Gulti 2016; Kassa and Alemu 2016). However, given the complexities to coordinate research nationally, this has remained rather nominal and is yet to function as planned.

Period	Program/event	Objectives/highlights	Remarks
1985–1986	Research–Extension Liaison Committee (RELC) was established under Research–Extension Division (RED) of IAR	The RELC was established at zonal and national levels: at zonal level to review and approve research proposals and extension recommendations, identify training needs for SMSs, and oversee research–extension and farmer linkages; at national level to give overall policy direction and capacity building.	RELC was largely ad hoc—that is, it did not have any legal status, which affected its decision-making power and institutionalizing accountability among members; farmers were passive participants.
Late 1990s	Research–Extension–Farmers Linkage Advisory Council (REFAC)	REFAC was organized at national, regional, and research center levels and was run by the RED of the Ethiopian Agricultural Research Organization (EARO). The main objective of REFAC was providing overall guidance to research and extension programs, and oversight of the linkage between the two activities. It was mainly funded by the World Bank Agricultural Research and Training Project (ARTP).	REFAC did not produce strong linkages as expected, mainly due to lack of clarity on actors' responsibilities.
2008–2009	Agriculture and Rural Development Partners Linkage Advisory Council (ARDPLAC)	ARDPLAC was organized at national, regional, zonal, and *woreda* levels and was run by the Ministry of Agriculture or Bureau of Agriculture. Later the name was changed to Agriculture Development Partners Linkage Advisory Council (ADPLAC). Like REFAC, it was mainly funded by the World Bank.	ADPLAC can be considered as a first attempt to institutionalize the linkage through allocation of regular finance and accountable institutional setup within the Ministry of Agriculture, even though there is still an ad hoc nature to the planning and execution of linkage activities.

Source: Authors' compilation from various sources, primarily Deneke and Gulti (2016); Kassa and Alemu (2016); Demekech et al. (2010); and Ethiopia (1999).

In sum, as discussed earlier, poor research–extension–farmers linkages are cited as the main reason for limited impact of the agricultural research system (Deneke and Gulti 2016; Kassa and Alemu 2016; Davis et al. 2010). A recent study by Kassa and Alemu (2016) points out the following: (1) the lack of integration and coordination between research and extension has created ambiguity as to who should conduct pre-extension trials before making technologies available to farmers; and (2) farmers and extension agents provide little or no feedback to research institutes about disseminated technologies, mainly due to the weak link between research and extension. Thus, given these challenges, the gains from the research and development investments made to run geographically spread national and regional agricultural research centers throughout the country would hardly be worth it.

6.5 Evidence of Effectiveness and Impact

Access to the Agricultural Extension Service System

Several improved technology sets and a great deal of modern inputs have been distributed to farmers through the extension system. According to the annual Agricultural Sample Survey of the CSA of Ethiopia, between 2004/2005 and 2010/2011 the proportion of farm households that had access to the extension system jumped from 33 percent to 71 percent. At the same time, the number of farm households with access to advisory services increased from 3.6 million to 10.8 million (Figure 6.4). Furthermore, the number of farm households covered by the extension services package, including advisory services, increased from 2.5 million to 6.6 million, while coverage increased from 1.5 million to 3.9 million hectares. Note that farmers are considered covered if they have applied one or more of the elements of the extension services package, including the use of fertilizers and improved seeds.

Promoted Technologies

According to the DAs interviewed, a wide range of technologies are being promoted in terms of land preparation, seed selections, row-planting, fertilizer applications, crop management, postharvest handling, and natural resource conservation. Many DAs (75 percent) reported having promoted market linkages, although one-quarter of development agents have yet to do so. It is interesting to note that market linkages are among the top requests by farmers but are among the least promoted extension services technology (Table 6.6). Ayenew (2016) documented that the Improving Productivity and Market Success (IPMS) project, which integrated productivity-enhancing technological schemes with the market development model, and contributed to input and output market integration and/or market-oriented agricultural production.

The topics requested by farmers are several, with the major ones relating to seed selection, crop management, market linkages, fertilizer application, and land preparation. According to the 2008 DA survey, the most overwhelmingly requested topic is improved seed and planting materials. This demonstrates improvements in the diversity of knowledge and information provided by the extension system beyond modern inputs promotion in more recent years. However, Table 6.6 shows that a major disconnect still exists as well as an inconsistency in the focus of development agents to establish such needs.

FIGURE 6.4 Farm households and area covered by the public agricultural extension system (millions) and farm households receiving advisory services, 2004/2005–2013/2014

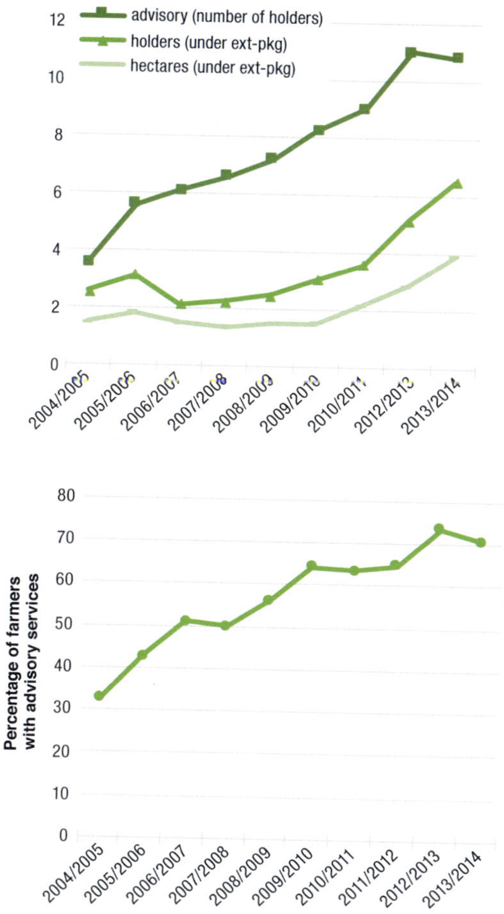

Source: Author's computation using Central Statistical Agency (CSA)'s annual reports (CSA, Volumes I, II, III, 2005–2014).
Note: Ext-pkg = extension services package.

Satisfaction with Agricultural Extension Services

Almost all households surveyed were satisfied (59 percent, strongly satisfied; 40 percent, satisfied) with the individual visits made by DAs (Table 6.7). Fewer than 2 percent indicated that they were not satisfied with the answers development agents provided or that the information given by DAs was incorrect or irrelevant. This shows overall high satisfaction with DA advice.

TABLE 6.6 Technologies or practices promoted by development agents

Technology/practices	DAs who promoted technology in 2015/2016 (%)	Topic or technology requested by farmers? (% yes)
Land preparation	98.6	57.0
Seed selection	97.0	60.0
Row-planting	98.0	53.0
Fertilizer application	98.2	57.4
Crop management	97.2	58.4
Postharvest handling	96.0	57.4
Natural resource conservation	96.4	49.2
Climate-smart practices	85.2	53.3
Market linkages	75.5	57.7

Source: Digital Green DA Survey (2017).

TABLE 6.7 Recipients' assessment of quality of agricultural extension services provided in the past 12 months, 2011 and 2013, percent

Assessment of quality	2011	2013
The service provided by the development agent was satisfactory?		
Strongly agree	62.5	58.7
Agree	35.5	39.6
Disagree	1.1	1.3
Strongly disagree	0.7	0.4
Total number of observations: 7,381		

Source: AGP survey conducted in 2011, 2013, two rounds.

Another study by Elias et al. (2016), although with a small sample size (150 beneficiaries in northwest Ethiopia) paints a different picture. It shows that about 55 percent of interviewees were satisfied, whereas 45 percent of them were dissatisfied with the extension services received, implying that the program still has a lot of room for improvement. Focus group discussions conducted with farmers show that lack of quality and diversified improved seeds, limited technology choices, high price of inputs (chemical fertilizer), and an inconvenient loan system are the main reasons for dissatisfaction.

Farmers also indicated serious issues with the quality, diversity, and availability of improved seeds in the study area. They have experienced crop loss due to seed quality. In addition, while they have a great need to access different types of crop seeds, they are only able to access improved seeds for maize, wheat, and teff crops. Their opinion is consistent with previous literature

(Spielman, Mekonnen, and Alemu 2012). Although this may not relate to the quality of extension services per se, these farmer responses highlight the inability of the extension system to communicate and respond to these constraints. In other words, supply-driven, instead of demand-driven, extension services limits farmers' satisfaction. Therefore, it is important that packages provided farmers should be of a high standard and be easily accessible, affordable, and needs-based to enhance overall satisfaction.

Yet other studies have highlighted caution in the interpretation of farmers' satisfaction due to overreporting and serious social desirability bias with responses obtained in some settings and country contexts (Ragasa and Niu 2017 in the case of Malawi). Moreover, satisfaction can be correlated with the promotion and provision of inputs, as highlighted by Elias et al. (2016) and Ragasa and Mazunda (2018). Therefore, one should exercise caution when interpreting such responses while learning from the insights. In addition to asking about farmers' satisfaction, it is often insightful and useful to ask farmers how the advice was used and how it changed their behavior and practices.

Agricultural Extension Services, Technology Adoption, and Productivity Growth: Empirical Evidence

Several studies suggest that the success of extension services programs in Ethiopia has been mixed. The most recent expansion of extension services has yet to be evaluated, although Davis et al. (2010) provide a careful review of system operation prior to 2010 and identify a series of weaknesses. A study by Dercon et al. (2009) shows that receiving at least one visit from a DA increases consumption growth (a measure of income) by 7 percent and reduces poverty by 10 percent, based on a panel dataset from 1999 to 2004. The authors caution readers on the implications of the results for extension services, as some of the effect may represent transfers of technology or knowledge, while some may reflect the influence that extension agents have on the increased use of fertilizer and other inputs.

Given the success of Ethiopia to put in place one of the largest public extension systems in Africa, and the associated quality challenges thereof, Berhane et al. (2018) (1) study the roles this system has played in the adoption of specific technologies; (2) identify direct effects of the system in improving agricultural productivity; and (3) infer and outline the challenges ahead to further improve agricultural productivity in Ethiopia. They control for many important sources of heterogeneity, including household wealth and weather-related factors. The following are the main findings:

- Access to the extension system significantly increases adoption of modern inputs (mainly chemical fertilizers, improved seeds, herbicides, and irrigation).
- Extension services do not directly increase the level of productivity other than through the indirect effect on increased adoption of these modern inputs. Use of modern inputs—including fertilizers, improved seeds, irrigation, and row-planting—contributes to productivity increases. This result is consistent across several indicators for extension services and advisory services directly provided by DAs as well as those provided through farmer-to-farmer interactions. These results are also in line with recent evidence suggesting that agricultural productivity increases in Ethiopia in the past decade are mainly associated with expansion in cultivated area of land and increased use of chemical fertilizers (Bachewe et al. 2018). The results suggest that agricultural productivity increases in Ethiopia are not knowledge-driven as yet and that achieving additional productivity increases will be difficult without investing in knowledge-based extension services.
- Berhane et al. (2018) do not find significant adoption and productivity variances among younger and older farmers as a result of access to extension services.
- There are variances in wealth, rainfall levels and agroecology, land, labor, and other critical farm inputs such as oxen, the last of which are associated with significant productivity growth. Both Berhane et al. (2018) and Ragasa et al. (2013) find that access to production-related information delivered by radio does raise productivity. Innovating on alternative methods for information dissemination such as radio and other mass media may prove to be a low-cost and effective way of increasing productivity and incomes.

These results show that the technical support currently provided through extension services is not directly linked with increases in productivity (other than through channeling inputs) and is consistent with a recent assessment by the government of Ethiopia, documented in GTP II. What is not clear is the mechanism that drives these low achievements in recent years.

The explanation for this is simple. Development agents, in practice, spend a substantial amount of their time on promoting and channeling fertilizer and improved seeds to farmers and increasingly less time on technical assistance (Figure 6.3). In fact, being at the receiving node of government parastatals involved with input supply, the development agent is viewed more as the sole link to communities than an extension agent whose expertise is in need. Recent studies that link extension services and productivity show a major effort to push for fertilizer and improved seed use (Abay et al. 2017; Berhane et al. 2018), considered the major contributors to rapid productivity growth and agricultural transformation in Ethiopia (Bachewe et al. 2018). Beyond this, studies show no other direct effects on productivity (Abay et al. 2017).

These results are largely consistent with an earlier study by Krishnan and Patnam (2014), which used a panel dataset from 1999 to 2009 to specifically review the adoption of fertilizer and improved seeds. However, their study shows that the impact of DA extension wears off over time as farmers learn more from other farmers. It shows that adoption is mainly a result of knowledge shared by neighbors, thus spreading the technologies during this period. DAs had a significant impact on adoption in 1999; however, by 2004 and by 2009 their role had become almost irrelevant in this process, despite a vast increase in the number of DAs throughout rural Ethiopia. Learning from neighbors is significant and stable throughout the study: an increase of one standard deviation in the average adoption of improved seeds by neighbors (corresponding to local diffusion rates increasing by 22 percent) raises the probability of one's own adoption by 11 percentage points. For 1999 the results by Dercon et al. (2009) are confirmed by Krishnan and Patnam (2014) in that the role of extension services does matter. Learning from extension services ceases to be relevant after 1999, however, and despite further extensive public investment in extension services in subsequent years, there is no evidence of return.

Another study by Nisrane et al. (2011) provides further data that is consistent with the results of Krishnan and Patnam (2014); they find a significant impact of extension services on output for the years 1994–1999 and 1999–2004, although no effect on output after 2004. These results are consistent with various studies showing that extension services in 1999 had the largest bearing on adoption—and hence potentially on output growth in the subsequent period—although this effect wears off in more recent years.

Ragasa et al. (2013) find no significant direct effect of extension services on crop productivity beyond their impact on fertilizer and improved seed use. In addition, their study shows that the perceived usefulness of DA advice is statistically significant in the productivity models, implying that "quality" extension services matter. They show that radio can contribute significantly to increasing productivity. These may reflect the situation in Ethiopia, where the large cadre of development agents is focused on fertilizer and improved seed promotion; and other information channels like radio are more likely to promote other forms of nonfertilizer soil fertility management practices (such as manure and other organic fertilizers, crop rotation, and soil conservation techniques) than DA visits (Ragasa et al. 2013).

They highlight the need to pay close attention to women's access to extension services and which delivery methods can best reach them. Male heads are more likely to be visited by and to receive advice from development or extension agents than female heads (Ragasa et al. 2013). Female-headed households appear to have lower productivity; but when extension services and land access are given equally to them, they are as, if not more, productive as male-headed households (Ragasa et al. 2013). Fortunately, more recent study by Buehren et al. (2019) shows that improvement in the extension services, particularly under the Rural Capacity Building Project (RCBP) in targeting women farmers. Women farmers' access and satisfaction with extension services increased significantly immediately after the start of the RCBP, but that effect did not last into the medium term (Buehren et al. 2019).

These findings have several implications, as follows:

- There is a need for both dynamic processes of innovation and regular reviews of the technologies promoted. As farmers become aware of and try these technologies, the issue of low adoption becomes less an issue of weak extension and information systems and more an issue of the nature and relevance of the technology. This calls for an urgent need to review the usefulness of different technologies currently promoted.
- There is need to look at the quality of advice. Again, this can be a product of irrelevant technologies being promoted or a reflection of some DAs' incompetence.
- The capacities and specific tasks of development agents need to be reviewed to understand how they can best support farmers within the limits of their resources and time. Despite recent efforts to clarify the specific roles and tasks of DAs as extension agents (Ethiopia, MoA 2011), the amount and type of work they are expected to perform is daunting and increasing with each intervention taking place, leaving them overburdened and poorly prepared to perform regular extension services activities.

In addition, the poor technical training of DAs and the deficient quality of facilities at FTCs means that farmers are not tempted to spend time around FTCs. DAs are challenged to demonstrate to farmers whatever knowledge or experiences they hold. Added to this is the fact that development agents often need to traverse difficult terrain to reach many farm communities, which are scattered over a wide geographical area, with poor infrastructure and (or even nonexistent) transportation facilities. This suggests that further gains in agricultural productivity will have to come from institutional innovations in the existing input supply-led extension system, upgrading it to one that is more focused and knowledge-driven to address complex problems in scattered and heterogeneous geographical areas. DAs will need a more holistic knowledge base, with more facilitation and knowledge-brokering skills in addition to technical skills. A stronger engagement with research institutes and knowledge sources will need to be institutionally cultivated with an incentive system accordingly in place.

6.6 Conclusion and Recommendations

Extension is critical to promote improved farm technologies and increase productivity. Ethiopia has heavily invested in its massive public extension system and has one of the highest agent-to-farmer ratios in the world. Ethiopia has registered substantial economic progress in recent years, largely attributable to growth in agriculture. This chapter has dealt with the extent of Ethiopia's investment in agriculture—particularly the extension system that is linked to this growth; the extent to which existing extension services structures maintain current growth levels; and what needs to be done to sustain current growth levels. In addition, while drawing on diverse sources of primary and secondary data, this chapter has documented the historical evolution and state of Ethiopia's extension system, providing suggestions on the way forward.

Results from data analyses at the household and development agent levels indicate that access to the extension system increased the adoption of modern inputs such as chemical fertilizers (primarily), improved seeds, herbicides, and irrigation. Moreover, the use of these modern inputs, as well as row-planting, increased productivity levels significantly. However, extension services do not directly increase productivity levels other than indirectly through their effects on the increased adoption of modern inputs. These results are plausible, given that Ethiopia's extension system is geared toward conveying these inputs to farmers and has limited capability to convey critical knowledge-based support to farmers.

Adoption of these fundamental inputs was instrumental to recent productivity increases and will continue to be important insofar as Ethiopia's agricultural production system starts from a rather low base. To that extent, the essential role that DAs place in channeling inputs to farmers will remain critical. However, further gains in agricultural productivity will have to come through significant improvement of the existing input supply–led extension system, upgrading it to one that is knowledge driven and able to address some of the complex issues that arise due to, among others, the dynamics of farming systems that arise from heterogeneous agroecology, as well as soil nutrient deficiencies (in the face of population growth and climate change).

The analysis in this chapter indicates three key constraints that play against the greater contributions of extension services to productivity growth and agricultural transformation. First, with limited institutional innovations and poor coordination with research centers—hence the limited injection of new knowledge into the system—development agents are left with little leverage to convince lead and other farmers. Their operation under poorly resourced work conditions (given the amount and diversity of work they are tasked with) implies that the link between research and extension services remains weak, which leads to diminishing returns to the technical support of development agents on technology adoption (Krishnan and Patnam 2014). Second, the fact that DAs are overburdened by activities beyond their regular mandates provides little time for them to search for additional knowledge and information. Although the current system can be commended for having one of the highest DA-to-farmer ratios, it is overly standardized (one-size-fits-all) and lacks the flexibility to adapt to local conditions. Third, the efficacy of FTCs is also constrained because they are generally underresourced and scattered, with little focus and scale. While evidence suggests that the number of farm households reached with extension services has substantially increased, these constraints negate sustaining future gains. It is unlikely, therefore, that the increased farmers' access to the system, as it is now, can be translated into productivity gains.

As such, significant reforms of the extension system are critical to Ethiopia's agricultural transformation. Reforms will need to extricate the system away from single-minded, top-down, package approaches of cereal intensification to more dynamic, context-oriented, responsive, and knowledge-based service provision. This will obviously require some major policy choices between a system that covers a wider area of the country with thinly spread resources (given Ethiopia's topography and limited resources) and a more focused but well-resourced system that is capable of addressing critical knowledge bottlenecks to proceed with transformation. Such choices are not straightforward, particularly given the political economy attractiveness of reaching everyone (even if that means keeping the thinly resourced current system) at the expense of not focusing and prioritizing where the investments are yielding the optimal development results.

These analyses also highlight the heavy work burden of development agents and the need to balance expectations against support they are provided with. In the short to medium term, ways to ease transportation time (such as providing more motorcycles or vehicles) and the facilitation of technologies and tools (such as radio, video, and smartphones) to make their extension services delivery more efficient would be useful, given their heavy involvement in numerous other activities.

More important, institutional innovations require the channeling of new knowledge to extension agents, with a strong link between research and extension services to remain country specific. Given the complexities associated with the size of farm communities to be served, and the physical and infrastructural constraints of Ethiopia, recasting the extension system as one that will be responsive to farmers' demands and to knowledge sources is the most pressing agenda for policymakers at this time. A first step toward addressing these challenges would be for policymakers to appreciate the extent to which the existing extension services are unable to adequately respond to current and emerging challenges and, as elsewhere (see, for example, Tsakok 2011), understand that decades of continued investments are required to achieve agricultural transformation. This chapter highlights some easier and more manageable changes that can be strived for in the short and medium term, including (1) more investments in the extension structures and human resources, where gaps are identified here; (2) more focus on knowledge-based extension services to complement the heavy focus on modern inputs promotion; (3) expanding tools and activities on promoting demand articulation of communities in various locations and catering the extension services accordingly; and (4) more effective institutional arrangements for systematic and sustainable research and extension linkages and integration into the agricultural innovation system.

Finally, extension services are very much tied to access to productive inputs, and thus there is a need not only to continue public investment to promote fertilizer, seed, credit, and other complementary inputs and extension services, but also to support private-sector development. This is of course the most difficult to achieve and an aspiration in the long term as the political economy and priorities transform over time. These findings reinforce other studies conducted in the region relating to the need for complete, rather than half-hearted, liberalization of input supply markets to support smallholders to intensify commercialization of their production. The findings recognize the necessity of continued public engagement in input and output markets and extension services, while carving out new space for private investment by providing goods and services for smallholders in an efficient manner. By being more flexible in how inputs and services are provided and ensuring a greater degree of choice for smallholders, new market and technological opportunities will emerge in Ethiopia's agriculture sector.

References

Abate, T. 2006. "Focusing Agricultural Research to Address Development Needs: The Way I See It." In *Successes with Value Chain, Proceedings of Scaling Up and Scaling Out Agricultural Technologies in Ethiopia*, edited by T. Abate, 1–20. Proceedings from conference, Addis Ababa, Ethiopia, May 9–11.

Abay, K. A., G. Berhane, A. S. Taffesse, K. Abay, and B. Koru. 2018. "Estimating Input Complementarities with Unobserved Heterogeneity: Evidence from Ethiopia." *Journal of Agricultural Economics* 69 (2): 495–517.

Abesha, D., A. Waktola, and J. B. Aune. 2000. *Agricultural Extension in the Drylands of Ethiopia*. Report. Oslo, Norway: Drylands Coordination Group.

Anderson, R. J., and G. Feder. 2004. "Agricultural Extension: Good Intentions and Hard Realities (English)." *The World Bank Research Observer* 19 (1): 41–60.

Ayenew, H. Y. 2016. "Production Efficiency and Market Orientation in Food Crops in North West Ethiopia: Application of Matching Technique for Impact Assessment." *PloS One* 11 (7): 0158454.

Bachewe, F. N., G. Berhane, B. Minten, and A. S. Taffesse. 2018. "Agricultural Transformation in Africa? Assessing the Evidence in Ethiopia." *World Development* 105: 286–298.

Belay, K. 2003. "Agricultural Extension in Ethiopia: The Case of Participatory Demonstration and Training Extension System." *Journal of Social Development in Africa* 18 (1): 49–84.

Belay, K., and A. Degnet. 2004. "Challenges Facing Agricultural Extension Agents: A Case Study from South Western Ethiopia." *African Development Review* 16 (1): 139–168.

Benin, S. 2014. *Identifying Agricultural Expenditures within the Public Financial Accounts and Coding System in Ghana: Is the Ten Percent Government Agriculture Expenditure Overestimated?* International Food Policy Research Institute (IFPRI) Discussion Paper 1365. Washington, DC: IFPRI.

Berhane, G., C. Ragasa, G. T. Abate, and T. W. Assefa. 2018. *The State of Agricultural Extension Services in Ethiopia and Their Contribution to Agricultural Productivity*. Washington, DC: IFPRI.

Bernard T., S. Makhija, K. Orkin, A. S. Taffesse, and D. Spielman. 2016. *Video-Based Agricultural Extension: Analysis of a Pilot Project in Ethiopia*. Project Note. Washington, DC: IFPRI.

Birkhaeuser, D., R. E. Evenson, and G. Feder. 1991. "The Economic Impact of Agricultural Extension: A Review." *Economic Development and Cultural Change* 39 (3): 607–650.

Birner, R., K. Davis, J. Pender, E. Nkonya, P. Anandajayasekeram, J. Ekboir, A. Mbabu, D. J. Spielman, D. Horna, S. Benin, and M. Cohen. 2009. "From Best Practice to Best Fit: A Framework for Designing and Analyzing Pluralistic Agricultural Advisory Services Worldwide." *Journal of Agricultural Education and Extension* 15 (4): 341–355.

Buehren, N., M. Goldstein, E. Molina, and J. Vaillant. 2019. "The Impact of Agricultural Extension Services on Women Farmers' Access, Adoption and Productivity: Evidence from Ethiopia." *Agricultural Economics* 50 (4): 407–419.

Davis, K., B. Swanson, and D. Amudavi. 2009. "Review of Agricultural Extension in Ethiopia." Draft Report for the Bill and Melinda Gates Foundation and the Government of Ethiopia, IFPRI, Addis Ababa.

Davis, K., B. Swanson, D. Amudavi, D. A. Mekonnen, A. Flohrs, J. Riese, C. Lamb, and E. Zerfu. 2010. *In-Depth Assessment of the Public Agricultural Extension System of Ethiopia and Recommendations for Improvement.* IFPRI Discussion Paper 1041. Washington, DC: IFPRI.

Demekech, G., F. Moges, G. Zeleke, K. Tesfaye, and M. Ayalew. 2010. "Multi-Stakeholder Linkages in Rural Innovation Processes in the Amhara Region, Ethiopia." Working document series 137. International Centre for Development-Oriented Research in Agriculture, University of Bahir Dar, Amhara Region Agricultural Research Institute and Bureau of Agriculture and Rural Development, Wageningen/Bahir Dar.

Deneke, T. T., and D. Gulti. 2016. "Agricultural Research and Extension Linkages in the Amhara Region, Ethiopia." In *Technological and Institutional Innovations for Marginalized Smallholders in Agricultural Development,* edited by Franz W. Gatzweiler and Joachim von Braun, 113–124. Cham, Switzerland: Springer.

Dercon, S., D. O. Gilligan, J. Hoddinott, and T. Woldehanna. 2009. "The Impact of Agricultural Extension and Roads on Poverty and Consumption Growth in Fifteen Ethiopian Villages." *American Journal of Agricultural Economics* 91 (4): 1007–1021.

Dorosh, P., and S. Rashid. 2012. *Food and Agriculture in Ethiopia: Progress and Policy Challenges.* Philadelphia, US: University of Pennsylvania Press.

Elias, A., M. Nohmi, K. Yasunobu, and A. Ishida. 2013. "Effect of Agricultural Extension Program on Smallholders' Farm Productivity: Evidence from Three Peasant Associations in the Highlands of Ethiopia." *Journal of Agricultural Science* 5 (8): 163.

———. 2016. "Farmers' Satisfaction with Agricultural Extension Service and Its Influencing Factors: A Case Study in North West Ethiopia." *Journal of Agricultural Science and Technology* 18 (1): 39–53.

Ethiopia, CSA (Central Statistical Agency). 2004–2016. *Agricultural Sample Survey: Report on Farm Management Practices (Private Peasant Holdings,* Meher *Season).* Volume III. Addis Ababa.

Ethiopia. 1999. *Agricultural Research and Training Project: Ethiopian Research—Extension-Farmer Linkages Strategy.* Volume I. Addis Ababa.

———. 2002. *Food Security Strategy.* Addis Ababa.

Ethiopia, MoA (Ministry of Agriculture). 2011. *Guideline for Agricultural Development Partners' Linkage Advisory Council (ADPLAC): Organizational Structure, Functions and Procedures.* Addis Ababa.

———. 2014. *Crop Variety Register.* Issue No. 17. Plant Variety Release, Protection and Seed Quality Control Directorate. Addis Ababa.

Ethiopia, MoFED (Ministry of Finance and Economic Development). 2002. *Ethiopia: Sustainable Development and Poverty Reduction Program.* Addis Ababa.

———. 2006. *A Plan for Accelerated and Sustained Development to End Poverty (PASDEP).* Addis Ababa.

———. 2010. *Growth and Transformation Plan: 2010/11–2014/15.* Volume 1, *Main Text.* Addis Ababa.

Gebremedhin, B., D. Hoekstra, and A. Tegegne. 2006. *Commercialization of Ethiopian Agriculture: Extension Service from Input Supplier to Knowledge Broker and Facilitator.* Addis Ababa: International Livestock Research Institute.

Hill, R., G. Inchauste, N. Lustig, E. Tsehaye, and T. Woldehanna. 2016. "Fiscal Incidence Analysis for Ethiopia." In *The Distributional Impact of Taxes and Transfers: Evidence from Eight Developing Countries,* edited by G. Inchauste and N. Lustig, 79–111. Washington, DC: World Bank.

Kassa, B., and D. Alemu. 2016. "Agricultural Research and Extension Linkages: Challenges and Intervention Options." *Ethiopian Journal of Agricultural Sciences* 27 (1): 55–76.

Krishnan, P., and M. Patnam. 2014. "Neighbors and Extension Agents in Ethiopia. Who Matters More for Technology Adoption?" *American Journal of Agricultural Economics* 96 (1): 308–327.

Nisrane, F., G. Berhane, S. Asrat, G. Getachew, A. S. Taffesse, and J. Hoddinott. 2011. *Sources of Inefficiency and Growth in Agricultural Output in Subsistence Agriculture: A Stochastic Frontier Analysis.* Ethiopia Strategy Support Program Working Paper 19. Addis Ababa: IFPRI.

NPC (National Planning Commission). 2016. *Growth and Transformation Plan II (GTP II) (2015/16–2019/20): Main Text.* Addis Ababa: NPC.

Purcell, D. L., and J. R. Anderson. 1997. *Agricultural Extension and Research: Achievements and Problems in National Systems.* World Development Sources (WDS) 1997. A World Bank Operations Evaluation Study. Washington, DC: World Bank.

Ragasa, C., G. Berhane, F. Tadesse, and A. S. Taffesse. 2013. "Gender Differences in Agricultural Extension, Technology Adoption and Productivity: Evidence from Ethiopia." *Journal of Agricultural Extension and Education* 19 (5): 437–468.

Ragasa, C., and J. Mazunda. 2018. "The Impact of Agricultural Extension Services in the Context of a Heavily Subsidized Input System: The Case of Malawi." *World Development* 105: 25–47.

Ragasa, C., and C. Niu. 2017. *The State of Agricultural Extension and Advisory Services Provision in Malawi: Insights from Household and Community Surveys.* Technical Report. Washington, DC: IFPRI.

Schmidt, E., and M. Kedir. 2009. *Urbanization and Spatial Connectivity in Ethiopia: Urban Growth Analysis Using GIS.* Ethiopia Strategy Support Program—2 Discussion Paper 3. Addis Ababa: IFPRI.

Spielman, D. J., D. K. Mekonnen, and D. Alemu. 2012. "Seed, Fertilizer, and Agricultural Extension in Ethiopia." In *Food and Agriculture in Ethiopia: Progress and Policy Challenges,* edited by P. Dorosh and S. Rashid, 84–122. Philadelphia: University of Pennsylvania Press.

Tsakok, I. 2011. *Success in Agricultural Transformation: What It Means and What Makes It Happen.* Cambridge, UK: Cambridge University Press.

World Bank. 2007. *World Development Report 2008: Agriculture for Development.* Washington, DC.

———. 2015. *Ethiopia Poverty Assessment 2014.* Washington, DC.

Wubneh, F. K. 2007. "Realizing the Dream: Agricultural Extension for Rural Livelihoods Development in Ethiopia." Institute of Social Studies (ISS) Graduate School of Development Studies. MSc thesis, The Hague, Netherlands.

Zhou, Y., and S. C. Babu, eds. 2015. *Knowledge Driven Development: Private Extension and Global Lessons.* London, UK: Academic Press. http://ifpri.worldcat.org/oclc/911268908.

Chapter 7

MALAWI

Catherine Ragasa and Catherine Mthinda

7.1 Introduction

Malawi has made some progress in increasing its agricultural production and economic growth and in reducing food insecurity in recent years, although there is much yet to be done. Malawi has not yet achieved the 6 percent target in agricultural growth, despite having exceeded the agricultural investment target of 10 percent under the Comprehensive Africa Agriculture Development Programme (Malawi, MoAIWD 2016a). Undernutrition and food insecurity are still widespread, with 37 percent of children under five being stunted, according to the 2015/2016 Demographic and Health Survey (DHS), and 6.7 million people estimated to be in need of food assistance in the 2016/2017 crop year due to severe drought (Malawi, MoAIWD 2016a).

Despite the early successes of the government's flagship agricultural program, the Farm Input Subsidy Programme (FISP), agricultural productivity has stagnated and food insecurity conditions in many areas of the country remain. For instance, in the case of maize, since the 2010/2011 season, productivity has been around 2 metric tons per hectare, remaining below the Agriculture Sector Wide Approach (ASWAp) target of 3 metric tons per hectare (Malawi, MoAIWD 2016a). This necessitates bold actions to revisit the design and implementation of FISP and, at the same time, rethink the other complementary services and systems within agriculture and other sectors that need to be strengthened.

Agricultural extension and advisory services are one of those critical complementary services and systems that remains underfunded (Ragasa et al. 2017). In early 2015 extension services were highlighted as the most important priority area for increasing agricultural productivity during the extensive consultation process on the content of the National Agriculture Policy (NAP), which involved representatives from a broad range of agricultural stakeholder groups coming from 28 districts. Snapp et al. (2014) suggest that lack of information among farmers on the appropriate use of hybrid maize seed and fertilizer due to ineffective extension services may have been a factor in the observed low nutrient use efficiency observed among beneficiaries of the FISP, thus limiting the productivity and development impact of Malawi's flagship agricultural development program. A recent analysis of the Malawi Integrated Household Panel Survey (2010, 2013) by Ragasa and Mazunda (2018) suggests that access to agricultural advice does not necessarily lead to greater crop productivity and food security. However, it is the quality of extension services, measured in terms of the farmer's perceived usefulness and relevance of the advice, that is a significant predictor of agricultural productivity and household food security. These findings and observations suggest a need to look more closely at the extension system in Malawi to identify the issues and constraints that currently prevent it from making a substantial contribution to the country's development objectives.

A number of initiatives are being implemented in response to this identified need. Among them, the United States Agency for International Development (USAID) implemented a five-year project (Strengthening Agricultural and Nutrition Extension [SANE]), and the European Union currently funds a five-year project (KULIMA), both of which primarily focus on extension services and their links to agricultural research. Malawi's Ministry of Agriculture, Irrigation and Water Development (MoAIWD) has also requested a study to look closely at the state of extension services provision, with the intent to further strengthen it to contribute to food security, economic growth, and other development goals. This has led to the three-year project (Assessing and Enhancing the Capacity, Performance, and Impact of the Pluralistic Agricultural Extension System in Malawi), funded by the government of Flanders and the German Agency for International Development, led by the International Food Policy Research Institute (IFPRI). This chapter summarizes the assessment data collected from this IFPRI project, synthesizes its research papers, and extensively reviews relevant literature on the evaluation of extension services approaches in Malawi. The chapter is structured following the best-fit framework discussed in Birner et al. (2009) and presented in Chapter 2.

Section 7.2 presents the sources of data analyzed and the analytical methods used. Section 7.3 describes the enabling environment, while Section 7.4 summarizes the structure and landscape of the extension system, management systems of organizations involved in extension services provision, capacity and incentives of extension services providers, and coordination and certification systems. Section 7.5 summarizes the various extension services approaches applied and the evidence of the effectiveness and impact of these. Finally, Section 7.6 concludes with major recommendations from the assessment.

7.2 Methods and Data Sources

This chapter utilizes various surveys and interviews conducted in Malawi between August 2016 and March 2017 by IFPRI, Wadonda Consult Limited, Lilongwe University of Agriculture and Natural Resources students and other partners in Malawi (detailed in Table 7.1). Figure 7.1 presents a map of Malawi, showing all districts (except Likoma) included in the main household survey datasets used in this chapter; the shaded districts where the census of extension service providers were conducted, and dotted points where two rounds of the focus group discussions with farming communities were conducted.

7.3 Governance Structures and Policies

The provision of extension services in Malawi has been guided by the National Agricultural Extension Policy (NAEP) put in place in 2000. The vision in this policy is that "all farmers are able to demand and have access to high quality extension services from those best able to deliver them" (Malawi, MoAIWD 2000: 16). Thus the main aim for introducing the policy was to ensure that farmers are able to demand and have access to high-quality extension services. The government of Malawi introduced a policy that promotes the provision of decentralized, demand-driven extension services and encourages the participation of many extension services providers. The policy sets out nine guiding principles to help extension services actors operationalize this policy, which includes demand-driven extension services, accountability, equalization, and decentralized coordination, among other principles.

The implementation of this policy was linked to other subsectoral policies and strategies whose effectiveness depend on sound provision of extension services and therefore support the implementation of extension policy. These include the Food Security Policy; Crop Production Policy; Livestock Development Policy; the Agricultural Research Master Plan; Gender, HIV, and AIDS Strategy as well as AIDS Agricultural Sector Policy; National Fertilizer Strategy; National Irrigation Policy and Development Strategy; and Land Resource Conservation Policy; among others. Thus the effectiveness of these policies depends on the sound provision of extension services.

TABLE 7.1 Datasets used in this chapter

Dataset and year of data collection	Description and sampling	Extension focus	Analytical method
IFPRI household and community surveys (2016, 2018)	Nationally representative panel household surveys covering 3,001 households in 299 randomly selected communities; community surveys answered by 4–8 knowledgeable representatives in the community: all districts (except Likoma) are covered. In each district, sections were randomly selected based on probability proportional to the population size of the district. In each selected section, one community is randomly selected. Dataset also contains 5,069 female and male primary adults within these 3,001 households, which enables intrahousehold analysis, particularly looking at women's empowerment and gender gaps in these dimensions within dual-headed households.	The survey focuses on agricultural extension and technology adoption. The survey also contains many other data, such as socioeconomic and production data, to allow linking extension services access to outcome indicators.	For the 2016 dataset: both descriptive analysis and econometric modeling conducted by authors presented for the first time here in this chapter or cited from published papers of the authors using same datasets. For the 2018 dataset, only descriptive analysis is done so far since the data is still being analyzed more in-depth at the time of the last revision of this chapter.
IFPRI interviews with extension service providers (2017)	Census of 217 extension service providers in 15 districts, which were selected by the research team in consultation with key stakeholders to represent the different regions, agroecological zones, agricultural development divisions (ADD), and farming systems (see Ragasa et al. 2017).	Three-page questionnaire mainly on general extension services provision.	Descriptive analysis conducted by authors and presented for the first time here in this chapter or cited from authors' published papers using same datasets.

The government of Malawi launched its National Agricultural Policy in 2016, which has become the umbrella policy for the agriculture sector. The main agenda for NAP is the transformation of the agriculture sector, which is expected to lead to significant growth in agricultural production, productivity, and real farm incomes (Malawi, MoAIWD 2016b). To achieve this, NAP has set out nine priority areas for realizing this transformation, one of which is sustainable agricultural production and productivity. This priority area cites weak agricultural extension services delivery as a key constraint. In addition, the provision of extension services is found in all the priority areas identified in the NAP, especially in the areas of agricultural productivity, crop diversification, and commercialization. NAP therefore promotes innovative, demand-driven, and pluralistic extension services as it is considered critical to the achievement of its goal.

Dataset and year of data collection	Description and sampling	Extension focus	Analytical method
IFPRI interviews with heads of organizations (2017)	Dataset covers 30 organizations, selected from the census of extension service providers above; each organization selected to represent and provide extension services provision data across diverse types of service providers.	Eight-page questionnaire that asked more details on extension services provision, staffing, budget, and others (to provide more details on extension services provision by type of providers).	Descriptive analysis conducted by authors and presented for the first time here in this chapter or cited from authors' published papers using same datasets.
IFPRI focus group discussions (2017, 2019)	Linked to the household and community surveys above, one remote community and one central community are randomly selected from the selected districts to represent the different regions, ADD, agroecological zones, and farming systems.	Six-page guide questions on feedback on extension services and technologies being promoted.	Coded and analyzed using Nvivo; descriptive analysis conducted by authors and presented for the first time here in this chapter or cited from authors' published papers using same datasets.
Strengthening Agricultural and Nutrition Extension project (SANE) (2016)	Census of all Area Stakeholder Panels (ASP), District Stakeholder Panels (DSP), and District Agricultural Extension Coordination Committee (DAECC) in 10 of 28 districts that the USAID Feed the Future zones of influence operate. This census is part of the monitoring and evaluation of the SANE project.	Semistructured questionnaire is mainly on extension services and functioning of these structures.	Descriptive analysis conducted by authors and presented for the first time here in this chapter or cited from authors' published papers using same datasets.

Source: Authors' compilation.

NAP was necessitated due to the existence of the many policies and strategies that were incoherent and limited the ability of the agricultural sector to respond to changing economic opportunities and challenges. The government is hopeful the NAP will provide a comprehensive policy framework for each subsector to positively contribute to agricultural development and the economy in a coherent manner. Subsectors are expected to respond with relevant strategies in line with the new agricultural policy. As such, the national agricultural extension policy was recently reviewed and a new national agricultural extension strategy is being drafted. Experts agree that Malawi has an appropriate policy environment for implementing extension services, although the policies are not fully operationalized due to a number of factors, including government funding and unclear implementation guidelines (MEAS 2012; Chowa, Garforth, and Cardey 2013; Ragasa, Mazunda, and Kadzamira 2015; Masangano 2017).

FIGURE 7.1 Malawi focus districts for the study of extension service providers and focus group discussions

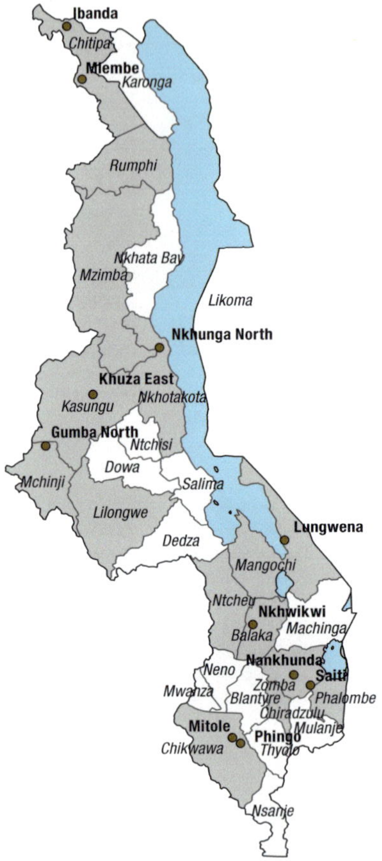

Source: IFPRI interviews (December 2016–March 2017).

Note: All districts are covered in the household and community surveys; those shaded gray are the 15 focus districts of the census of extension services providers; and those with dots are the locations of the focus group discussions.

7.4 Organizational Capacity and Management

There is a broad set of different providers for extension services in Malawi, indicating quite a strong degree of pluralism. Extension service providers identified by key informants in the focus districts include the government; trusts; international nongovernmental organizations (NGOs); local organizations including church-based organizations, farmer-based organizations, community or youth organizations; private companies including banks, agrodealers, traders, and commodity exchange; and mass media operators including radio, mobile application, and call centers.

Public Extension System

Even though the Malawi extension system has become increasingly pluralistic, the public extension system is still the largest extension service provider in Malawi. In a district there are roughly 82 government frontline workers and a total of 41 frontline workers for all nongovernment service providers on average (2:1 government to nongovernment ratio).[1] Moreover, almost all nongovernment service providers work with extension agents to implement their activities and compensate them with fuel allowance and training opportunities (Ragasa et al. 2017). In addition, the community survey shows that 99 percent of group villages have been visited by government agents; while only 15 percent of group villages have been visited by nongovernment agents (Ragasa and Niu 2017). Therefore, there still remains limited spatial coverage in the provision of agricultural extension and advisory services by nongovernment actors when viewing the national scope.

MoAIWD operates through its own structure as well as the decentralized local government structure (Figure 7.2). The ministry headquarters is in Lilongwe, the capital city. The technical services, such as extension services, crops, research, and land resources in the ministry are headed by the Controller of Agricultural Extension and Training Services (CAETS), who supervises the directors. For agricultural planning and implementation purposes, the country is divided into eight agricultural development divisions (ADDs). The ADDs are further divided into districts and extension planning areas (EPAs). The EPAs, which are the basic agricultural operational units, are divided into sections and then villages. There are total of 28 districts, 185 EPAs, and 2,880 sections (Kamkwamba 2015). It is in the districts that the ministry's structure is aligned along the local government structure. That is, for local government the structure begins with the district, which is divided into traditional areas. These areas are further divided into group villages and then villages, where farm families are located. Kamkwamba (2015) estimated the number of farm families to be about 4 million.

The ADDs are headed by the program manager and cover two to five districts. Program managers coordinate all agricultural activities from the

1 There is also a possible interexchange or mixing of what subject matter specialists and frontline workers are within the nongovernment service providers, so the figures are our best interpretation of the data available.

FIGURE 7.2 Organizational structure of the MoAIWD

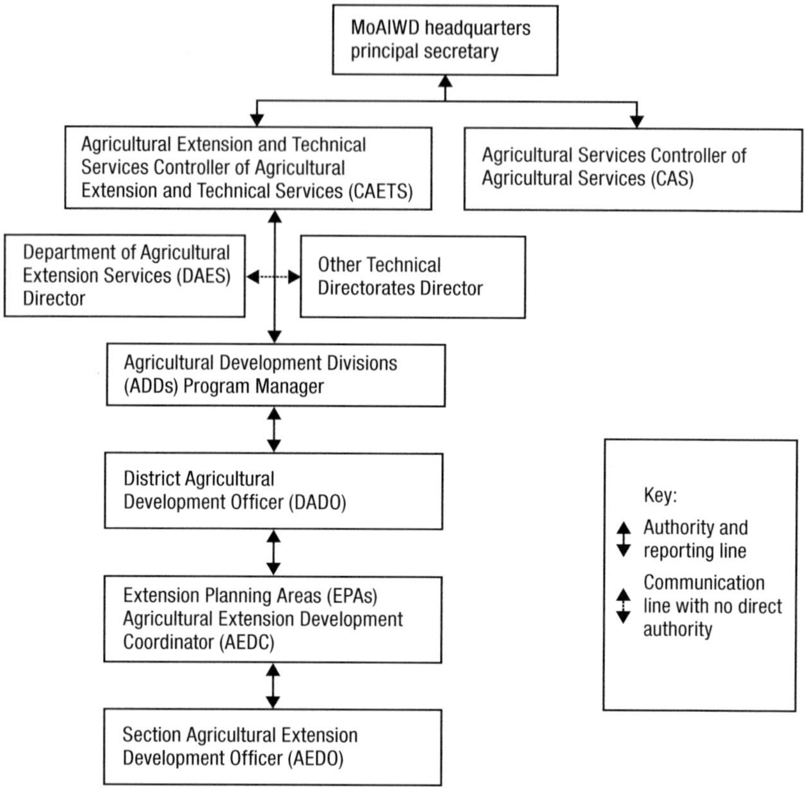

Source: Adapted from Masangano and Mthinda (2012).
Note: MoAIWD = Ministry of Agriculture, Irrigation and Water Development.

ministry's technical departments, such as crops, land resources, and conservation, livestock, fisheries, irrigation, research, and extension. With decentralization, MoAIWD works through the district assemblies headed by the district commissioner. In each assembly, extension services fall under the responsibility of a district agricultural development officer (DADO), supported by a team of subject matter specialists (SMSs) from various technical departments who backstop the frontline extension agents, who are referred to as agricultural extension development officers (AEDO) in Malawi. The DADO is responsible for all agricultural programs in the district, including extension services. Similarly, the agricultural extension development coordinators (AEDC) head the EPAs and are coordinators of all agricultural activities in the Ministry of Agriculture. However, the AEDOs manage the sections and are responsible for implementing all agricultural activities in the sections. Ideally, it should be one AEDO per section, although due to inadequate staff, some sections have no AEDO. For consistency throughout this chapter, AEDOs will be referred to as "extension agents."

Within MoAIWD, the public extension services are provided and coordinated through the Department of Agricultural Extension Services (DAES) structure. At the national level, DAES has five technical branches with subject matter specialists: namely, Extension Methodologies and Systems, Food and Nutrition, Agricultural Communication, Agribusiness, and Gender and Extension Support, headed by the director of agricultural extension services. These branches are represented at the ADD and district levels by one or two subject matter specialists. The chief agricultural extension officer (CAEO) heads the extension team at ADD level and assists the program manager and district commissioner (DC) in supervising all extension activities at ADD and district levels respectively.

Figure 7.2 shows that agricultural staff at the ADD and district level report extension information to DAES through the supervisors at their respective levels. On a day-to-day basis, the staff report to their direct supervisor at their duty station. However, they are all technically accountable to their supervisor in DAES, and this arrangement is typical with all technical departments in the ministry. Staff recruitment, training, performance management, and promotions are all centralized in the technical department, which poses a challenge of unity-of-command principle in management between the central technical departments and decentralized services, including extension (Chiweza 2010). However, there is hope for improvement as government is moving toward completing the sector devolution process. This move can encourage nongovernment service providers to become more demand-driven.

Nongovernmental Service Providers

There are also about 120 nonstate extension service providers active in the 15 sample districts covered. The number of service providers per district averages 14, and ranges from 6 in Chiradzulu to 25 in Balaka and 35 in Lilongwe. Two-thirds are local NGOs, farmer-based organizations, or private companies (including input dealers, banks, local radio stations, and mobile telephone operators). One-third are international NGOs. With all nonstate providers combined, each district has roughly 16 nonstate specialists (a 1:1 state-to-nonstate ratio) and 47 nonstate frontline workers (a 2:1 state-to-nonstate ratio).

Farmers' organizations are also active as extension services providers to their members. According to the extension policy, the roles of farmers' organizations include the provision of extension services to their members, adequately representing farmers' interests, contributing to policy formulation, and capacity building of members. The Farmers' Union was established in 2003 as the umbrella body of farmers' organizations in Malawi, with 256-member organizations of about 1 million smallholder farmers in 2016. The National Association of Smallholder Farmers of Malawi (NASFAM) is the largest independent, smallholder-owned membership organization in Malawi. Unlike other farmers' organizations, NASFAM has a fully-fledged extension system. The extension services is headed by the farm services manager. As of 2013, there were 1,571 farmer trainers who were being used by 62 association field officers (AFOs) to facilitate the delivery of extension services (Kumwenda 2013). In addition, NASFAM has three farm services specialists and 14 association business managers, and the national association also uses these lead farmers. NASFAM commissions its own research, which is done through government research stations, the university, and international research centers. It has a business entity that facilitates marketing of produce for its associations. However, NASFAM works in partnership with the government extension system as its field staff is located at the EPA level. Hence, the national association works with government AEDOs and lead farmers in the villages. NASFAM reaches its farmers through radio and TV programs as well as through newsletters and leaflets. It conducts its own demonstrations and field days. It supports its staff with training and necessary working materials such as stationery, transport, and communication.

Media-based extension services providers largely focus on market information provisions. Most commonly, they collect and disseminate information on prices for key agricultural commodities in rural and national markets as well as on technologies that would best suit farmers' local conditions. Such information is also provided online, through short message services and radio programs. Some extension services providers (for example, startup mobile application developer Geek Bit) have gone a step ahead to develop and roll out mobile applications that would help farmers with information on farming practices, comparative market prices for seeds, farm products, and farming resources. Most information displayed on the application is collected from seed manufacturing companies, agrodealers, and research stations (as per a key informant from Geek Bit). However, other extension services providers such as Human Network International—in partnership with mobile phone operators, the government of Malawi, and some key development organizations—launched 3-2-1 and 7-1-1-1 service to increase access to information for farmers.[2] Farmers on the Airtel or TNM mobile network can access the application through 3-2-1 or 7-1-1-1 service, respectively, by dialing and following prompts to listen to information of interest.

Radio operators, including Farm Radio Trust, design radio programs to spread messages or technologies to wider masses of the farming community. Such programs can be 30-minute broadcasts repeated weekly over some defined period of time. These can cover a wide range of topics or content, depending on what would be of interest for farmers. For example, the topics may touch on issues of soil fertility improvement; effects of climate change on the community; climate change solutions; nutrition; or accessing market information. Such broadcasts could take the form of studio interviews, mini-dramas, testimonials, phone interviews, panel discussions, subject matter specialist presentations, live phone-ins, on-farm recordings, or feedback from listeners through recorded discussions in community listener groups (Chapota, Mthinda, and Fatch 2014).

Coordination through the District Agricultural Extension Services System

Adding to the already complex setup is the District Agricultural Extension Services System (DAESS), which mirrors the decentralized local government structure (Malawi, MoAIWD 2004, 2006) and the main structures envisioned to implement NAEP. The DAESS is the cornerstone in the Agricultural Extension Policy Implementation Guide developed in 2004. The DAESS is composed of interlinked structures—from the village to the district and national levels—as a system for demand articulation from the bottom up, to improve coordination and responsiveness among different service providers. The DAESS aims to (1) reduce information asymmetry between users and service providers, (2) provide platforms for demand articulation and aggregation, (3) coordinate and harmonize the activities and messages of extension service providers, and (4) improve accountability among various service providers to provide better-quality extension services. Put differently, with an emphasis on improving coordination and making agricultural extension services more demand-driven, the government of Malawi promoted the creation of various connected structures at various levels, starting with farmers' involvement at the village level. These different structures are model villages and village

2 These are the names of the mobile apps; the numbers represent the numbers that farmers dial to access the apps.

agriculture committees (VACs) at the village or group village level; area stakeholder panels (ASPs) at the extension planning area level; district stakeholder panels, district agricultural extension coordination committees, and district agriculture committees (DACs) at the district level; and a national stakeholder panel, the Malawi Forum for Agricultural Advisory Services (MaFAAS) at the national level (Figure 7.3).

Within the DAESS, each structure performs key functions. The model villages are the entry point for agricultural development programs and use participatory approaches to plan, implement, monitor, and evaluate the progress of those programs at the village level. Also, at this level the VACs, which are subcommittees under the village development committees (VDCs), serve as forums for farmers and other relevant stakeholders such as NGOs and agribusinesses to express their demands and agree on common issues that require action. ASPs perform a similar function at the EPA level by serving as a discussion space to identify agricultural priorities, ensure representation of varied stakeholders, and aggregate villages' agricultural demands to obtain quality responses. At the district level, the DSP collects the agricultural demands expressed at the ASPs and, with extension service providers, coordinates responses to the prioritized agricultural needs. The DAECCs are responsible not only for ensuring that quality extension services are provided throughout the district but also for advising the DACs on agricultural development issues and for mobilizing resources in support of agricultural extension services. The DACs also connect the multilevel extension structure with other institutions of the local government by being composed of elected members of the district councils.[3] Last, DSPs and DAECCs are linked to the national stakeholder panel (MaFAAS) at the top, which connects to the higher levels of the Ministry of Agriculture, Irrigation and Water Development. Altogether, this system aims to create pluralistic, demand-driven, and decentralized extension services that assist farmers in an efficient and effective way.

In addition, the establishment of the National Agriculture Content Development Committee (NACDC) in 2014 presents another opportunity for enhancing coordination. The purpose of NACDC is to improve quality of agriculture extension to avoid conflicting technical messages. This committee was initiated by the service providers interested in disseminating extension messages to farmers using platforms that are based on information and

3 The DACs work with the district councils to, for example, make recommendations on extension service policy, support local agricultural institutions and development, assist in resource acquisition, and encourage community participation.

FIGURE 7.3 Agricultural extension governance structures in Malawi

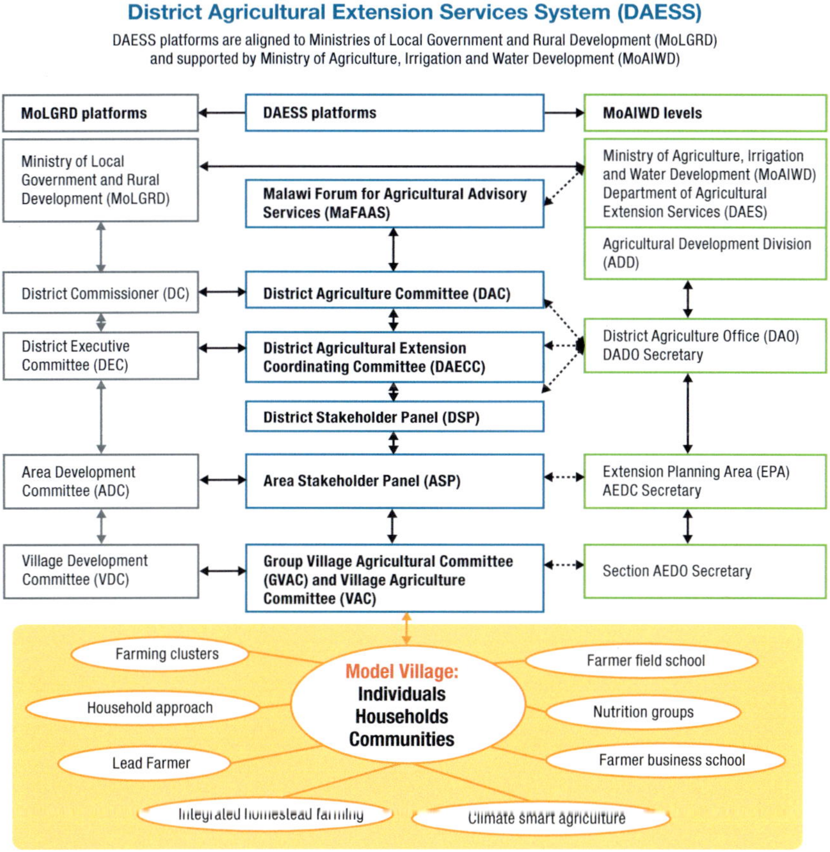

Source: Modified from Malawi, MoAIWD (2006).

Note: Straight lines reflect reporting lines from the DAESS platforms to the MoLGRDs. For instance, the ASPs report to the area development committees on agricultural development at the area level and the functioning of the ASPs. A similar structure applies to the other levels (village, district). The dashed lines connect MoAIWD with the DAESS platforms in terms of facilitating the existence of these platforms, being part of them (for example, the district agriculture development officer chairs the DAECC), and building their capacity so that these platforms perform their key functions. MoAIWD is also an advocate for the role of MaFAAS at the national level. The DAES is part of MoAIWD, together with the other technical departments concerning fisheries, livestock, crop development, land resources, and agricultural development. The DAESS structure encompasses all of these technical departments, so the heads of the agricultural technical departments are represented on the DSPs and DAECCs.

communication technologies (ICTs). However, the committee's mandate has since been broadened to include all agricultural extension messages to ensure that farmers receive messages that are validated and are of good quality. Membership of NACDC is open to public and private service providers in the agriculture sector who are cautious on the quality of messages disseminated and is chaired by a nongovernment service provider. As the committee is relatively new, its capacity and workings would need strengthening to help harmonize and coordinate agricultural extension service provision in Malawi. In addition, since the committee is only at the national level, an opportunity exists to test it at the district level through the DAEC.

Human Capacity and Training

QUANTITY INDICATORS

In a district there are roughly 18 technical staff or subject matter specialists from nongovernment and 16 SMSs from government (1:1 ratio). As mentioned previously, there are roughly 86 government frontline workers and a total of 47 frontline workers for nongovernment extension services providers in a district, on average (2:1 government to nongovernment ratio).[4] In addition, in a district there are roughly 1,520 lead farmers on average, although this can be overestimated since the lead farmers of all government and nongovernment extension services providers are summed, except that these can be the same lead farmers. This shows that there are a substantial number and size of nongovernment extension services providers, and the need for coordination and harmonization in this pluralistic system has never been more essential and urgent.

In particular, the number of farming households is not consistent in Malawi, with the Agricultural Production Estimates Survey (APES) providing much higher numbers of households than the Agricultural Census. Farmer-to-government agent ratio is roughly 2,240 or 3,316, depending on whether one is using APES or the Agricultural Census. The farmer-to-all agents ratio is roughly 1,568 or 2,232, depending on whether one is using APES or the Agricultural Census. Given that there are no good data for benchmarking, it is hard to establish where Malawi is in terms of farmer-to-staff ratio. Given available data from other countries, these ratios are much higher than some countries in Africa south of the Sahara, including Ethiopia, are comparable to Tanzania, and are much lower/better than Nigeria or India (see Figure 6.2, Chapter 6). While there is much attention on farmer-to-agent ratios, without funds for operating costs, additional people in the payroll will not make an impact. Therefore, there should be a balance and prioritization of

4 There is also a possible interexchange or mixing of what SMSs and frontline workers are within the nongovernment extension providers, so the figures are the best interpretation of the data available.

additional funding in order to make a positive impact in terms of coverage and agricultural outcomes.

QUALITY INDICATORS

Based on the interviews with 71 extension workers, the majority (64 percent) of frontline workers have at least a two-year diploma, most common for government and nongovernment organizations. A fifth have a one-year certificate/diploma, dominated by government. Interestingly, 16 percent have only a primary or high school education (19 percent from nongovernment) and do not have at least a certificate or diploma. In the case of AEDOs, further investigation was carried out, since there appears to be an error, given the requirement of two-year diploma to be an AEDO. The one-year certificate course was phased out to become a diploma course in the 1980s. At one point, government-recruited secondary-level candidates were to be trained on the job and later trained to diploma level at Natural Resources College (NRC) on a part-time basis following employment. Those AEDOs with no certificate or diploma may represent the remnants that did not make it to NRC.

AEDO trainees were recruited toward the end of 2016 under the Sustainable Agriculture Promotion Programme (SAPP). These are individuals who have generally completed secondary education. They are expected to periodically attend NRC for short training courses and return to their designated EPAs for field work. Once they complete the required training, they will become fully fledged AEDOs. The focus was on individuals remaining in the rural area so that they can stay and work in the same area upon becoming an AEDO. Besides formal education, the majority of frontline workers (85 percent) reported that they ever had received professional training, similar to that of government and nongovernment (Figure 7.4). This means that 15 percent have never received training or retraining.

About half of the frontline workers for government and nongovernment received some training in 2016 (Table 7.2). It can be assumed that roughly 50–60 percent received training in the past three years, which means that roughly 40–50 percent have never received any training and retraining in at least three years. This has serious implications on the skills upgrading among frontline workers in government and nongovernment. This, ultimately, has implications on the quality of advice and extension services being provided to the farmers.

GENDER COMPOSITION

Female subject matter specialists are roughly 19 percent of all government SMS. There is a higher proportion of female frontline workers (25 percent

FIGURE 7.4 Frontline workers who received professional training (%)

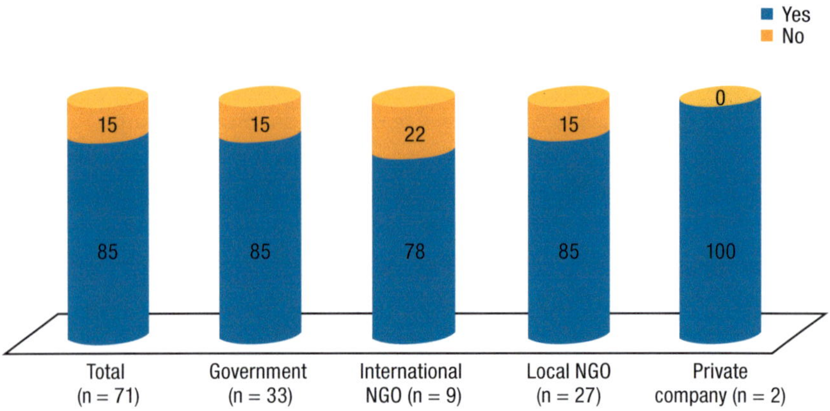

Source: IFPRI in-depth interviews of 71 frontline workers (December 2016–February 2017).

TABLE 7.2 Proportion of frontline workers by last time they received professional training, 2010–2016 (%)

Latest year of training	Total (n = 60)	Government (n = 28)	International NGO (n = 7)	Local NGO (n = 23)	Private company (n = 2)
2016	47	50	57	43	0
2015	7	4	0	13	0
2014	7	0	14	9	50
2013	7	7	0	9	0
2012	5	4	0	9	0
2011	2	0	14	0	0
2010	8	11	0	9	0

Source: IFPRI in-depth interviews with 71 frontline workers (December 2016–February 2017).

in the government and 32 percent in nongovernment). There is also a greater proportion of female lead farmers (35 percent). These figures are higher or better than some countries (for example, 5 percent in the Democratic Republic of the Congo), although considering the proportion of female farmers to the farming population in Malawi, the proportion of female technical staff and extension services workers may still be low.

There is evidence showing that gender equality at the level of policy and extension services providers affects gender equality in accessing extension services (see Ragasa [2014] for a review). Evidence shows that extension services from female agents may be better targeted to female farmers. For example, the World Bank and IFPRI (2009) find that female extension services workers serve a higher proportion of female farmers than do male agents (the average ratio of female farmers to male farmers who were provided with extension services by agents is 1.3 for female agents and 0.53 for male agents for the countries under study: Ethiopia, India, and Ghana). This shows that female agents are more likely to work with a greater proportion of female farmers, while male agents are more likely to work with a greater proportion of male farmers. If the aim is to ensure that female farmers have equal likelihood of having access to extension services to male, and to address the earlier reported bias in their access to these services, efforts will have to be exerted to achieve more gender balance at the level of the extension services provider and frontline worker. Gender-based constraints, such as social norms that limit women's school attendance or mobility, also limit their opportunities and willingness to work as extension agents. It may be difficult at a practical level for a married woman to work in a rural area away from her husband and family or to find appropriate housing and schooling for her children.

TIME ALLOCATION

It is essential to understand the workload of frontline workers, since it affects the quality of their work. Possible reasons for underperformance among extension agents may be because they are unmotivated or demoralized by the lack of support to the extension system or that they are overworked and that the expectations from them are too high compared to the amount of resources they are given to perform their work. Some also question the added responsibilities of government extension services workers in FISP or APES, which, according to some critics, are not really a part of training and supporting farmers with knowledge and skills, and may crowd out the scarce time and energies among extension services workers, keeping them from undertaking more knowledge-brokerage and technical assistance.

The data show that among AEDOs, only 37 percent of their time was spent on working with farmers and 38 percent is spent on FISP and APES-related activities (Figure 7.5). We are not sure if AEDOs' involvement in FISP might have reduced following the change in the program implementation; identification of beneficiaries is now done centrally, after we had already collected our data. The time allocation proportions were much better for nongovernment frontline workers who spend half of their time helping and supporting farmers. Still, it seems that expectations are too high for frontline workers, who are not guided and not supported enough, especially the state AEDOs.

If this time-use and expectations were to continue, it would be necessary to support extension workers to perform all these roles, including the following: (1) provision of motorbikes and fuel allowance, which should be a relatively low-cost investment to ensure better mobility of extension workers; and (2) use of low-cost mass media (radio, mobile phone/text messaging, videos) to ensure that more farmers are reached with information and options to create greater demand for extension services.

INCENTIVES

An average monthly salary for an AEDO is Malawi kwacha (MWK) 79,440 (US$110) to MWK 113,260 (US$160) for a nongovernment frontline worker. The monthly salaries of nongovernment extension services workers are 40 percent higher than those of the government. The field allowance per year is minimal (MWK 33,500 to 34,500 per year, or US$4 per month) and similar for government and nongovernment employees. Despite not having major differences in monthly salaries and yearly allowances between government and nongovernment agents, one major difference is the type of mobility. Almost all frontline workers from nongovernment are provided with a motorbike and fuel allowances, while most frontline workers in the government are provided with only a pushbike. This is exacerbated whereby an AEDO is requested to cover two sections due to staff shortages, made worse in hilly areas. Only 32 percent reported having access to a motorbike. The fuel and maintenance allowance per month for nongovernment frontline workers is MWK 40,000 (US$55) compared to MWK 1,900 (US$1.50) for AEDOs.

This poor mobility, especially among government extension services workers, is clearly obvious in the responses of frontline workers when asked about their constraints and motivations. The overwhelming majority of AEDOs (85 percent) reported better mobility as their top motivating factor. Others cited good remuneration packages, support for activities (funding, protective gear, and quick responses to issues raised by communities), frequent and more refresher courses, provision and maintenance of good housing, and timely provision of field allowances. Among nongovernment frontline workers, responses were more diverse and less concentrated, although the most common were reliable means of mobility, good remuneration packages, support for activities, and frequent and more refresher courses. Though noted by very few frontline workers, the last five factors seem to suggest that some nongovernment frontline workers look beyond for such things as work relationships, collaboration, and the sort of feedback they get from the communities they serve.

FIGURE 7.5 Time allocation of frontline workers during planting season (%)

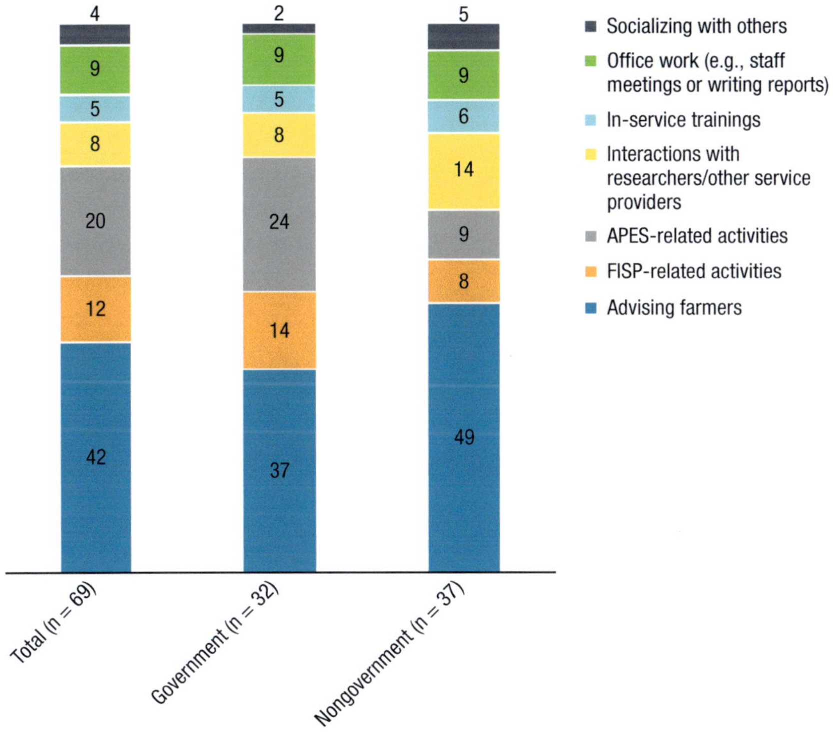

Source: Authors' in-depth interviews of 71 frontline workers (December 2016–February 2017).
Note: APES = Agricultural Production Estimates Survey; FISP = Farm Input Subsidy Programme.

Similarly, the most pressing general constraint was the poor mode of mobility for frontline workers, which is more pronounced for government frontline workers. The other major constraint most common for both frontline workers was inadequate or lack of resources for activities, including funds and materials. The other constraints for AEDOs were poor housing conditions; low salaries; limited recognition in terms of promotion, rewards, or allowance; and lack of skills upgrading and retraining. In-depth interviews pointed to poor housing conditions as a major constraint hindering AEDOs in performing better. For government and nongovernment, some also cited constraints such as the lack of interest, cooperation, and commitment from farming communities.

Monitoring and Accountability

Monitoring of performance and evaluation of outcomes and impacts are poor, particularly in the government system. Under both state and nonstate systems, targets are usually limited to inputs and outputs (such as the number of households trained) and do not reach the level of outcomes or impacts. Very few (13 percent of frontline workers and 10 percent of all service providers) reported having outcome indicators, such as crop yield performance or number of households that are food secure, as their performance targets. While much data are collected through APES, it is often not linked and utilized to inform how extension services are performing or how they can be improved. At a national scale, the main indicator still used is farmer-to-agent ratio, which can potentially be counterproductive, putting more strain on scarce resources without clear impact if no operating funds are added for extension work. Moreover, there are few incentives for good performance and no system to evaluate government extension workers and hold them accountable for their performance.

Financial Capacity

Government funds for extension services are mainly for personnel compensation (73–83 percent), with operating funds (known as "other recurrent transactions") being limited (17–27 percent).[5] The latter are mainly for administrative expenses, and funds to do actual extension work and farmer support are extremely limited. The estimated public funds for extension services are roughly 740,000 Malawian kwachas (MWK) (US$1,000) per AEDO per year, or MWK 250 (US$0.33) per farmer per year, which is very little.[6]

To put this into context, salaries for agricultural research are roughly 40–60 percent of the budget in most African countries, and in cases such as Ghana—with more than 80 percent of total research budget for personnel—salaries are considered very alarming (Flaherty, Essegbey, and Asare 2010). One would assume that the operating fund requirements for extension services would be similar, given that technologies and knowledge are disseminated to widely spread rural communities.

---

5 These figures include development projects, such as ASWAp-SP and SAPP, that enter the government accounting system, but they do not include any agriculture-sector funds given to NGOs.

6 It was difficult to collect reliable data from both government and nongovernment service providers in the 15 sample districts (see Figure 7.1). The figures summarized here were from the government district offices (DADOs) and verified with approved budgets from the Ministry of Finance and the National Local Government Finance Committee.

Numerous projects and programs provide extension services. However, these are largely uncoordinated efforts. One would think of these efforts as "projectized," which generally are ad hoc, irregular, and infrequent. Focus group discussions also confirmed issues of duplication and concentration of the efforts, which result into conflicting and inconsistent messages that tend to confuse farmers. Survey results also reveal existence of gaps in many areas such as disadvantaged groups (females, uneducated farmers, and youth) being often left out. Together, all this mirrors the tendency of checking boxes to achieve quick results in many of the short-term projects. Ultimately, such development efforts tend to be unsustainable and deepen the sense of dependence and dole-out mentality among rural communities.

7.5 Evidence of Effectiveness and Impact of Extension Services

Access to Extension Services

Three-quarters of households reported having received some agricultural advice in the past two years (Figure 7.6). This mirrors major efforts in extension service delivery overall. However, half of households received advice in the past 12 months, indicating that provision of advice is irregular and infrequent as we see a large difference between access to advice within a two-year duration (panel a) versus a one-year duration (panel b). About 48 percent of the households reported access to nutrition advice in the past two years and 27 percent in the past 12 months, respectively—not quite as high (Figure 7.6). Moreover, provision of advice is heavily focused on crop production; with only 27 percent reportedly receiving information on market access or marketing and 17 percent reporting receiving information on livestock in the past 12 months (Figure 7.6).

Comparing 2016 and 2018, we see some improvements in extension service delivery. In panel b, 54 percent of the households reported access to agriculture advice in 2018—a 1 percent increase from 2016. Improvements are mainly in access to marketing/agro-processing and environment/climate change. There was a decline in advice on nonfarm livelihoods. In nutrition advice we see a significant increase in the proportion of households reporting accessing nutrition advice, with 27 percent of the households reporting access to nutrition advice in 2018—up from 21 percent of households in 2016.

While the National Extension Policy envisions equity and states that "the public sector must make sure that the poorest segments of the population,

FIGURE 7.6 Households receiving agriculture or nutrition advice from any source, 2016 and 2018 (%)

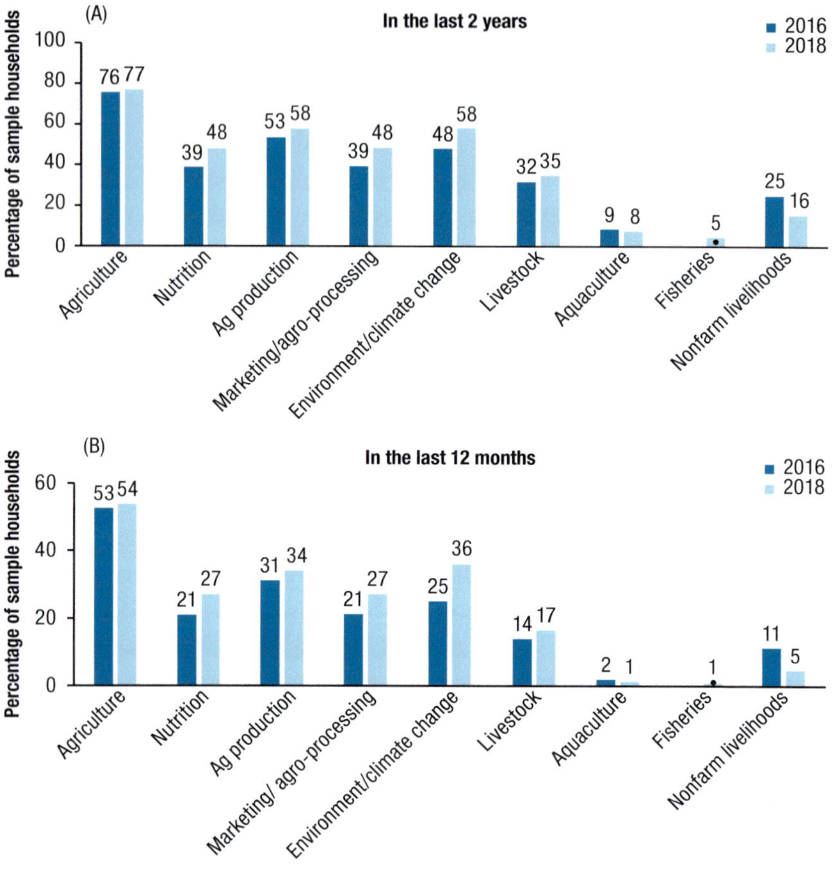

Source: IFPRI household surveys (2016, 2018).

Note: Agriculture combines all non-nutrition–related advice, including crop production, market/agro-processing, environment/climate change, livestock, aquaculture, fisheries, and nonfarm livelihoods. The dotted bar (fisheries) means only 2018 data were collected on this subject. Survey question: "In the last 2 years [In the last 12 months], did you receive any information or advice on any of these topics?"

women, youth, and people with disabilities are not left out of the development process" (Malawi, MoAIWD 2000: 25), our dataset shows that access to extension services is lowest among the more disadvantaged segments of the farming population (see Ragasa and Niu 2017). The poorest households, households with young heads, those adults with no or only a few years of primary education, female producers, and those in more remote areas are less likely to access agricultural advice (see Ragasa and Niu 2017; Ragasa and Comstock 2019).

The issue of youth is of particular interest. Youth (defined here as younger than 35 years of age) are less likely to access advice on most of these topics. In general, youth are less likely to participate in village and less likely to demand or request advisory or extension services (Ragasa and Niu 2017). As a result, youth are less likely to be aware of, to try out, and to adopt improved technologies than their older counterparts. Youth are less likely to receive extension services and advice compared with their older counterparts (Figure 7.7). This is a topic that needs further investigation since these results contrast with claims and hopes that youth are the drivers of agricultural innovation and rural transformation.

Males are more likely to receive agriculture-related advice than females, although both are equally likely to receive advice on other livelihoods and or health/nutrition. Older men are more likely to receive agricultural advice than younger men and women of both age groups (Figure 7.7). Among females, female members in male-headed households have a lower likelihood of receiving agriculture advice than females who are household heads. This may be due to the persistent focus on household heads as recipients or beneficiaries in many extension, training, and capacity-building programs. A study by Ragasa, Aberman, and Alvarez-Mingote (2019) find that reducing the gender gap and targeting both female and male adults in the same household can increase productivity and food security of the households.

Quality of Extension Services

Farmers gave very high ratings and are satisfied with the advice that they receive overall: 76 percent of farmers are very satisfied; 77 percent said they found the advice very useful; 86 percent said they followed the advice; 92 percent said the advice was something they wanted or needed (Table 7.3). This is largely similar for 2016 and 2018 data. Ratings are similar to those reported in the Integrated Household Survey (IHS3) (see Ragasa and Mazunda 2018; Ragasa and Comstock 2019). Spielman et al. (2019) and World Bank and IFPRI (2009) also reported high ratings and satisfaction of extension services, but this may be due to the free provision of these services—from the perspective of farmers, something free is better than none, and better to report as useful than not get any information at all. Either the extension services are really of high quality or our measurement and survey questions are not capturing the issues in terms of quality and content of extension services. One implication is that self-reported satisfaction ratings should be interpreted

FIGURE 7.7 Access to extension services, 2016 and 2018 (%)

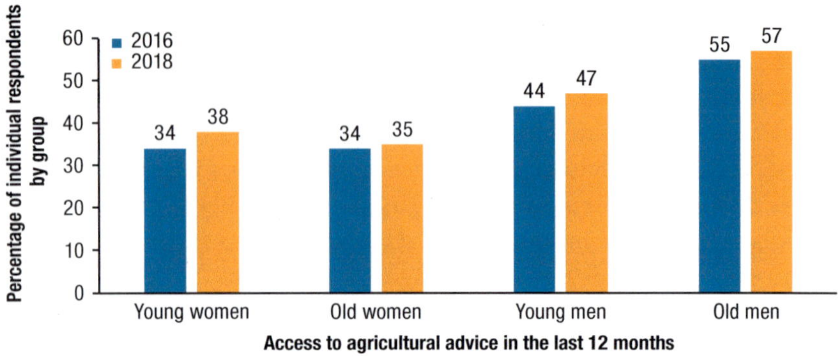

Source: IFPRI household surveys (2016, 2018).

Note: "Youth" is officially defined in Malawi to cover those under 35 years.

with caution; and the focus could be on understanding the barriers to uptake and adoption of extension messages. Satisfaction questions should be accompanied by questions and assessment of actual adoption or use of the messages and practices being promoted by the extension services.

Demand-Driven Extension

Most farming households do not participate in meetings or processes where they would be able to articulate their demands for agriculture- or nutrition-related advice. Only 14 percent of those receiving advice actually demanded or requested it (see Table 7.3), indicating that not many farmers are requesting or articulating their demand for extension services. This indicates that the provision of nutrition and extension services is still heavily supply-driven, rather than demand-driven, as envisioned in the National Extension Policy.

The majority of respondents are either those who received advice—despite the fact that they did not request it (41 percent of respondents)—those who did not seek advice, or those who did not receive agriculture-related advice (half of respondents). Within the latter, one-third of them reported having particular needs or demands for advice, which implies that there is still scope for extension services providers to improve their reach to farmers. In general, the majority of respondents neither adopt nor demand advice on particular technologies being promoted. This may be a reflection of the projectized extension service provision, characterized by lack of coordination and limited awareness and information campaigns on options available to farmers to support their demand for services and inform their choices and decisions.

TABLE 7.3 Farmers' feedback on the latest advice received from agricultural extension service providers

Topic	Were you satisfied with the advice on topic [. . .]?				Was the advice on topic [. . .] useful?				Did you follow advice on the topic?	Was it something that you needed?	Was it something you requested?
	Not satisfied	Somewhat satisfied	Satisfied	Very satisfied	Not useful	Somewhat useful	Useful	Very useful			
Agricultural production	1	1	19	79	1	2	19	79	85	89	12
Livestock	1	1	19	79	1	2	21	76	79	92	14
Aquaculture	3	14	43	40	1	19	37	36	24	67	12
Marketing	1	2	17	80	1	4	14	80	66	90	5
Postharvest	1	0	19	80	0	1	17	82	88	95	13
Sustainable land management	1	1	29	70	1	1	26	73	86	93	20
Environment	1	1	27	71	0	2	25	73	84	89	13
Health or nutrition	0	1	23	77	0	0	22	78	94	93	15
Total	**1**	**1**	**22**	**76**	**1**	**2**	**21**	**77**	**86**	**92**	**14**

Source: IFPRI household survey (2016). Similar results for 2018 household survey.

Promoted Technologies

In terms of specific types of agricultural technologies and practices being promoted (see Ragasa and Niu 2017 for the list), awareness rates range from 14 percent to 49 percent for soil fertility and land management technologies; we see improvements in terms of proportion of households being aware and knowledgeable about these technologies (Figure 7.8). The only exception is on multiple food groups (dietary diversity), where fewer household heads reported having knowledge or awareness of this advice. Of the new questions asked in 2018, there was widely reported adoption of orange-fleshed sweet potato (59 percent of households). Awareness of Purdue Improved Crop Storage (PICS) bags (currently at 26 percent of households), fall army worm control (21 percent of households), and aflatoxin control (13 percent) can be further improved.

Looking at plot-level management practices and production data, we see major improvements in technology adoption from 2016 to 2018 (Figure 7.9). In 2018 more households reported adopting the agricultural technologies being promoted. Major improvements were seen in the adoption of bunds/ridges, crop residue incorporation, proper plant spacing, water management, row planting, and intercropping. We see higher usages of pesticides, maybe due to fall army worm. We do not see improvements in soil cover or mulching, minimum tillage, and pit planting.

Comparing awareness and adoption, we see major gaps (Figure 7.10). Aside from intercropping, there are large gaps in the awareness and actual adoption of practices/technologies being promoted. It seems that extension services are not able to fill the gap between awareness and adoption or that there are major bottlenecks beyond information constraints that limit technology adoption.

Impact of Extension Services on Productivity and Food Security at the National Level

Ragasa and Mazunda (2018) examine the interplay between Malawi's input subsidy and access to extension services, and the impact of both on farm productivity and food security using Malawi's Integrated Household Panel Survey. A correlated random effects (CRE) device is used, and consistency and robustness of results are checked using various other estimation models. The receipt of fertilizer and seed subsidies is shown to have an inconsistent impact on farm productivity and food security; at the same time, access to agricultural advice is consistently insignificant in explaining these. Further analysis, however, shows a statistically significant and strong association with farm

FIGURE 7.8 **Household heads aware or knowledgeable of specific technologies, 2016 and 2018 (%)**

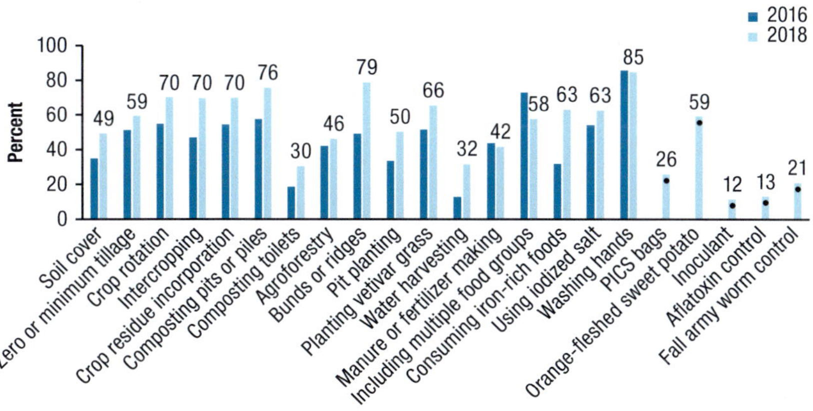

Source: IFPRI household surveys (2016, 2018).

Note: Dotted bar (PICS bags, orange-fleshed sweet potato, inoculant, aflatoxin control, fall army worm control) means only 2018 data were collected.

FIGURE 7.9 **Household heads adopting specific technologies, 2016 and 2018 (%)**

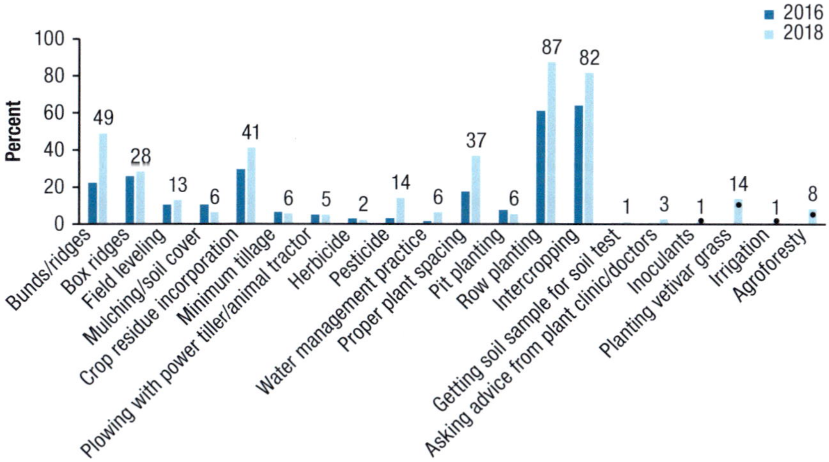

Source: IFPRI household surveys (2016, 2018).

Note: Dotted bar (Inoculants, Planting vetivar grass, Irrigation, Agroforesty) means only 2018 data were collected. "Getting soil sample for soil test" and "Asking advice from plant clinic/doctors" are not specific technologies per se, but added here as they are actually the messages by extension officers for advice on soil quality and crop pests and diseases.

FIGURE 7.10 Gap in household heads awareness versus adoption of specific technologies, 2018 (%)

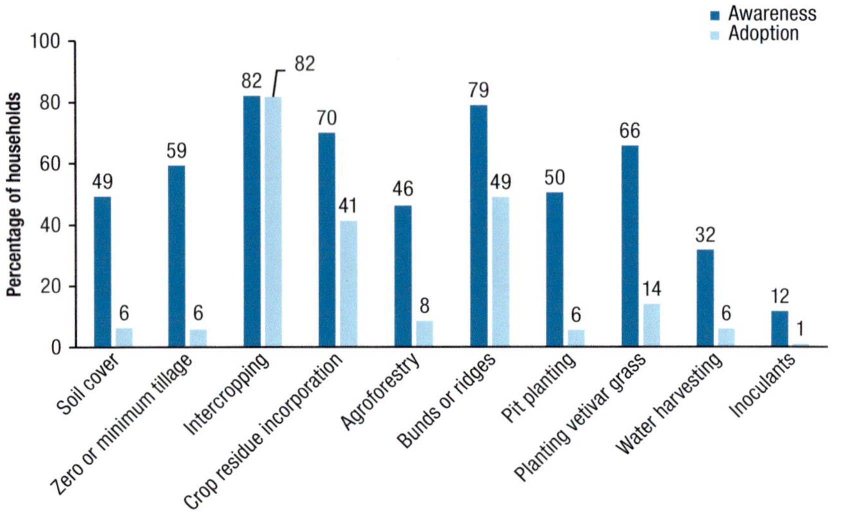

Source: IFPRI household surveys (2016, 2018).

productivity and food security when access to extension services is unpacked to include indicators of usefulness and farmers' satisfaction.

For each crop studied, households receiving "very useful" agricultural advice saw the largest boost to computed plot-level crop productivity and household food security; while those reporting "not useful" advice are strongly associated with lower levels of productivity and food security. More so than the type of advice or its method of communication to farmers, the relevance of extension advice, often measured by the satisfaction of farmers with agricultural outreach, is a signal of quality and relevance of the information provided and reliable predictor of improved farm-level and household-level outcomes. After all, to farmers, agricultural services are only as helpful as they are useful. Providing generalized outreach is not enough; advice to farmers must be locally tailored to be effective, and there is need to ensure the provision of relevant and useful agricultural advice. We also found positive impact of agricultural extension, coupled with market access information. This is consistent with Andarge and Ragasa (2019), in which they found strong effects of market extension on commercialization and crop diversification using the MIHS (2010–2016).

When it comes to effective agricultural extension services, content and quality of information are important to achieving agricultural development outcomes. These results suggest the following ways to better target extension services: (1) get feedback from farmers as consumers of agricultural outreach products; and (2) use the findings to create timely, reliable, relevant outreach materials that go beyond the current focus on crop production (mainly chemical fertilizer and modern seed varieties) to include markets, agro-processing, other livelihoods, nutrition, and climate resilience—all tailored to local conditions.

Impact of Extension Programs or Approaches

Evidence of the effectiveness of extension programs or approaches is scanty. A relevant assessment of successful approaches is the effort carried out by MaFAAS to document innovative extension services in Malawi. Selection of these approaches was based on the Framework for African Agricultural Productivity principles. The assessment criteria provide a useful framework for extension services providers to assess extension services approaches and methods for their effectiveness and innovativeness, which is lacking. Mthinda (2015) summarizes the selected eight documented innovative approaches: farmer-to-farmer extension (by National Association of Smallholder Farmers of Malawi, NASFAM); host farmer demonstrations (Agricultural Research and Extension Trust, ARET); radio-based extension delivery (Farm Radio Trust); long-term extension programs for significant poverty alleviation (World Vision Malawi); farmers' clubs (Development Aid from People to People, DAPP); farmer voice radio (Malawi Broadcasting Corporation); lead farmer approach (Development Fund of Norway); and farmer cluster and the *Ulimi wa Mndandanda* model (Department of Agricultural Extension Services, DAES).[7] All these approaches seem to have the commonality of putting farmers at the center of the process. For some, the long duration of the program, the long-term approach to institutional and human capacity building, and the keen attention to monitoring and reporting on their development objectives were key. However, these are based on self-reported achievements of the project teams, with limited validation, and no impact evaluation. Other studies have attempted to evaluate different extension approaches used in Malawi. Here we draw the main lessons from these studies.

First, combining approaches has worked better than relying on a single delivery tool. One of the successes of Farm Radio Trust is the ability to

7 Regarding the farmer cluster and *Ulimi wa Mndandanda* model, farmers in a stretch of not less than 1 kilometer agree to implement similar recommended and innovative agronomic practices in their field, more like collaborative large-scale demonstrations.

combine radio programs with call-in features and text messaging (Chapota, Mthinda, and Fatch 2014). Ragasa and Niu (2017) highlight the huge potential of radio and other mass media in technology awareness campaigns to create demand for extension services. Ragasa (2019a) found that radio has the widest coverage of all extension delivery methods, reaching 48 percent of households in 2017/2018 (Figure 7.11). More than half of women (52 percent) reported listening to radio at least once a week. Younger women and men used radio more than other sources for their agricultural information needs. Younger and older men used radio more than other sources for nutrition education. Radio seems to be a critical delivery platform for nutrition education for men, circumventing strong gender norms on women's role in domestic work and nutrition while men are usually teased and laughed at when attending meetings/trainings related to nutrition. Ragasa (2019a) shows the significant effect of interactive radio programming on both women's and men's empowerment scores. Relative to other sources, access to radio had a greater effect on improving the empowerment of women than men. Mechanisms for this gendered outcome come from lower time burden and awareness campaigns and messages on gender equality. Access to radio by women and by men improved crop diversification and legume harvested area and improved dietary diversity.

Niu and Ragasa (2018) also stressed the importance of more intensive training and face-to-face interactions for complex agricultural management practices. Follow-up and continued mentoring by extension agents, facilitators, or field officers of both lead farmers and other farmers are necessary. For instance, Ward et al. (2018) and Fisher, Holden, and Katengeza (2017) detail the complexities of conservation agriculture and show that encouraging the adoption of the whole conservation agriculture package leads to very low compliance. Rather, it may be more effective to find "leverage" points and prioritize and sequence the promotion of specific dimensions of conservation agriculture based on a farmer's current practice (see specific suggestions in Ward et al. 2018). For some technologies, reassessment and reevaluation will be critical to know if they are really beneficial to farmers and what is constraining their adoption.

Second, an evaluation of farmer business schools (FBS) implemented by the government, using the case of Dedza, shows no follow-up of graduates and poor monitoring of project outcomes. A third of FBS participants did not complete the program; the most common reason was that the facilitator or the program did not continue for various reasons; some initial participants mentioned that they just lost interest. Chilemba and Ragasa (2018) show no impact of FBS participation or graduation; and only a few FBS

FIGURE 7.11 **Household heads participating in various methods or approaches, 2016 and 2018 (%)**

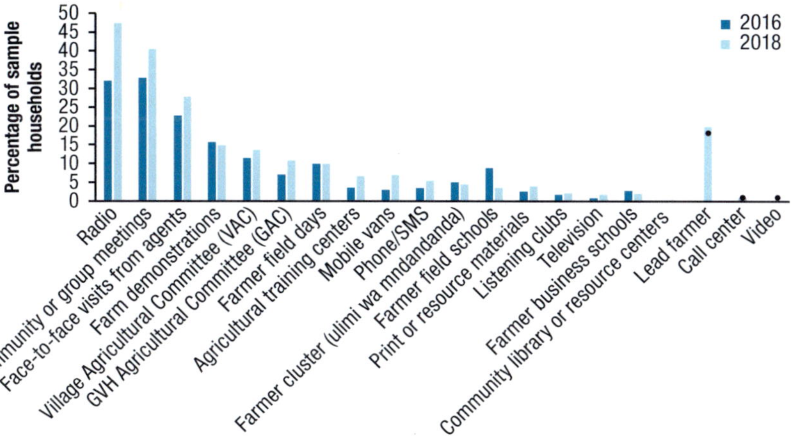

Source: IFPRI household surveys (2016, 2018).

Note: Dotted bar (Lead farmer, Call center, Video) means only 2018 data were collected. "Village Agricultural Committee (VAC)" and "GVH Agricultural Committee (GAC)" are not extension approaches or methods per se, but respondents mentioned them as sources of agricultural advice so they are included here.

participants experienced (small) increases in income. Malawi has implemented various extension approaches in the past. They involve good principles and are well-intentioned and effective in delivering information to communities. Oftentimes, it is not the extension approach that is the source of concern. Limited access to information or low awareness and adoption of promoted technologies is often due to lack of funds to scale up proven good approaches. When projects finish, so does much of the extension service provision.

Third, another potentially low-cost extension approach being heavily promoted in Malawi is the lead farmer approach. Musopole et al. (2013) show that transparent community selection of lead farmers, as well as incentives and support for lead farmers, can be critical to the success of the approach. Fisher, Holden, and Katengeza (2017) and Niu and Ragasa (2018) stress the complementarity of the lead farmers approach to other approaches. In fact, experiments by Beaman et al. (2018) show that most farmers are convinced to adopt a new technology only if they receive information about it from multiple sources. The challenge would be to use local contextual knowledge to find entry points or opportunities where most farmers would have multiple contacts or sources of information.

Ragasa (2019b) shows that while lead farmers play a crucial role in supporting and assisting AEDOs' activities in the communities (farm demonstrations, community and group meetings, etc.), their current coverage or interactions with farmers is limited (7 percent of households claimed that they have received advice from a lead farmer in the past 12 months and 20 percent claimed to have ever interacted and participated in any activity or meeting organized by lead farmers). Ragasa (2019b) shows no statistical association of farmers' exposure, access to, and interactions with lead farmers on awareness and adoption of most technologies being promoted.

When the type and heterogeneity of lead farmers were unpacked, Ragasa (2019b) shows that access to quality lead farmers, adoption behavior of lead farmers, and regular training received by lead farmers have strong and consistent effects on awareness and adoption of most technologies being promoted. The implications are the need for regular training of lead farmers; ensuring an inclusive and transparent selection of quality lead farmers, including greater sensitization of communities; supporting and incentivizing quality lead farmers, including periodic orientation and joint stakeholder supervisory visits by extension agents of local leaders on the lead farmer approach; and mobilization of the communities and local leaders to take ownership of the process and the approach.

Another implication is that activities of lead farmers seem to depend on how active and motivated the AEDO or NGO extension worker is that they are supporting. If the AEDOs or NGO agents are not active in the community, lead farmers are also not active. Serious issues in terms of building capacity, motivation, and accountability at the service provider level need to be addressed first so that farmer-to-farmer extension can be more effective. So equipping them with proper training and support is crucial. Also needed is to go beyond using lead farmer-to-farmer ratio as a main indicator of extension service provision. Instead, indicators should be on the number of households that have adopted the technologies and that are better-off because of the knowledge imparted by the lead farmers. Similar to the conclusions of Behaghel et al. (2018), the lead farmer approach should be supported by a well-working extension system and come as a complement to that system, not a substitute.

Fourth, the surveys and interviews highlight the need to revisit the expectations placed on the decentralized structures for demand articulation and coordination and to be realistic about what they can really do. Only a third of the communities have a village or group villages agricultural committee (VAC/GAC) set up; and participation by households in these VAC/GAC/VDC is low (only a third of those farmers aware of these committees were actively participating). There is generally low awareness of these VAC/GAC/VDC and much confusion about their roles and functions. Only a fifth of the randomly selected communities have implemented the model village concept, being promoted as an integrated approach for solving communities' challenges. The model village concept is not associated with improved community outcome indicators based on econometric models; and its implementation should be reviewed and improved in order to improve development outcomes (Ragasa, Alvarez-Mingote, and McNamara 2018). Moreover, only a few VAC, ASPs, and DAECC are active and responsive; almost all district stakeholder panels (DSPs) are not working; and many of them do not have the institutional setup, resources, and capacity to coordinate, monitor, and harmonize service provision at the district level (Ragasa, Alvarez-Mingote, and McNamara 2018). Current capacity strengthening of these structures seems to be done mechanically, focusing on their setup but without providing adequate support for their long-term functionality such as designing monitoring practices and tracking of key objectives.

Last, several studies highlight that in many instances information may not be the only binding constraints—access to inputs and capital is also crucial. Chilemba and Ragasa (2018) show that the lack of positive impact of farmer business schools on farm incomes in Dedza district is likely due to limited resources of FBS participants to apply the FBS lessons and start profitable ventures. A study by Ambler, de Brauw, and Godlonton (2018) of NASFAM farmers in Ntchisi and Dowa districts shows that financial support to farmers (cash and in-kind transfers) lead to shifts away from tobacco toward groundnut and soy production and increases in legumes production; no difference in impact was found among different extension approaches used. Therefore, a holistic approach to production and livelihoods is necessary, including information access. A holistic approach highlights putting farmers as the center, and all technical departments and service providers contributing in a coordinated matter and not working in silos.

7.6 Conclusion and Recommendations

This section makes concluding remarks and provides recommendations to improve the state of agricultural extension in Malawi.

Governance Structures and Policy Environment

The policy environment for Malawi's extension services can be considered enabling. The National Agricultural Policy was launched in 2016 with a focus on strengthening key drivers for the transformation and commercialization of the agriculture sector, with a prominent role for extension services. A national extension services strategy is being developed in response to the new NAP. Once this is finalized, the Malawi MoAIWD should sensitize stakeholders at all levels so that they know what is expected of them. The ministry and its partners should mobilize resources for it to become operational. In addition, the ministry should develop detailed and clear implementation guidelines for the different levels of service provision. As other subsectors develop their strategies, the DAES and MaFAAS should ensure these strategies address extension and advisory issues.

More funding from government for extension services (with greater balance between operating versus salary costs) is needed, especially to fulfill its function in terms of coordination and addressing the gaps in the provision of extension services, where nongovernment providers are not filling (for example, sustainable production systems, natural resources management, and disadvantaged groups). Without funds for operating costs, adding staff to the payroll will not help and would be counterproductive, crowding out other investments that can better deliver development outcomes. Moreover, the system has to move beyond farmer-to-agent ratio as its main performance indicator and should monitor farmers' access to useful information and their feedback on the messages and technologies being promoted.

Organizational Capacity and Management

PUBLIC SECTOR

This chapter has demonstrated that the public extension system operates through a dual decentralized and centralized system. This is because the agriculture sector has not fully decentralized its system, allowing both structures to operate simultaneously. As such, DAES at the national level and the ADD level continue to directly influence priorities in the districts and which farmers' demands are responded to. This has resulted in an unhealthy unity of command and dual accountability that stifle the full implementation of the extension system. In addition, it continues to implement the extension policy through the District Agricultural Extension Services System with limited success. The DAESS is generally weak with limited resources and support from its stakeholders, which renders it ineffective for the articulation of farmers' demands and for the provision of a coordinated response by extension services providers. Although DAESS staff participates in developing work plans and budgets and monitors activities through these structures, not much is being done on the monitoring of outcomes, impacts, and lessons learned.

To improve the organization and management of extension services, Malawi's MoAIWD can do the following:

1. Abide by the decentralization act and complete the decentralization process by empowering farmers to articulate their demands and for stakeholders at the district and national levels to organize a quality response.

2. Thoroughly review the DAESS to establish the key bottlenecks and opportunities, as well as generate lessons from best practices to effectively implement the pluralistic demand-driven extension policy.

3. Ensure that field staff report activities they do on behalf of or jointly with other service providers. Incentives systems can be aligned not only on the activities organized but also on the level and quality of coordination and collaboration with other extension services providers.

4. Strengthen its monitoring system by going beyond activities/inputs to include outcomes, impacts, and key lessons.

PRIVATE SECTOR AND NGOS

The number of nonstate actors has increased as a result of the introduction of the pluralistic extension policy, which is a welcome development. These are present in every district in varying degrees, offering a wide range of extension services. Their increased presence has resulted in unhealthy competition for resources and clients In some districts, resulting in unsustainable tactics such as free handouts to farmers. Few farmers' organizations participate in extension services provision and most of them are young. For the nonstate actors to provide more effective extension services, there is a need for a more coordinated and regulated extension services. Malawi's MoAIWD should therefore establish a mapping system of extension services providers by district to determine their strengths and opportunities as well as how best they can be supported to enable them to provide quality extension services. The ministry should strengthen its efforts in empowering farmers' organizations to participate in the provision of extension services as well as help organize their demand for the same.

DECENTRALIZED COORDINATION

Since the pluralistic extension policy was launched in 2000, the government of Malawi has made efforts to establish the DAESS with limited success. Some efforts by the donor community were made to strengthen the DAESS structures. There is some evidence that many of these structures are active and functional because of these new efforts. However, participation of households in these structures remains very low, and ratings in terms of their responsiveness are generally low. Constraining factors include unwillingness of the nonstate extension services providers to own the DAESS structures, viewing them as still heavily linked to the national level Department of Agricultural Extension Services and to the centralized decision-making process.

It therefore is recommended that Malawi's MoAIWD should critically review the DAESS to fine-tune it. Particular attention should focus on the DAESS structure, operational procedures, performance, knowledge management, efficiency, and sustainability. In addition, service providers should adhere to the decentralization act and complete devolution of decision-making powers to the districts to enhance coordination and linkages among extension services providers in the agriculture sector. The MoAIWD should walk the talk and set a good example to nonstate actors. In addition, the MoAIWD should continue to build capacity at the lower levels to understand and use the structures for organizing and responding to farmers' demands.

PROFESSIONALIZATION, REGISTRATION, CERTIFICATION OF EXTENSION AGENTS

There is a substantial number, size, and coverage of nongovernment extension services providers, and the coordination and harmonization of messages in this pluralistic system has never been more important and urgent. The government of Malawi can focus on this role through the District Agricultural Extension Coordinating Committee (DAECC) at the district level and national structures, such as MaFAAS, at the national level. The NACDC is a good start and should continue to be funded and strengthened.

The MoAIWD with MaFAAS should promote professionalism in the provision of extension services by doing the following:

- Establishing acceptable qualifications of professional extension agents,
- Monitoring the conduct of members of the profession,
- Acting as a regulatory body and licensing extension organizations, and
- Advising and influencing curricula of higher learning institutions as well as its students.

FOCUS ON HARMONIZED AND RELEVANT CONTENT OF EXTENSION SERVICES

A critical step for harmonization and certification is relevant content. First, research, other technical departments, and extension should collaborate on assessment of the productivity, profitability, and usefulness of technologies and extension messages being promoted. A common assumption is that the "improved" technologies promoted are productive and profitable, when in fact various studies (summarized in Ragasa and Mazunda 2018) highlight that this may not always be the case. There are wide variabilities in soil types, farming systems, and local conditions; highly productive and profitable technologies on average may not be productive and profitable to many conditions. Evidence presented in this chapter reinforces the need to further review and evaluate the technologies being promoted to determine how they really impact different farming communities.

CAPACITY AND TRAINING

Regular training of extension services workers to upgrade technical and facilitation skills will be required. Nongovernment extension agents are being trained by projects more often, although several of these capacity strengthening efforts are largely uncoordinated. Coordination of these and a feedback system to and from the agricultural colleges and training institutions will be useful to ensure that demand and supply of skills and expertise meet. Quality training and follow-up training should also extend to lead farmers and community volunteers. Most important, there is need to strengthen the agricultural education and training institutions that are critical in molding future extension services workers in the country. Exchange forums on how to strengthen these systems and reform the curricula and training modules should be initiated. Hands-on training of students and programs for internships and apprenticeships should be supported.

References

Ambler, K., A. de Brauw, and S. Godlonton. 2018. *Agriculture Support Services in Malawi: Direct Effects, Complementarities, and Time Dynamics*. IFPRI Discussion Paper 1725. Washington, DC: IFPRI.

Andarge, T., and C. Ragasa. 2019. "Can Information Induce Farmers to Adopt Riskier Livelihood Strategies? Evidence from Malawi." Paper selected for presentation during the Development and Sustainability conference, Ann Arbor, MI, US, October 14–18.

Beaman, L., A. BenYishay, J. Magruder, and A. M. Mobarak. 2018. "Can Network Theory-Based Targeting Increase Technology Adoption?" National Bureau of Economic Research, Cambridge, MA, US.

Behaghel, L., J. Gignoux, M. K. Beraho, J. Kugonza, K. Macours, M. N. Mangheni, and J. Oduol. 2018. *Dissemination of New Agricultural Technologies in Africa: Making Extension Work: Impact Evaluation of a Farmer Trainer Program in Uganda.* Final project report. New Delhi: 3ie; Cambridge, MA, US: J-PAL; Paris: Paris School of Economics.

Birner, R., K. Davis, J. Pender, E. Nkonya, P. Anandajayasekeram, J. Ekboir, A. Mbabu, D. J. Spielman, D. Horna, S. Benin, and M. Cohen. 2009. "From Best Practice to Best Fit: A Framework for Designing and Analyzing Pluralistic Agricultural Advisory Services Worldwide." *Journal of Agricultural Education and Extension* 15 (4): 341–355.

Chapota, R., C. Mthinda, and P. Fatch. 2014. *The Role of Radio in Agricultural Extension and Advisory Services—Experiences and Lessons from Farm Radio Programming in Malawi.* MEAS Case Study # 8. Urbana-Champaign, IL, US: University of Illinois, MEAS. www.farmradiomw.org/publications/MEAS-CS-Malawi-Farm-Radio-Chapota-Fatch-Mthinda-Feb-2014.pdf.

Chilemba, J., and C. Ragasa. 2018. *The Impact of a Farmer Business School Program on Incomes of Smallholder Farmers: Insights from Central Malawi.* MaSSP Working Paper 23. Washington, DC, and Lilongwe, Malawi: International Food Policy Research Institute (IFPRI).

Chiweza, A. L. 2010. *A Joint Study of the Ministry of Local Government and Rural Development and Concern Universal.* Lilongwe, Malawi: Ministry of Local Government and Concern Universal.

Chowa, C., C. Garforth, and C. Cardey. 2013. "Farmer Experience of Pluralistic Agricultural Extension, Malawi." *Journal of Agricultural Education and Extension* 19 (2): 147–166.

Fisher, M., S. Holden, and S. Katengeza. 2017. *Adoption of CA Technologies among Followers of Lead Farmers: How Strong Is the Influence from Lead Farmers?* Centre for Land Tenure Studies Working Paper 7/17. Ås, Norway: Norwegian University of Life Sciences.

Flaherty, K., G. O. Essegbey, and R. Asare. 2010. *Ghana: Recent Developments in Agricultural Research.* ASTI Country Note. Washington, DC: IFPRI.

Kamkwamba, S. 2015. "Agricultural Extension Services in Malawi: Past, Present and the Future." Paper presented at MEAS Symposium, Washington, DC, June 1–5.

Kumwenda, W. 2013. "NASFAM Extension System." Paper presented at the Workshop and General Assembly for the Malawi Forum for Agricultural Advisory Services (MaFAAS), Malawi Institute of Management, Lilongwe, May 8–9.

MaFAAS (Malawi Forum for Agricultural Advisory Services). 2015. *Report on the Technical Support Visit by the Malawi Forum for Agricultural Advisory Services (MaFAAS) to District Agriculture Extension Services (DAESS) Structures in Nkhatabay, Ntcheu and Mulanje Districts.* Lilongwe.

Malawi, MoAIWD (Ministry of Agriculture, Irrigation and Water Development). 2000. *Agricultural Extension in the New Millennium: Towards Pluralistic and Demand-Driven Services in Malawi*. Lilongwe.

———. 2004. *Agricultural Extension Policy Implementation Guide*. Lilongwe.

———. 2006. *District Agricultural Extension Services System Implementation Guide*. Lilongwe.

———. 2016a. *Agriculture Sector Performance Report 2015/2016 Fiscal Year*. Draft final report from the joint sector review. Lilongwe.

———. 2016b. *National Agricultural Policy (2016–2020)*. Lilongwe.

Masangano, C. 2017. "Presentation of Issues Paper for the Development of Extension and Advisory Services Strategy for Malawi." PSI Consult International, Lilongwe, Malawi, April 12.

Masangano, C., and C. Mthinda. 2012. *Pluralistic Extension System in Malawi*. IFPRI Discussion Paper 1171. Washington, DC: IFPRI.

MEAS (Modernizing Extension and Advisory Services Project). 2012. "Strengthening Pluralistic Agricultural Extension in Malawi." A MEAS Rapid Scoping Mission, MEAS, University of Illinois, January 7–27.

Mkwambisi, D., S. Khaila, C. Mthinda, A. Mulwafu, and M. Chikowi. 2013. *Baseline Draft Report for the Agriculture Technology Transfer of the Malawi Agricultural Sector Wide Approach Support Project (ASWAp-SP)*. Lilongwe, Malawi: Millennium Centre for Research and Development.

Mthinda, C. 2015. *Innovative Agricultural Extension and Advisory Services Approaches: A Synthesis Report*. Lilongwe, Malawi: MaFAAS.

Musopole, R., F. Nkana, A. BenYishay, M. Mobarak, M. Jones, and F. Kondylis. 2013. *Communicating with Farmers through Social Networks*. Innovations for Poverty Action (IPA) Policy Brief. Lilongwe, Malawi.

Niu, C., and C. Ragasa. 2018. "Selective Attention and Information Loss in the Lab-to-Farm Knowledge Chain: The Case of Malawian Agricultural Extension Programs." *Agricultural Systems* 165: 147–163.

Ragasa, C. 2014. "Improving Gender-Responsiveness on Agricultural Extension." In *Gender in Agriculture and Food Security: Closing the Knowledge Gap*, edited by A. Quisumbing, R. Meinzen-Dick, T. Raney, A. Croppenstedt, J. Behrman, and A. Peterman, 411–430. Rome: Food and Agriculture Organization of the United Nations (FAO); Dordrecht, Netherlands: Springer.

———. 2019a. "Knowledge Is Power: Impact of Interactive Radio Programming on Women's Empowerment and Agricultural Development Outcomes in Malawi." Selected paper presented during the Seeds of Change Conference, Canberra, April 4.

———. 2019b. *Modeling the Effectiveness of the Lead Farmer Approach in Agricultural Extension Service Provision: Nationally Representative Panel Data Analysis in Malawi*. IFPRI Discussion Paper 1848. Washington, DC: IFPRI. http://ebrary.ifpri.org/cdm/ref/collection/p15738coll2/id/133285.

Ragasa, C., N. L. Aberman, and C. Alvarez-Mingote. 2019. "Does Providing Agricultural and Nutrition Information to Both Men and Women Improve Household Food Security? Evidence from Malawi." *Global Food Security* 20: 45–59.

Ragasa, C., C. Alvarez-Mingote, and P. McNamara. 2018. *Is There a Role for Agricultural Committees and Stakeholder Panels in Improving Information and Accountability in Extension Services Provision?* Evidence from Malawi, IFPRI Discussion Paper 1726. Washington, DC: IFPRI.

Ragasa, C., and A. Comstock. 2019. *Dynamics in Agricultural Extension Services Provision in Malawi: Insights from Two Rounds of Household and Community Panel Surveys*. IFPRI Discussion Paper. Washington, DC: IFPRI.

Ragasa, C., and J. Mazunda. 2018. "The Impact of Agricultural Extension Services in the Context of a Heavily Subsidized Input System: The Case of Malawi." *World Development* 105: 25–47.

Ragasa, C., J. Mazunda, and M. Kadzamira. 2015. "The National Extension Policy of Malawi—Lessons from Implementation." *Malawi Country Strategy Support Program (Massp) Policy Notes* 23. Washington, DC: IFPRI.

Ragasa, C., D. Mzungu, E. Kaima, C. Kazembe, and K. Kalagho. 2017. *Capacity and Accountability in the Agricultural Extension System in Malawi: Insights from the Survey of Service Providers in 15 Districts*. IFPRI Discussion Paper 1673. Washington, DC: IFPRI.

Ragasa, C., and C. Niu. 2017. *The State of Agricultural Extension and Advisory Services Provision in Malawi: Insights from Household and Community Surveys*. MaSSP Technical report. Washington, DC: IFPRI.

Snapp, S. S., T. S. Jayne, W. Mhango, T. Benson, and J. Ricker-Gilbert. 2014. "Maize-Nitrogen Response in Malawi's Smallholder Production Systems." Paper prepared for National FISP Symposium, Lilongwe, July.

Spielman, D., C. Azzarri, N. Beintema, S. Makhija, and G. Nico. 2019. "Internationally Comparable Metrics for Agricultural Extension Data, Indicators, Challenges, and Opportunities." DLEC presentation, Addis Ababa, June.

Ward, P., A. Bell, K. Droppelmann, and T. Benton. 2018. "Early Adoption of Conservation Agriculture Practices: Understanding Practical Compliance in Programs with Multiple Adoption Decisions." *Land Use Policy* 70: 27–37.

World Bank and IFPRI. 2009. *Gender and Governance in Rural Services*. Washington, DC: World Bank.

Chapter 8

DEMOCRATIC REPUBLIC OF THE CONGO

Catherine Ragasa and John Ulimwengu

8.1 Introduction

This chapter analyzes the provision of agricultural extension and advisory services in the Democratic Republic of the Congo (DRC) in the context of a postconflict country and the experiences and challenges in reconstituting an extension system. Since 1996, the DRC has experienced a succession of wars and lower-scale conflicts that caused more than 5 million deaths over the 1998–2008 period (Maystadt et al. 2014). These have had a devastating effect on the country's infrastructure, technology, and institutions, creating a vacuum that has led to a laborious process of restoration. As a result, the DRC faces serious food insecurity and human development challenges. In 2011 through 2013 the DRC was ranked as the country with the most severe food insecurity, based on the *Global Hunger Index* of the International Food Policy Research Institute (IFPRI). It remains so according to current alarming hunger indicators (IFPRI 2016). At least 50 percent of the population is deficient in key nutrients, including iron, vitamin B12, calories, riboflavin, vitamin E, folate, and zinc (Ulimwengu, Roberts, and Randriamamonjy 2012). Life expectancy in the DRC is around 47 years for men and 51 years for women; 1 in 5 children die before reaching age 5 (Pfingu 2011; UNDP 2010).

Although the Democratic Republic of the Congo has significant potential to be the food basket of Africa (Tollens 2004; World Bank 2006), it has failed to create an enabling environment for the investments and policy changes that it needs. The key challenge faced by the decisionmakers is how to reform the DRC's extension and advisory services, including other necessary key services and activities to support the efforts of the Ministry of Agriculture, Livestock and Fisheries (MINAGRI) to support the transformation of the agriculture and food sectors (World Bank 2018). This chapter explores the delivery process of extension services to provide a better understanding of strategies and approaches to improve the provision of extension services in the country. It aims to contribute to the transformation of agricultural production and the increase in incomes. Specifically, the chapter aims to do the following:

1. Provide a diagnostic assessment of their structure, capacity, activities, and performance;
2. Characterize and compare the capacities and performance of various extension organizations and agents as well as analyze contributing factors;
3. Synthesize the evidence and lessons of various effectiveness analysis and impact evaluation studies of extension approaches; and
4. Identify feasible entry points to strengthen and reform the national extension system.

The guiding principle in this chapter is the "best-fit" (conceptual) framework that is described in Chapter 1 of this book (Birner et al. 2009). Section 8.2 discusses the methods and data sources for this chapter, while Sections 8.3 and 8.4 include the main findings summarized based on the various components of the best-fit framework (items 1 and 2 above). Section 8.5 summarizes the analysis of the performance of extension organizations and agents. Section 8.6 summarizes the evidence and lessons on effectiveness and impact of extension programs and approaches (item 3 above). Last, Section 8.7 provides some concluding remarks and policy implications.

8.2 Data Sources and Methods

This chapter synthesizes a series of literature on agricultural policy processes and extension services in the Democratic Republic of the Congo as well as various data collection approaches and analytical methods. Table 8.1 presents the diversity of methods used. First, an extensive literature review of the status of extension services was conducted. Second, a mapping was carried out of the institutional process network as well as interviews with key informants to better understand the role of agriculture sector players in general, with an emphasis on extension services (Ragasa, Babu, and Ulimwengu 2014). Third, surveys were carried out in August to October 2011 on extension organizations, extension workers, rural producer organizations, and Agricultural and Rural Management Council (Conseil Agricole et Rural de Gestion, CARG) management teams. These included three key provinces: Bandundu, Bas-Congo, and Kinshasa (Table 8.1) (Ragasa, Babu, and Ulimwengu 2014; Ragasa, Badibanga, and Ulimwengu 2016; and Ragasa, Ulimwengu et al. 2016).

The selection of these three western provinces is mainly based on the focus of the government at that time in piloting an effort to strengthen and reengage with the CARGs in relation to its National Agricultural Investment Plan (NAIP). Given the continuing turmoil in the eastern region of the DRC, the research team was unable to carry out its field work there. Since much of the literature focuses on the eastern part of the country, primary data collection for the western area was combined with the literature reviews relating to the eastern part, in order to draw an overall representation of DRC's extension system.

Sample territories and villages were randomly selected in the three western provinces, consistent with the National Statistics Institute's national 1-2-3 household survey, where each number refers to a separate phase: (1) employment, (2) informal sector, and (3) consumption.[1] In Table 8.1 we indicate the method that was applied to ensure that the selection of respondents was not biased. The various datasets were analyzed using various qualitative and quantitative approaches. An institutional analysis and process-network mapping were conducted (Ragasa, Babu, and Ulimwengu 2014). In addition, comparative and econometric analyses were employed to analyze the factors affecting the performance of CARGs, extension organizations, extension agents, and rural producer organizations (Ragasa, Badibanga, and Ulimwengu 2016; Ragasa, Ulimwengu et al. 2016). These provide rich datasets informing and assessing performance of meso-level organizations, which are providing and facilitating extension and other services to the rural communities. Last, we collected household-level data in the three provinces in line with the midline of the USAID's Food Production, Processing and Marketing Project (Ragasa, Nkonya et al. 2016).

8.3 Governance Structures and Policies

Institutional Landscape

The institutional landscape of the agriculture sector is presented in Figure 8.1. This is based on interviews with 45 high-level officials and representatives of the MINAGRI, the Ministry of Rural Development (MINRD), CARGs, parliament, nongovernmental organizations (NGOs), the private sector, and

(text continues on page 270)

1 Territories are lower administrative units than province and are composed of several *secteurs* (sectors). In terms of size, the following depicts the administrative units in the Democratic Republic of the Congo: "province" is larger than "territory," which is larger than "sector," which is larger than "groupment," which is larger than "village."

TABLE 8.1 Methods used to collect data for this study

Method and purpose	Date	Respondents	Selection of respondents	Analytical methods
Exploratory visit and key informants' interviews to understand policy and institutional context; including focused institutional-process-network-mapping exercises	May 2010	45 key informants (interviews were mainly conducted in Kinshasa, but experts are from various provinces)	Purposive, snowball approach	Descriptive analysis
Literature review	Mainly from May 2010 to May 2012; updates done in November 2015; April 2017; and May 2019 corresponding to the times this chapter was revised	National	Illustrative rather than exhaustive	Descriptive analysis
Review of meeting minutes and reports to look for the diversity and depth of topics discussed and problems raised, indications of follow-up or outcome of the previous discussions, indications of participation/ attendance	February–March 2011	30 meeting minutes from 10 different CARGs were collected (national)		Descriptive analysis
Survey of CARGs to understand functioning and assess performance	April–October 2011	Three provinces: 55 CARGs in 23 randomly selected territories*	(1) Randomly selected 23 territories, consistent with the National Statistics Institute implementation of the national 1-2-3 household survey; (2) interviewed all CARGs in the target territories; (3) acquired full listing of the CARG leaders and members (stratified by government and nongovernment); (4) randomly selected 3 from leadership; 1 member from government strata, 2 members from nongovernment strata; (5) conducted face-to-face group interviews lasting between two and three hours	Descriptive analysis and econometric analysis to identify factors explaining CARG performance

Method and purpose	Date	Respondents	Selection of respondents	Analytical methods
Survey of rural producer organizations (RPOs)	April–October 2011	Three provinces: 181 community organizations in 23 randomly selected territories	(1) Randomly selected 23 territories; (2) interviewed all RPOs (with some agriculture related activities) in the target territories; (3) interviewed a group of four to six opinion leaders from these RPOs; (4) conducted face-to-face group interviews lasting between two and three hours	Descriptive analysis and econometric analysis to identify factors explaining RPO performance
Survey of agricultural extension organizations (AEO)	April–October 2011	Three provinces; 107 extension organization heads in 23 randomly selected territories	(1) Randomly selected 23 territories; (2) interviewed all AEO in the target territories; (3) interviewed the head or representative from these AEO; (4) conducted face-to-face individual interviews lasting between two and three hours	Descriptive analysis
Survey of agricultural workers	April–October 2011	Three provinces: 162 randomly selected agricultural workers in 23 randomly selected territories	(1) Randomly selected 23 territories; (2) interviewed all AEO in the target territories; (3) acquired list of workers in each AEO and randomly selected four to eight workers; (4) conducted face-to-face interviews lasting between two and three hours	Descriptive analysis and econometric analysis to identify factors explaining extension agents' performance
Household survey	February–August 2014	Three provinces: household survey and midline assessment of an agricultural program (FPPM)	Total of 3,110 households in randomly selected treatment communities and randomly selected comparable/control communities	Descriptive analysis and difference-in difference method to assess indicative impact of the interventions

Source: Authors.

Note: * Three provinces include Bas-Congo, former Bandundu, and Kinshasa. FPPM = food production, processing and marketing project. CARG = Agricultural and Rural Management Council.

universities, guided by a Net-Map approach and a semistructured questionnaire.[2] MINAGRI and MINRD were the two key ministries responsible for leading the agricultural strategy and policymaking process in the Democratic Republic of the Congo. The political system—including the president, prime minister, and parliament at the national level and elected government administrators at the local level—determines the financial resources and creates the enabling environment for the agricultural sector. Various government ministries (each in its own jurisdiction) are responsible for policymaking and planning, implementation, and monitoring of development programs. The DRC has 26 provinces (including Kinshasa), 30 districts, 145 territories, and 800 sectors (formerly called *collectivities*); within the decentralization process, the provincial services of the ministries are being revitalized to provide service provision to the rural population.[3] Staff recruitment to support the decentralized units is ongoing.

Farming communities and their organizations are critical players in the agricultural sector. The three federations of NGOs—the Federation of Laic and Economic NGOs (FOLECO), the National Council of Development NGOs (CNONGD), and the Regional Council of Development NGOs—so far have not gained much influence in the agricultural policymaking process. For a more systematic engagement of civil society organizations into the policymaking and planning process, the government initiated in 2008 the creation of CARGs, which are the platforms for multistakeholder discussions, policy dialogues, and information sharing. Private-sector participation in the agricultural policymaking process in DRC has not been strong in the past years, except for some advocacy efforts by the Federation of Private Enterprises in Congo (FEC). This national federation of private companies has to date only a few agriculture-based, large-scale companies as members. Similarly, the Confederation of Small and Medium Enterprises of the Congo (COPEMECO), of which numerous small-scale enterprises are members, has not gained a strong presence in the agricultural policymaking process. Despite the presence of these CARGs and confederations, weak linkages and

2 MINRD was first created in the 1980s, but it was absorbed twice into MINAGRI and then recreated as an independent ministry. CARGs are platforms for consultation, dialogues, problem-solving, articulation of demand of rural services, and monitoring of programs at the territory and local levels and an avenue for engagement in policymaking processes at the provincial and national levels (Ragasa, Badibanga, and Ulimwengu 2016 for details).

3 Based on Decree Law 081 (July 2, 1998) on the territorial and administrative organization of the DRC. This figure includes three districts in Kinshasa (excluding a chief of district). Most of the districts are to become provinces (except the newly created district of Plateaux), according to Article 2 of the Congolese Third Republic Constitution (www.presidentrdc.cd/constitution.html).

FIGURE 8.1 Organizational structure of the agriculture policymaking process in the Democratic Republic of the Congo

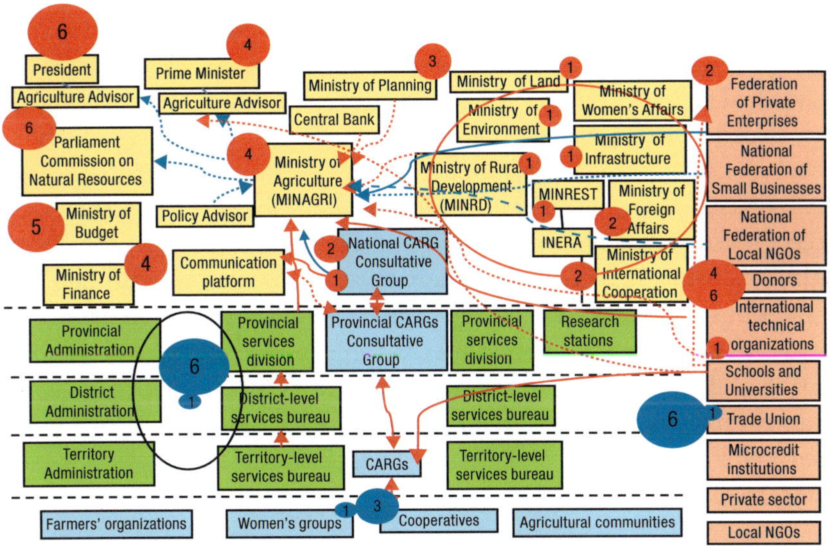

Source: Adopted from Ragasa, Babu, and Ulimwengu (2014), based on Net-Map interviews with a high-level official of the Ministry of Agriculture, CARG representatives, and a staff member of an international organization who has worked in the DRC for several years. These Net-Maps were complemented by interviews with several key actors in the agriculture sector.

Note: CARG = Agricultural and Rural Management Council; INERA = National Agronomic Research Institute; MINREST = Ministry of Research, Science, and Technology. The structure and placement of the actors do not represent political ranking or authority. The purpose is to inventory the various actors involved in the agricultural policy process at the national and local levels; to illustrate links among the various actors; and to identify who has influence in the actual agricultural policies being designed, funded, and implemented. The numbers in circles represent the influence level (measured on a 0–6 scale) of these actors, with red circles (and darker numbers) for those at the national level and blue circles (and lighter numbers) for those at the local level. The size of the circle represents the influence score, consistent with the number indicated. For some actors (donors, local government and trade union), two influence scores mean a range, indicating that the score varies across provinces. The different actors (in rectangles) are represented by different colors by type of organization: (1) national-level public institutions (yellow); (2) private sector, donor community, NGO, and academe (orange); (3) CARG and producer groups (blue); and (4) local-level public institutions (green). The lines connecting key actors represent the extent to which links and information flow between the actors (red lines represent information flow while blue lines represent the flow of policy advice; solid lines represent strong links while broken lines represent weak links). Organizations without an influence score were given a zero by respondents.

communication exist among various actors in the agricultural landscape (illustrated by the broken lines in Figure 8.1).

Interviews conducted and summarized in Ragasa, Babu, and Ulimwengu (2014) show that overall organizational and human capacities are also particularly weak in the DRC, with human capacity feeble at the MINRD and MINAGRI (the national and local levels). Of all the platforms and associations, CARG has the most inadequate human capacity. Other areas of capacity (including facilities and infrastructure, organizational systems, monitoring and evaluation systems [M&E], coordination, and communication within organization) were rated weak or very weak, based on interviews with 45 stakeholders using a 5-point Likert scale (see Ragasa, Babu, and Ulimwengu 2014).

Of particular interest are the universities and research organizations that provide evidence-based analyses and planning for the ministries. Interviews conducted and summarized in Ragasa, Babu, and Ulimwengu (2014) show that the university system faces numerous challenges, including a disconnect between teaching and research by faculty and classroom training of students on the one hand and policy problems and realities in the field on the other. The stakeholders interviewed highlight decaying human capital, the lack of investment and funding in university, agricultural research, education and training, and extension systems that result in insufficient facilities, computer centers, and training materials. As suggested by respondents, training of staff should be accompanied by upgrades in infrastructure and equipment in these systems. It appears that there is need for systemic infrastructure investment and human capacity building in the agriculture sector, including extension services.

As indicated in Figure 8.1, according to the key stakeholders interviewed and summaries in Ragasa, Babu, and Ulimwengu (2014), the most influential agricultural policy decision-makers are the high-level policymakers and lawmakers (the president and the members of parliament) because they set national priorities which, in turn, influence the budget allocation and disbursement to the different sectoral ministries and public services. The next most influential is the Ministry of Budget, which decides on actual budget allocation for sector ministries, followed by the Ministry of Finance, which sets priorities on actual budget disbursement. The prime minister has strong influence in setting national priorities and agricultural policies. Donors are considered influential in the agricultural policy process. These are the common ratings of the key stakeholders interviewed (see Ragasa et al. [2013]).

The patterns of influence illustrated in Figure 8.1 suggest that greater attention is needed to ensure that compelling evidence from research and dialogues reaches the influential decisionmakers and funders. This highlights the required strengthening of policy dialogue and negotiation skills, among many other skills and competencies, at the key ministries. It may also require more systematic mechanisms of linking MINAGRI and these decisionmakers, such as initiating regular meetings with the agriculture advisers of the president and the prime minister, and to examine the reestablishment of the socioeconomic research committee, with emphasis on the agriculture and rural sectors, at parliament. It will require a more systematic and innovative means of communication among these organizations. This implies that donors can play a major role in influencing policies in the DRC, taking care not to undermine the existing institutional structures and local ownership of these policies. It follows that institutional and policy reforms and funding for extension services will require similar strategies.

Numerous agricultural policies and programs were developed, although the episodes of political turmoil and conflict have disrupted their implementation. Table 8.2 shows the major agricultural policies and programs that began during the postcolonial period. As pointed out by Tollens (2004), many actions set forth by different governments to revive the agricultural sector and induce rural development since the DRC's independence have produced some positive, albeit insignificant, results. Most of the policies failed to be comprehensive and lacked internal coherence on which of the various interventions could have been articulated. Moreover, the lack of commitment from the government with respect to funding and implementation of these programs is evident across time.

The most notable is the Five-Year Plan (1986–1990), which was an ambitious long-term program, implemented to induce economic and social development and to achieve food self-sufficiency. It targeted agricultural development through the increase in size and productivity of peasant farms, promotion of entrepreneurship in the sector, intensification of extension and applied agronomic research, maintenance of feeder roads, and job creation and increase in incomes in rural areas. It is characterized by the move toward liberalism and the promotion of private initiative; promotion of partnership including the private sector, NGOs, peasants, small and medium agricultural firms, and donors; and presence of a detailed five-year investment plan supporting the general policy. Extension services were one of the major themes in this program. However, the implementation suffered, as did the previous ones, from a shortage of funds (that is, the government defaulted on its financing commitments).

To support the government of the Democratic Republic of the Congo, many externally funded programs operated in the 1970s and 1980s, and promoted agricultural technology transfer and advice to small producers, among others. Most notable among these programs are the World Bank–funded projects that were based on the training-and-visit method. In 1988 the National Extension Service (*service national de vulgarisation,* or SNV) was created to harmonize approaches and methods previously developed for various extension services and to integrate them into the ministry's day-to-day activities. However, starting with the unrest of 1991, which *(text continues on page 276)*

TABLE 8.2 Major policies after colonial period affecting agriculture and extension system

Major policies and programs	Year	Objectives and major components	Prominence of extension	Outcomes
Zairianization	1973–1976	Aimed at agricultural development, improvement of social conditions of farmers, and food self-sufficiency	Not prominent	Difficult transition; enormous abandonment of fields; plunge in agricultural production; selection of the new acquirers of the agro-industrial units was based on the political criteria rather than on the management skills of new owners
Radicalization	1974–1976	Aimed at transferring strategic agro-industrial units and large fields mismanaged by unskilled acquirers either to the state or to more skilled new acquirers; new government agencies were created to provide technical support to the sector and to ensure marketing of agricultural products	Not prominent	Failed to revive the agricultural sector as expected; agricultural research and extension were almost nonexistent; several agencies created to support the sector were nonoperational; most agro-industrial units were bankrupted; fall in production of all commercial crops except for coffee
Retrocession	1976–1978	Aimed at rehabilitation of bankrupt firms by focusing on providing financial and technical support		Limited success. Many owners of nationalized agricultural firms declined the invitation; productivity and production stagnation continued
Economic Revival Plan or Plan Mobutu	1978–1982	Aimed to revive the agricultural sector and other ones (mining, transportation, and energy), specifically, aimed at improving food security, increasing the production of exportable crops and of crops used as inputs by local industrial firms, and funding projects contributing to rural development. This was with support from IMF	Prominent	Limited success even with its 26 projects during the first phase (1979–1981) and 22 (1981–1983) during the second phase mainly because of funding shortage from the government

Major policies and programs	Year	Objectives and major components	Prominence of extension	Outcomes
Development Conventions	1979–1980	Aimed at inducing industrial and commercial large firms to contribute to agricultural development through the increase in agricultural productivity and the supply of basic inputs to the local agro-industry; the policy used several instruments including a special tax	Prominent	Limited success; the policy was selective and thus benefited somehow a few rural areas where the selected firms chose to operate; taxes led to increased prices of agricultural products, which was a disincentive among many firms to reinvest in the agriculture sector.
Agricultural Revival Plan	1982	Aimed at improving the partnership between the government and the private sector for a more rational management of the agricultural sector	Not prominent	Some success, but small farmers and other stakeholders of the agricultural sector were ignored
Interim Economic Rehabilitation Program	1983–1986	Aimed as a preparatory stage to the five-year plan (see below); first time that donors were seen as stakeholders; supported by the World Bank and the IMF; the involvement of the donors in the management of the sector included a diagnosis of previous agricultural policies and the identification of the causes of their failure		Some success; public investment plan under this program were funded and included 68 agricultural projects (including forest, livestock, and fishery); government defaulted on its financial commitments
Five-Year Plan (Plan Quinquennal)	1986–1990	Aimed at agricultural and rural development through the increase in size and productivity of peasant farms, promotion of entrepreneurship, job recreation in rural areas, intensification of extension and applied agronomic research, and maintenance of feeder roads	Prominent role of extension	Limited funds; government defaulted on its commitments
Master Plan	1990–2002	Aimed at achieving agricultural and rural development through participatory approaches; rehabilitation of rural infrastructure, improvement of the supply of inputs and credit; funding the agronomic research and extension; and improvement of agricultural markets and agro-processing	Very prominent role of extension	Political instability prevented its implementation

(continued)

TABLE 8.2 Continued

Major policies and programs	Year	Objectives and major components	Prominence of extension	Outcomes
Agricultural Policy Code (Note de Politique Agricole)	2009–2013	Aimed at reducing poverty and hunger by 50% by 2015 (MDG goal); specific objectives: 10% of the government budget going to agriculture to meet the Maputo commitment; MINAGRI restructuring; decentralized agricultural service provision; improvements for financial services; and promoting agricultural entrepreneurship	Very prominent role of extension	Targets not achieved. Funding for agriculture remains low and government commitment continues to be lacking
National Agricultural Investment Plan	2013–2020	Its overall objective is to stimulate sustained annual growth in the agricultural sector of at least 6%, which is required to reduce poverty, to ensure food and nutritional security in the DRC, and to generate sustainable jobs and income. Main components are as follows: • Promote agricultural value chains; • Develop and disseminate research products to users; • Improve sector governance; • Promote gender mainstreaming, and strengthen human and institutional capacities; • Reduce the vulnerability of the agricultural sector to climate change.	About 12.9 percent of the budget was allocated to research and extension	The government failed to mobilize required resources for the program. As a result, the program has been implemented piecemeal and is difficult to assess.

Source: Authors' compilation based on Tecsult International (2009) and DRC, MINAGRI (2013).
Note: MDG = Millennium Development Goals; MINAGRI = Ministry of Agriculture.

brought a nationwide crumbling of institutions and withdrawal of most donor support, SNV institutional capacities were weakened and ceased to function almost immediately. This gave way to an emergency model of extension services through NGOs stepping in to fill the gap, often without the adequate experience and capacity previously brought by SNV. As a result, the system was plagued with inefficiencies, redundancies, and confusion due to conflicting messages to farmers on a specific issue or technology highlighted by the interviews with farmers' organizations and CARG members (see also Tecsult International 2009; Ragasa et al. 2013).

The political instability that has characterized the country since 1990 has not allowed a full implementation of any agricultural policy. In 1990 an agricultural strategy, known as the Master Plan, was developed although never implemented. The Master Plan was a comprehensive and coherent agricultural and rural development policy involving all agricultural sector stakeholders. It represented the first time the DRC implemented the decentralization of services and restructured the functions and staffing of the MINAGRI. Improvement in the provision of extension services, through SNV, was a major component of this policy. Despite its coherence, the program was never implemented due to political and social unrest.

In 2003, after the war, new political leadership expressed interest in implementing the Master Plan. However, the context was totally different politically and economically for this policy to fit. New policies were implemented, although at a rather slow pace. Moreover, as part of the billions of dollars in the form of international assistance security, humanitarian, stabilization, peace-building, and economic recovery interventions, many donor-supported projects have been implemented targeting the agriculture and rural sectors. In 2009 a new policy document, referred to as the Agricultural Code, was drafted. It outlined planned reforms and takes into account the new challenges and the environment faced by the agricultural sector. The code aimed to reduce poverty and hunger by half by 2015 (Millennium Development Goal) with the specific objective to achieve 10 percent of the government budget going to agriculture to meet the Maputo commitment. Among other things, the three major components of this code were the decentralization of agricultural services, the restructuring of MINAGRI, and the implementation of the CARGs. However, this code remained in draft form and has never been implemented.

Under the Comprehensive Africa Agriculture Development Programme (CAADP) initiated by the New Partnership for Africa's Development (NEPAD), a socioeconomic flagship Program of the African Union (AU), the DRC government developed National Agricultural Investment Plan (NAIP) for 2013 through 2020. After years of sluggish growth in the sector, accompanied by rapidly growing food imports that reached US$1.2 billion in 2012, the trend could be reversed thanks to a new political will, expressed by the commitment to the CAADP process, the signing of the CAADP compact, and the design of a US$5.7 billion NAIP to cover the 2013–2020 period.

The remaining challenge is to increase the share of the national budget going toward agriculture to reach 10 percent, per the Maputo engagement.

The DRC government has expressed a commitment to engage significantly more domestic resources as part of the CAADP compact and its NAIP 2013–2020, of which 10 percent of the funds will go to agricultural research, extension services, and education. Aside from this, extension services and training will be an essential component in major programs on value-chain promotion and climate change adaptation. The distribution of planned funds by subprograms is given in Figure 8.2. A focus of the NAIP is to encourage private-sector investment, including foreign investors and home-grown medium and large agribusinesses.

In particular, the vision to move the country toward a green revolution entails a two-track approach. The first track consists in transforming small-scale subsistence farmers into commercial farmers by (1) increasing production through access to improved genetic materials (seed breeds), fertilizers, finance, and technology; and (2) reaching markets more efficiently by enabling producer cooperatives to identify markets, pool resources, add value, negotiate contracts, and access feeder roads and storage facilities. This all while ensuring that development does not infringe on ecologically sensitive areas. The second track consists in restarting private investment in commercial agriculture using modern technologies geared toward agro-processing and exports. The government aims to attract private investment in agro-industrial parks, where it plans to provide basic infrastructure (transport, water, power, ICT) and a transparent administrative and fiscal framework (for example, land titles and special economic zone status). This safe haven approach is piloted in peri-urban Kinshasa. There is potential for combining small-scale agriculture and large-scale agro-industries through contract farming and by supporting mid-size operators in relevant value chains and regions. The productive alliance model will be tested in new operations to reinforce the commercialization of products and ensure market integration of small farmers. Extension services are strongly featured in both tracks. Pilot implementation of these are ongoing and yet to be evaluated.

Extension Policy and Structure

There is no existing national extension policy and strategy in the Democratic Republic of the Congo. The extension system described in the 1988 and 1993 decrees point to three systems involved in the provision of extension services in the DRC: (1) national extension services (SNV) for coordination with national headquarters and a coordinating team in each province; (2)

FIGURE 8.2 Available funds for the National Agriculture Investment Plan by subprogram (%)

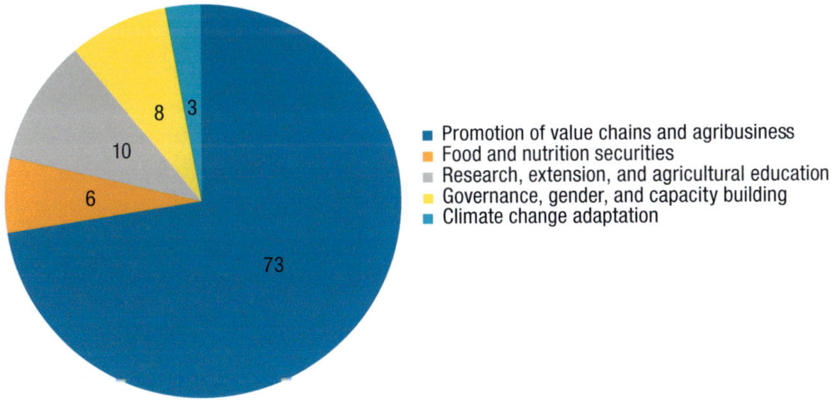

Source: DRC, MINAGRI (2013).

supporting structures, including researchers and specialists, for training and technical backstopping, as well as INERA (the National Agronomic Research Institute), SENAMA (the National Mechanization Agency), SENASEM (the National Seed Agency), and SENAFIC (the National Fertilizer Agency); and (3) agricultural inspection system within the MINAGRI, with an estimated 11,245 field staff, complemented by agricultural workers from NGOs for the actual delivery of extension services to rural communities (Figure 8.3).[4] The number of field staff in the system is still in the payroll, but most operations have ceased after the donors' financial support ended in 1997.

Coordination and Pluralism

In 1988 the SNV was created to coordinate, harmonize, and support rural extension services activities as an independent unit within the National Rural Development Department, with their own budget, to institutionalize the earlier National Extension Program. In 1993 another decree was passed to grant administrative and financial autonomy to the SNV under the umbrella authority of the MINAGRI. During its early implementation (in the late 1980s), the SNV had 121 staff at the national level, and 82 coordinators and officers assigned to the different provinces. The SNV was tasked to develop extension services materials and production guides and to regularly update

4 A census survey conducted in 2005 within MINAGRI reveals that the total number of staff was estimated at 18,500, of which 11,245 is said to be composed of qualified extension agents (DRC, MINAGRI 2005).

FIGURE 8.3 Agricultural extension system in the Democratic Republic of the Congo

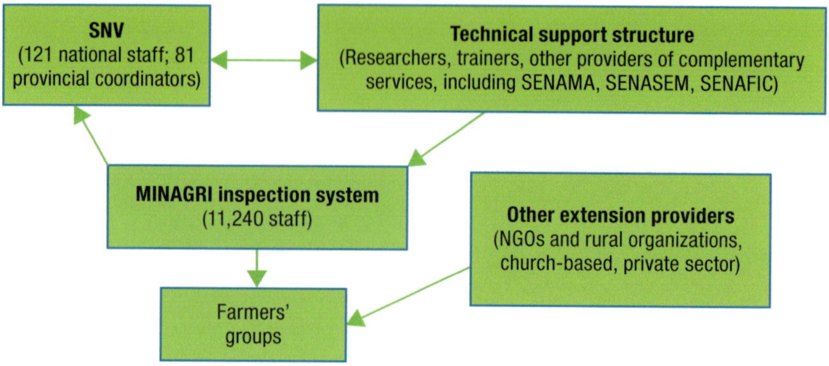

Source: Authors; 1988 and 1993 decrees creating SNV.

Note: MINAGRI = Ministry of Agriculture; SENAMA = National Mechanization Agency; SENASEM = National Seed Agency; SENAFIC = National Fertilizer Agency.

them, to provide training and backstopping, based on these extension services materials, and to coordinate and monitor extension services provision.

From 1989 to the mid-1990s, the SNV was financially supported by the DRC government as well as by international organizations including the World Bank, the French aid agency Agence Francaise de Developpment, the United Nations Development Program (UNDP), and the Food and Agriculture Organization of the United Nations (FAO). By the mid-1990s, donor support ended and government funds were substantially reduced. From 1997 to 2001, funding came solely from the government and mostly went to pay for salaries. Starting in 2002, there was no operating funding available from any source and staff salaries were not regularly paid. This made it difficult for the SNV to fulfill its mandate, given that staff of the agricultural inspection system does not fall under the supervision of the SNV; neither could the SNV train staff nor provide other relevant services without budget support. Despite preparing and submitting budgets each year, the SNV has not been able to garner any sort of financial support from within or outside the government. SNV staff has remained unpaid for years, although employees have kept their positions while looking for other means of economic survival. A sizable portion of staff remains untrained in newer approaches to transmitting improved technologies to farmers, and members have now passed retirement age.

The current extension system in the Democratic Republic of the Congo is characterized by a largely uncoordinated system, with a largely underfunded public system that has an extensive field staff still on its payroll. The system has numerous NGOs, church-based organizations, or producer organizations that are trying to fill the gaps, mostly with ad hoc and project-based funding. In terms of pluralism, results show that 55 percent of the surveyed extension organizations are government agencies (different locations of MINAGRI or MINRD) and 45 percent nongovernment (one-third of these are NGOs; private sector and producer associations represent about 6 percent each; church-based organizations account for about 3 percent) (Table 8.3). This indicates that coverage of NGOs is still limited, and government agents are still the dominant source of extension services in the rural areas in various territories and sectors in western DRC. Moreover, government agents are also sometimes employed as agents for NGOs to implement their projects. This indicates limited pluralism of extension services in the DRC.

Overall, linkages between extension organizations and between their agents and other stakeholders are rather limited. Absence of interactions with any rural bank and financial institution was reported by the extension agents, organization heads, and supervisors. Almost all of the sample extension agents have neither interacted in the previous year with central MINAGRI staff, financial institutions, other extension agents, nor input suppliers (Figure 8.4). Similarly, only 7–8 percent of supervisors of extension organizations have met with agro-processors or financial institutions only a few times a year. About 80 percent of surveyed extension organizations have never interacted in the previous year with input suppliers, traders or buyers, colleges and universities, and research institutes. The only set of actors with which a majority of sample extension agents met in the previous year was other extension agents (36 percent of sample extension organizations). Given that extension agents are the brokers of information and bridge the gap between users and sources of innovation, their lack of interaction or linkage with other organizations and stakeholders poses a serious issue in the fulfillment of their role in strengthening the capacity of farmers and achieving agricultural growth.

8.4 Organizational Capacity and Management

The majority of respondents (83 percent) reported having a planning process for their activities and operations. In terms of performance targets or indicators, only about 24 percent of extension agents are reporting any target set by supervisors or heads. About 20 percent of these targets relate to some kind of outcome (yield improvement), while the rest are associated with training and other supports. Performance indicators and targets remain at the level of input/output and only a few use outcome and impact indicators.

TABLE 8.3 Distribution of sample organizations and agents interviewed

Type of organization	Agriculture extension organizations (Number)	Agriculture extension organizations (%)	Agriculture extension (extension services) agents (Number)	Agriculture extension (extension services) agents (%)
Government (different locations)[1]	59	55	114	70
Nongovernmental organizations[2]	33	30	25	15
Private sector[2]	6	6	9	6
Church-based organizations[2]	3	3	6	4
Farmer-based organizations[2]	6	6	8	5
Total	107	100	162	100

Source: IFPRI survey (August–October 2011).

Note: 1. Different locations (territories or sectors) of MINAGRI or MINRD. 2. There is an unclear distinction between NGO, church-based organizations, private sector, and rural producer organizations. The classification retained here is what the respondents reported, although it is likely that respondents belong to the other types of organizations.

FIGURE 8.4 Distribution of extension agents based on frequency of interaction with other actors (%)

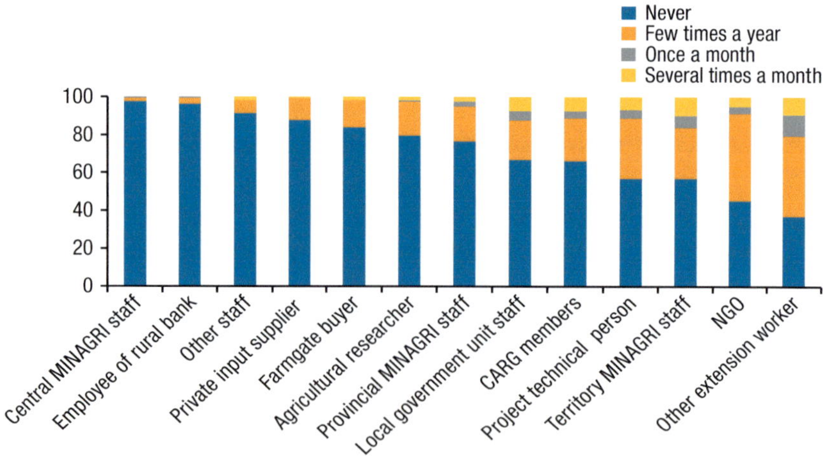

Source: IFPRI survey (August–October 2011).

Note: MINAGRI = Ministry of Agriculture; CARG = Agricultural and Rural Management Council.

Table 8.4 highlights a few key responses from extension agents concerning supervision and incentive systems. Two-thirds of extension agents claimed that their supervisors visit them in the field, and they reported that on average the frequency of supervisors' visits is four times during the past quarter. However, a third of extension agents were not visited by a supervisor. The enforcement of the incentive system (reward or sanction), based on

TABLE 8.4 Distribution of extension agents based on perception on supervision (n = 162 agents)

Statements	Percentage that agree or strongly agree (%)
Supervisor visits you in the field.	67.3
Supervisor is satisfied with your work.	95.1
Supervisor knows enough about your daily activities.	93.8
Supervisor is always around when needed.	87.0
Know someone in the office subject to sanctions for poor performance.	14.8
Have ever received an award for good performance.	10.5
Farmers never complain about the performance of the organization.	69.1

Source: IFPRI survey (August–October 2011).

performance, is also very weak: only 15 percent of extension agents know someone in his or her office who was subject to sanctions for poor performance or unprofessional conduct, and only 10 percent of extension agents have received or know someone who received a reward for good performance.

Human Capacity and Incentives

Field staff in the DRC totals 11,245, according to estimates, most of which is comprised of those referred to as agricultural monitors or frontline workers. These are mandated to work directly with rural communities. At the sector level they are supervised by sector agronomists and veterinarians who report to agricultural and veterinary inspectors at the territorial level, who in turn report to the provincial directors and inspectors.[5] In provinces where the decentralization policy is not fully implemented, there are also agricultural and veterinary inspectors at the district level.

Estimates suggest that there are about 6 million farming households, around 15 million farmers, and 39 million in the agricultural population in the DRC (FAO 2010). Thus the DRC has about 535 farmers per extension agent or 3,400 agricultural population per extension agent. These two ratios for the country are an improvement over those in India, Nigeria, and Tanzania (see Figure 6.2 in Chapter 6). Moreover, one of the two ratios is higher in the DRC than in China and Indonesia. Anderson and Feder (2004) show that the farmer-to-agent ratio in most developing countries is more than 1,000:1. This signifies that the number of extension agents or

5 A "sector" is an administrative level between a territory and its villages.

the farmer-to-agent ratio may not be the major issue; rather, it falls on how to manage the available human resources while providing them with a clear vision, incentives, and funding.

With respect to the profile of field staff, 78 percent has higher than a secondary or high school degree (Table 8.5). A large majority (57 percent) of field staff from government has vocational training or a three-year agricultural college degree. In contrast, most of field staff from NGOs have a university degree. Across all organization types, a substantive proportion of field staff (22 percent) has only a primary or secondary degree.

A full 61 percent of extension agents received other professional training besides formal education (Table 8.6). There are more extension agents (more than 80 percent) in NGOs and church-based organizations that have received training than in the government, private sector, or producer organizations. Across all respondents, a greater proportion of extension agents with lower formal education (primary or high school) did not receive on-the-job or skills development training compared to those with a higher education level. This reinforces the serious challenge of weak skill sets of a good proportion of extension agents due to a lack of formal education and technical training. For those who received on-the-job training, most of the training sessions were in areas of crop production, general education and communication, monitoring and evaluation, and general management.

An overwhelming majority of extension agents and managers/supervisors in extension organizations are trained and specialized in crop production. About 13 percent of extension agents specialize in livestock and veterinary services, and 3 percent specialize in fisheries and aquaculture. No extension agent reported specialization or training in postharvest or extension services marketing. Field staff should also be trained to have marketing skills so that they can provide extension services marketing among farmers. Furthermore, the skill set should include food preparation as well as nutrition and health advice to prepare mothers and adults on how best to utilize their existing food production and resources to maximize the nutrient intake of their children. Overall, extension agents should be trained in a diverse set of skills so that they can respond to the widely varying challenges and bottlenecks faced by farmers.

Numerous organizations have been providing ad hoc, short-term training to extension agents and rural development workers. Of all the training attended by the extension agents interviewed, more than half are offered by donors or international organizations; around 21 percent are given by government agencies such as MINAGRI, SNV, MINRD, and MINHEALTH,

TABLE 8.5 Distribution of agents by type of organization and level of education (%)

Organization type	Primary	Secondary	Vocational/ college	University
All respondents (n = 162)	**4**	**18**	**47**	**31**
Government (n = 114)	2	17	57	25
Nongovernmental organizations (n = 25)	8	16	20	56
Church (n = 6)	17	17	33	33
Private sector (n = 9)	11	33	11	44
Farmer-based organizations (n = 8)	0	25	50	25

Source: IFPRI survey (August–October 2011).

TABLE 8.6 Distribution of agents by training received and organization type (%)

Organization type and education level	Primary	Secondary and high school	Vocational and technical	University	Other professional training
All respondents	**33**	**59**	**52**	**80**	**61**
Government	0	53	51	79	57
Nongovernmental organizations	50	100	100	86	88
Church-based	100	100	50	100	83
Private sector	0	67	0	50	44
Farmer-based organizations	0	0	25	100	38

Source: IFPRI survey (August–October 2011).

and another 21 percent are from national NGOs and other local organizations. Only 5 percent of the training is from the education system, including the Rural Development Institute/Agro or Agro-veterinarian Studies Institute/College (ISDR/ISEA), and less than 1 percent is provided by the agricultural research stations (INERA). The issue with the training is that it has been provided on an ad hoc basis and in an uncoordinated manner, leaving the capacity of the agricultural training and education institutes largely weak and underfunded.

A major concern with skills development programs and the capacity of field staff is the serious weakness in the agricultural training and education system in the country. Within the Ministry of Education, Agro or Agro-veterinarian Studies Institute/College (ISEA/ISEAV) and the ISDR are the government institutions that offer agricultural training and education, which are responsible for training agricultural extension agents and rural development workers. ISEA/ISEAV trains agricultural technicians (agronomists and veterinarians) while ISDR trains rural development workers. Almost all territories have at least one ISEA or ISDR. In some cases, farmers' organizations also go to them directly to request training, oftentimes at a fee. Moreover, students taking their practicum to the villages is a useful way to disseminate technology packages and production techniques to farmers. In these cases, ISEA and ISDR are potentially useful pathways of technology dissemination and extension services delivery that can be explored more.

However, a closer look at some of these institutes (Ragasa et al. 2013) highlights the following challenges in agricultural education and training in the DRC: (1) lack of sustained funding; (2) no strategic planning that match the sector needs, and forward-looking mentality for ISDR/ISEA, in particular, and the education system in the DRC more generally; (3) outdated curricula; (4) problems of quality of education starting at the primary level; (5) lack of up-to-date training and skills development for staff; and (6) lack of linkages with the rest of the agricultural support system, including extension services, research, and universities.

The average age of extension agents in the sample is 52 years. About 60 percent of the extension agents is older than 50; 44 percent is older than 55. The extension agents in the government are relatively older, while those from NGOs and FBOs are relatively younger. About 51 percent of agents in government are older than 55, compared to only 16 percent in NGOs and 13 percent in FBOs. About 29 percent of agents from government are older than 60, while there are only 12 percent in NGOs and 13 percent in FBOs.

Only 5 percent of the random selection of extension agents is made up of women. Of all the extension services supervisors/heads interviewed, only 7 percent are women. There was no female supervisor or head from public-sector agencies. There is a limited dataset on female and male farmers' differentiated access to extension services, although evidence from the literature with regard to many countries suggests a strong correlation between the sex of the agent and the gender differential in access to extension services. Gender-based constraints, such as social norms that limit women's school attendance or mobility, also limit their opportunities and willingness to work as extension agents. It may be difficult at the practical level for a married woman to work in a rural area away from her husband and family or to find appropriate housing and schooling for her children.

In addition to capacity, incentives are critical for performance. In an attempt to measure financial incentive and compensation among extension agents, agents were asked about the various sources and types of compensation in exchange for their extension services work. About half of them receive a basic salary. They reported receiving 75,000 Congolese francs (CF) (roughly US$78) per month as a basic salary, on average, although it could range from 15,600 CF to 921,000 CF per month across agents (Table 8.7), way below the country per capita GDP of US$460.

NGOs can use stronger incentives than a government extension agent receives (see Table 8.8). The salary of public-sector extension agents is seriously low. To put this into context, the daily farm labor wage is from US$0.5 to US$3.5, which on average is higher than the salary of extension agents in the public sector. About half of the extension agents interviewed reported receiving some kind of bonus or commission in cash for their extension services work. Agents reported receiving 22,000 CF per month as a cash bonus per month, although this ranges from 800 CF to 78,000 CF per month across the agents. This indicates additional financial incentives for extension agents, although the total cash compensation reported is still not much compared to the DRC cost of living. Only 3–5 percent of extension agents interviewed reported any in-kind compensation in the form of a gift for their extension services work from farmers or other clients.

Financial Resources

Due to the irregularity of funds disbursed by the government, more than half (54 percent) of government-based extension organizations reported no funding at all since 2009 from government. Only 2 percent of government-based extension organizations obtained operating funds from government since 2009; and only 7 percent received capital funds since 2009. More than half (51 percent) of government-based extension organizations received no funds from any source since 2009. Only 23 percent of extension organizations (all types) received external funding (international NGOs, donors, private sector, or foundations) (Table 8.8). The greatest proportion of extension organizations that received external funding are NGOs. Among government-based ones, only 5 percent obtained external funding. None of the farmer-based organizations (FBO) received external funding. Still, a large proportion of NGOs, church-based organizations, private sector and FBOs do not get funding to do extension services. For government-based extension organizations, more than half do not get funds from government, and only 5 percent receive them from other sources. This still left about half of the government-based extension organizations without any type of funding for their operations and extension services work in 2009 or 2010. Respondents stated that this has been the case for several years.

TABLE 8.7 Descriptive statistics of financial compensation received for extension provision (Congolese francs)

Compensation type	Number	Mean	Standard deviation	Minimum	Maximum
Basic salary	86	75,497	105,527	15,600	921,000
Bonus	73	22,468	17,253	833	78,000
Commission	1	5,000	na	5,000	5,000
Total number of organizations	107	76,056	97,892	833	921,000

Source: IFPRI survey (August–October 2011). na = not applicable.

TABLE 8.8 Distribution of extension organizations by external funds received (%)

Organization type	Received external funds	Did not receive external funds
Total	23	77
Government (different locations)	5	95
Nongovernmental	61	39
Church based	33	67
Private sector	17	83
Farmer based	0	100

Source: IFPRI survey (August–October 2011).

Physical Resources

The dismal state of the DRC's transportation infrastructure also hinders the performance of extension services. Although 81 percent of extension agents believe that the workload is adequate, only 40 percent feel that travel time to farms is manageable, and 47 percent report that mobility to their operational area is difficult (Table 8.9). Supervisors estimate that, on average per month, 55 percent of extension agents' time is spent getting to the field. This discrepancy is likely to be not only due to poor road conditions but also to the fact that extension agents are not provided a travel allowance and bikes or vehicles as a result of insufficient funds. While a motorbike is a common mode of transport for extension agents in many African countries, survey results indicate that 91 percent do not have access to either a bike, motorbike, or vehicle for work.

Extension Approaches

The main approach is still heavily based on the training-and-visit system, remnants of the World Bank–supported program from the 1980s. The survey

TABLE 8.9 Distribution of extension agents by transportation challenges

Statement	Percentage agree or strongly agree (%)
Workload is adequate.	81
Farm distances from where you stay are manageable.	40
Mobility to your operational area is difficult.	47
Do not have access to a motorbike or vehicle for work.	91
Amount spent per month on mobility for job (CF)	21,600
Amount received in travel and transport per month (CF)	18,545
Estimation of extension agents' time spent for getting to the field in a month (%)	55

Source: IFPRI survey (August–October 2011).
Note: Total number of agents = 162.

of 162 field staff in western DRC reveals that the most common method of delivering extension services is field staff visits to farms or homes, followed by visits and information sharing at farmer-based organizations, training sessions, and demonstration farms. NGOs and farmer-based organizations make more frequent visits and meet with other FBOs, training sessions, and demonstration farms, while government, church-based organizations, and the private sector make more frequent visits to farmer's fields or homes than those from other organization types. The survey results indicate that 74 percent of field staff hold special meetings, such as planning meetings, where farmers can express which types of activities they expect to carry out. These meetings are usually held at the community or sector levels. Field staff activities also include the formation, mobilization, and support of FBOs.

Regarding the distribution of inputs, together with the provision of extension services, a quarter of surveyed field staff from government extension services in western DRC reported that inputs were distributed, while more than half of field staff from NGOs and all field staff from farmer-based organizations reported distributing inputs. This indicates that nongovernmental field staff often has a more integrated approach to service delivery and tends to focus more on the complementarity of inputs, technical knowledge, and extension services for farm management.

In eastern DRC, where numerous donor-funded projects have been implemented, the use of radio and innovation platforms (IPs) are common methods in disseminating information on agricultural management practices. IPs assemble stakeholders to share information, identify opportunities, discuss challenges, and agree on joint activities relating to a shared interest, often with a specific commodity focus.

Farmer-centered approaches have also been implemented, although at a small scale to date. The most common of these approaches is farmer field schools (FFS).[6] These approaches rely on the higher involvement of farmers and producer organizations to identify issues at the local level, define adaptive research and on-the-farm trial needs, implement trials, and disseminate appropriate technologies. In western DRC, half of field staff from government interviewed reported teaching or facilitating in farmer field schools; while 64 percent from NGOs and only 17 percent from farmer-based organizations use farmer field schools (or at least in theory). However, key informant interviews and visits to some farming communities suggest limited participation in FFS, and the commonly cited constraint is the lack of trained facilitators.

8.5 Performance of Extension Organizations, Producer Organizations, and Extension Agents

Ragasa, Ulimwengu et al. (2016) modeled the variation in performance among 107 extension organizations and 167 extension agents. The performance indicators collected were the number of farmers trained, number of villages visited, number of technologies disseminated, number of trainings conducted, number of demonstration plots organized, and number of training materials produced and disseminated. In the survey the research team used the French translation *technologie agricole*, and used examples such as improved varieties, fertilizer use, fertilizer timing and application methods, planting methods, disease and pest control techniques, and soil fertility management practices. Half of the sample extension organizations reported having disseminated at least one new technology or improved management practice, and 55 percent of extension organizations reported organizing farm demonstrations in the past two years. While 82 percent of extension agents reported having taught and disseminated at least one technology or improved management practice in the past two years, only 52 percent of agriculture extension agents reported having monitored farmers' adoption of these new technologies

6 A FFS is a group of 20 to 25 farmers with associated land (usually 0.5 to 1 hectare) who are trained in the field following the various stages of development of a given crop, based on field observations and analyses. The objectives are to strengthen the capacities of farmers themselves to identify/solve problems encountered during crop production, and reinforce their organizational capacities to further manage their own development. Main principles are as follows: FFSs are assisted through local organizations (farmers' organizations, women's groups, local NGOs) in which resources persons are identified and trained in the method; FFS facilitators (two farmers per FFS) receive specialized training each month in crop development and disseminate it in the field to other members; inputs given to conduct trials and FFS organizations are managed by an FFS committee; and FFSs should rapidly become financially self-sustained.

disseminated and only 19 percent of extension agents reported having monitored the impact of these technologies on farmers.

Ragasa, Ulimwengu et al. (2016) show some lessons that can be learned on improving the performance and effectiveness of extension organizations and agents. Significant factors explaining agent's performance include (1) external funding received; (2) linkages and interactions with others; (3) training received; and (4) presence of performance targets and enforcement of reward and sanction systems based on performance. Those explaining the performance of an organization are (1) external funding received; (2) linkages and interactions with others; (3) presence of performance targets and enforcement of reward and sanction systems based on performance; (4) proportion of female staff in the organization; and (5) proportion of time spent by field staff accessing farms and farmers. Ragasa and Golan (2014) analyzed the performance of 181 rural producer organizations (RPOs) in service delivery and found that management training has a positive effect on performance and increases the probability of the RPO providing advice and marketing to its members.

Results show that extension organizations with performance targets are more likely to perform well in terms of extension provision and monitoring technology adoption and impacts on farmers. Extension organizations with performance targets have 26 percent higher probability to conduct training and visits compared to those without targets. Extension agents in organizations with performance targets have 13 percent higher probability of being active in technology dissemination and rural education than those without performance targets. Results show strong evidence on the importance of linkages and interaction in organizational and individual agent's performance. Extension agents and organizations that interact and link more to other actors and organizations are 11–33 percent more likely to perform well in technology dissemination, training, farm demonstration and monitoring adoption, and impacts on farmers than those that did not have any interactions. Similarly, Ragasa and Golan (2014) found that the greater the interaction with external actors, the greater is the likelihood that services and information are provided to members through the RPOs. Interacting with external organizations increases directly the probability of providing inputs and advice by 17 percent, but there is no direct effect of external interaction on marketing of outputs. These findings are consistent with other studies (Bernard et al. 2008; Karami and Rezaei-Moghaddam 2005). This finding suggests that there is a need for greater recognition of the importance of linking with other actors that are potential sources of services, information, technical support, and market outlets. Policies and investments to help RPOs link more to each other and to others are critical. From the supply side, interventions to strengthen the capacity of service providers and external actors (government, NGO, church based, and private sector) will be needed to complement strategies supporting linkages.

Strong institutions and management systems are positively correlated with RPO performance. Especially in the context of weak capacity and institutions in fragile states, external support during setup appears crucial for performance, and it also increases an RPO's capacity to link to external actors. This is consistent with a general consensus in the literature that collective action may as well not emerge at all in the absence of external interventions (Varughese and Ostrom 2001). The widespread lack of capacity and high levels of illiteracy among the rural population in the DRC may make external assistance in setting up groups, raising community awareness, and mobilizing collective action central for the performance and viability of these RPOs.

There is no conclusive evidence that the number of staffs in the extension organizations is a statistically significant factor in explaining good performance in terms of extension service provision among extension agents and organizations. However, there is some evidence that training received by extension agents and gender balance in staffing in extension organizations matter in performance. Extension agents who received training are 6–18 percent more likely to be active and perform better than those who did not receive training. Extension organizations with female agents are more likely to be active and perform well in extension service provision compared to extension organizations without female agents. A 1 percent increase in the proportion of female agents in extension organizations is associated with a 1 percent increase in the probability of good performance in technology dissemination and farm demonstration among extension organizations.

The statistical analyses suggest that the availability of financial resources, especially in the form of external funding, is a strongly significant factor in explaining good or poor performance among extension organizations and agents. Extension organizations that received external funds are more likely to be active and perform well in extension service provision. Extension organizations that received external funds have 33–48 percent higher probability to conduct farm demonstrations and promote training materials in rural areas compared to the extension organizations that did not receive external funds. Extension agents that received external funds have 18–56 percent higher likelihood to be engaged in technology dissemination, rural education, monitoring farmers' technology adoption, and monitoring technology impacts on farmers. It will be therefore crucial to support knowledge provision and agriculture extension through external funds in the coming few years, but the government could start streamlining its human resources to free some resources from salary to operating and capital funds.

Based on these results, we conclude the following interventions can help improve the performance of extension agents and organizations:

1. First, training opportunities have been proven to be important in performance as confirmed by findings in several studies. Especially in the case of the DRC, where almost a quarter of agents only have primary education and the majority of government agents have no university degree, regular training is extremely important and needs to be institutionalized rather than being provided as an ad hoc activity. Also, agricultural extension system relies on the strength of the agricultural education system, and therefore any efforts to revitalize the former need complementary investments in the latter.

2. Second, linkages and partnerships have been shown to be important in extension performance. Interactions with other extension agents, NGOs, agrodealers, and agribusiness, and local political authorities are particularly important. This finding suggests that there is a need for greater recognition of the importance of linking with other actors that are potential sources of services, information, technical support, and market outlets. Policies and investments to help extension agents and organizations link more to each other are critical. For example, providing means of transportation, better roads, and reducing the time and transaction costs for extension agents to go to the field, visit farmers and interact with other actors has been shown to be important based on the analysis.

3. Third, clear direction and vision, coupled with measurable targets, are extremely important. With clear mandate and clear definition and communication of roles and responsibilities and with corresponding performance-based incentive mechanism, the weak extension system can be transformed into an efficient system that supports knowledge dissemination and technology transfer.

4. Fourth, with the absence of government funding, external funding has been crucial for the functioning of the extension system. External funders such as donor agencies and international NGOs also tend to put conditionality and more emphasis on monitoring adoption and

impact, and this has implications on the importance of their support for a transition economy such as the DRC. However, as has been proven in the past decades of neglect on extension in the DRC, the short-term support from external actors is extremely important but it needs to be supported and continued by the government for sustainability.

5. Fifth, the analysis suggests a strong correlation between having female agents in extension organizations and performance in service provision that confirms results of past studies. In places where female heads and agents are a few such as in the case of the DRC, focusing on girls' education and recruiting more women students and graduates to work on agriculture and on extension will be an important strategy.

8.6 Evidence of Effectiveness and Impact of Extension Services

Access to Extension Services

Since no national or subnational dataset is available, indications of access to extension services and extension performance presented here include (1) activeness in technology transfer and (2) visits made to the villages, as reported by the village representatives. Approximately half of the 107 extension organizations included in the survey reported that they had not disseminated or promoted any agricultural technology in the previous two years. An even greater proportion of government extension agencies had not promoted or disseminated technologies to farmers in the previous two years. This mirrors limited activity by a good proportion of extension organizations. Only 17 percent of the sampled villages reported having had visits from any extension agent or development worker (Table 8.10) in the previous two years (2009–2010). A large majority (83 percent) of villages reported not receiving any extension visits. Of those receiving visits, about 70 percent of the villages reported a single agent who visited and provided extension services, and 19 percent of the villages reported being visited by two extension agents, with the remaining 11 percent reporting having been visited by three to five extension agents. These agents come mostly from MINAGRI and NGOs. Of the 43 extension agents who visited the sample villages, only 9 percent were women.

This poor performance is not only for extension services but the agricultural sector overall. AU-DREA (2017) show that the Democratic Republic of the Congo is the lowest score among various countries in country

TABLE 8.10 Distribution of sample villages and extension visits by number of visits

Category	Number of villages	Percentage of total villages
No visit by extension agent	129	83
Visited by at least one extension agent	27	17
Number of agents who visited		
1	19	70
2	5	19
3	1	4
4	1	4
5	1	3

Source: Village-level survey implemented by IFPRI in 2011.

performance according to the Malabo Declaration. They recommend that the DRC should facilitate access of men and women engaged in agriculture to fertilizer, financial services, and agricultural extension services as one of the key interventions to enhance investment in the sector. They recommended that the DRC increase its spending in the agriculture sector toward these recommended interventions and to meet the CAADP target of 10 percent of the national budget. There are no national or subnational datasets available but there are smaller datasets and past studies that provide empirical evidence of effectiveness and impact of agricultural programs and extension approaches to help guide the investments in these areas.

Agricultural and Rural Management Councils

One of the major initiatives on agriculture and extension services is the Agricultural and Rural Management Councils (CARGs). The Belgian Technical Cooperation and various donors and NGOs have supported the establishment of more than 140 CARGs across the country since 2008. The CARGs are councils and forums of consultation that aim to improve planning and policy processes through the participation of stakeholders in the design, implementation, and monitoring of agricultural policies and programs, including demand articulation and coordination of agricultural extension services. CARGs operate at the local (territorial) level, which is equivalent to a district in many countries and is at a level of government just below a province (which reflects the level of decentralization in the DRC) (Figure 8.5).

In theory, CARGs are expected to (1) organize and facilitate dialogues for information sharing and problem solving, including such topics as harassment

FIGURE 8.5 Structure of the Agricultural and Rural Management Council

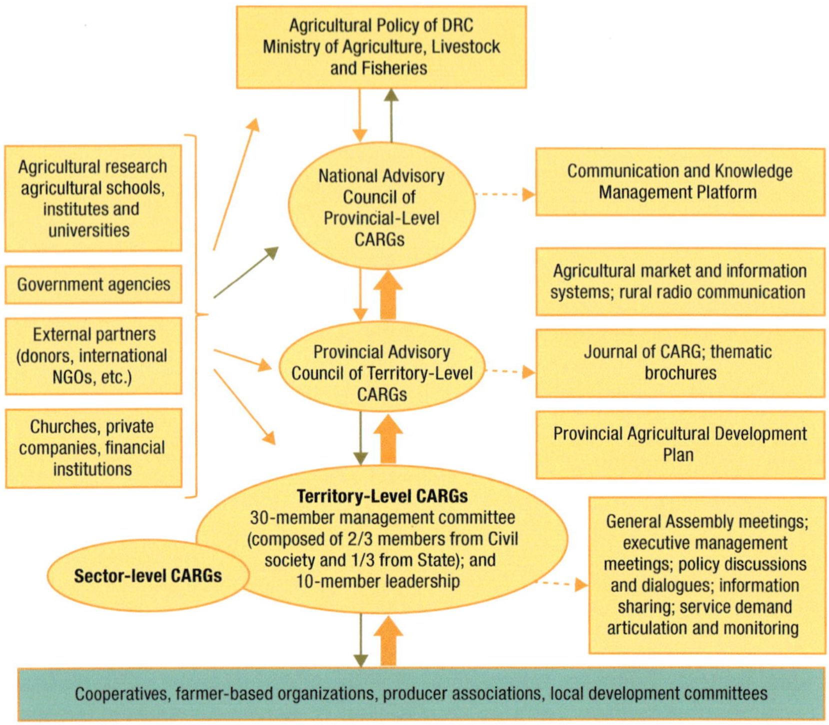

Source: Created by the authors, modified from Ragasa, Badibanga, and Ulimwengu (2016) and CARG (n.d.).

Note: CARG = Agricultural and Rural Management Council of the Democratic Republic of the Congo. Thin solid lines represent the flow of financial or technical support; thick solid lines represent the flow of feedback and information; broken lines represent outputs/products.

and land conflict; (2) document, aggregate, and communicate inputs from grassroots organizations for the design and implementation of provincial and national agricultural plans and programs and advocate actions to address their concerns and constraints; (3) coach, train, and strengthen farmers' associations; (4) act as brokers, bridges, or platforms for providing knowledge, advice, and innovative sharing among farmers; (5) inventory, coordinate, monitor, and evaluate agricultural development projects and activities; and (6) support farmer organizations in designing proposals and programs. Such CARG activities are thus more diverse and encompassing compared with other platforms and panels.

Ragasa, Badibanga, and Ulimwengu (2016) evaluate the effectiveness of 55 local-level Agricultural and Rural Management Councils (CARGs) in 23 randomly selected territories in the western DRC. All CARGs (territory, sector, city, or group) in the target territories were included in the sample. A full listing of the CARG leaders and members (stratified by government and nongovernment) were obtained and selected randomly to participate in the group interviews: three from leadership, one member from the government strata and two members from the nongovernment strata. Face-to-face group interviews lasted between two and three hours.

Ragasa, Badibanga, and Ulimwengu (2016) measure CARG performance using three indicators:

- The first measure of CARGs' performance is how consistent activities and outcomes are with reported objectives, which is a proxy for tangible output or outcome from the multistakeholder platform process. This is important because key informants and some members highlighted the fact that CARGs tend to pursue objectives other than their goals. The study results show that that only half of the surveyed CARGs achieved results consistent with at least one of their main goals, while the rest have not achieved any tangible outputs consistent with their objectives. Here, we see a mismatch between the goals and actual activities among many CARGs, which is a cause for concern as it signals a lack of focus.

- The second set of performance measures is related to the satisfaction and perception of relevant stakeholders (in this case, extension organization heads, agricultural workers, and members of rural producer organizations) about CARGs. Results show that although 79 percent of the heads of extension organizations were aware of the presence of CARGs, 42 percent of agricultural workers and 49 percent of rural producer organizations (RPOs) were not aware of CARGs. This shows limited outreach and sensitization about CARGs among key stakeholders. Although the majority of stakeholders interviewed were aware of CARGs, only 33 percent attended CARG meetings and perceived CARGs to be useful; and only 11 percent reported having benefited or knowing someone who had benefited from CARGs.

- The third set of performance indicators was the number of times CARGs interacted with service providers and other organizations and institutions; and we also used the CARG members' ratings of these meetings as a measure of the quality of the interactions. Since CARG is a conduit for finding solutions and services for rural communities, it is crucial to measure how they are performing in terms of their linkages and interaction

> with other institutions. External partners included MINAGRI, churches, NGOs, donors, farmers' organizations, other CARGs and research institutes and universities. All CARGs reported interaction with at least one of these external partners. CARGs met more often with RPOs, churches, and NGOs (average of five meetings per year); fewer with MINAGRI and donors (average of two meetings per year), and many fewer with other CARGs and research institutes and universities (average of one meeting per year). On average, 15 percent of CARGs found their meetings with external partners "very satisfactory," 70 percent "satisfactory," and 15 percent "not satisfactory." Meetings evaluated as "not satisfactory" were due to the lack of follow-up or limited activities or actions arising from the meetings.

Ragasa, Badibanga, and Ulimwengu (2016) show that CARGs' functioning and performance are diverse and have looked at factors that can help explain CARG performance level using probit regression analysis. The specific characteristics of CARGs that were strongly correlated with CARG performance are as follows: (1) effective coordination of CARG processes; (2) financial capacity; and (3) training received.

Coordination time (a proxy for the commitment of CARG leadership to CARGs) is correlated with CARGs performance. The leaders of the better performing CARGs devoted three times more to CARG activities than those in other CARGs. A substantial amount of time was allocated to the coordination of CARG activities by some CARG leaders, with 39 percent of CARGs in this group allocating between 100 and 286 hours to the coordination of CARG activities. Ways to incentivize CARG leaders to devote more time to management and activities could be part of the discussions on strategies to support CARGs. Borrowing from the literature on incentivizing public-sector service providers (see Musopole et al. 2013), opportunities for training and learning, reimbursement for CARG-related activities spent by CARG leaders, offering free bags of seeds of improved varieties, or vouchers for extension services are some of the incentives for CARG leadership that could be tested. Similarly, borrowing from the literature on IPs (Schut et al. 2016; Faysse 2006; Moellenkamp et al. 2010), investing in an effective broker or facilitator, either from the CARG leadership or an external expert, has worked in many cases to enhance commitment and management of the platform processes.

The financial capacity of CARGs is correlated with better performance. This is consistent with Sanyang et al. (2016), who found that incentives and access to resources are of high importance in the functioning of innovation platforms. A common element in all the successful platforms reviewed above is that they were funded and supported, which the regression analysis seems to support. The regression analysis provides an indication of strong correlation between training provided to CARG leadership and/or membership and CARG performance. All the platforms indicated training of its membership and executive committees. Areas for capacity strengthening that were implemented in the platforms are facilitation, conflict resolution, negotiation, organization, and management. Most of the platforms reviewed have supporting technical committees in addition to the executive committees that CARGs could also adopt. In addition to the membership and leadership committees, it may be worthwhile to facilitate the creation of innovation clusters (those with similar interests and working on the same value chains) to address specific issues in the priority value chains to encourage stronger market orientation and foster greater economic incentives for participation.

Overall, study results identify several problems and challenges in CARG implementation and the overall weaknesses in CARG formation. When it comes to their role in extension services, there is much controversy in terms of what they can do in extension services, which is partly explained by the lack of a clear definition of the CARGs' mission. While some stakeholders in the agricultural sector believe CARGs can handle any issue relating to agricultural strategies and policies, others do not see how a multistakeholder platform such as the CARG can deal efficiently with an issue that requires resource commitment and regular management operations, such as the provision of agricultural extension services. CARG can play a role as facilitator of access to agricultural extension services and as a platform for demand articulation and aggregation. It appears that CARGs are showing potential as an effective demand-side strategy and a bridging institution for demand and supply. However, they cannot play the role of service provider and therefore cannot function as a supply-side strategy.

To operate effectively as a service provider, an extension organization needs (1) to design and implement a coherent agricultural extension program with clearly defined targets; (2) to use different means and methods for providing agricultural extension services; (3) to use well-trained and experienced extension personnel; and (4) to have substantial financial and material resources to complete its goals. The CARGs do not meet these conditions. In fact, a CARG is a multistakeholder platform for consultation and thus differs from any permanent structure or organization that can design a program and commit the resources needed to implement it. Supply-side strategies, including capacity building and training of pluralistic extension service providers, will be crucial to complement investments and support to demand-side strategies such as CARGs.

Integrated Agricultural Research for Development and Innovation Platforms

The Lake Kivu region, capturing parts of eastern DRC, Rwanda, and Uganda, is one of three project sites of the Sub-Saharan Africa Challenge Program (SSA-CP), coordinated by the Forum for Agricultural Research in Africa (FARA). SSA-CP aims to develop technologies for sustainably intensifying subsistence-oriented farming systems; develop smallholder production systems that are compatible with sound natural resource management; improve the accessibility and efficiency of markets for smallholder and pastoral products; and catalyze the formulation and adoption of policies that would encourage innovation to improve the livelihoods of smallholders and pastoralists. It started in 2004 with the aim of introducing a new approach to promoting innovation and diffusion of innovations in African agriculture.

Integrated Agricultural Research for Development (IAR4D) is based on the innovation systems perspective and creates coalitions of stakeholders to identify and address local bottlenecks to agricultural development. Through this approach the program aims to promote agricultural innovations by using farmers' indigenous knowledge through a participatory framework and interactions between different stakeholders. The innovation systems perspective focuses on recognition of a wider, differentiated set of innovation suppliers; demands responsiveness and better connectivity of agricultural research with a wider range of innovation actors beyond extension agents and farmers; and expands the definition of the innovations being developed to include both economic and social applications (World Bank 2007, 2012).

A central concept of this approach is innovation platforms (IPs), which are decentralized local innovation systems. IPs are vehicles to bring stakeholders together. Each IP serves a group of villages and theoretically chooses representatives from different stakeholders via a participatory process. These representatives of farmers' associations, traders, researchers, extension workers, NGOs, and government policymakers regularly meet at the platforms, articulate their views and negotiate joint strategies for action. To provide proof of concept, the IAR4D program and its IPs were rolled out as a large experiment whereby some communities received IPs and others did not. In addition, baseline and midline surveys were conducted. These processes ensure rigorous monitoring and evaluation of the concept.

Available studies show mixed outcomes of these IPs (Pamuk, Bulte, and Adekunle 2012, 2014). Even within eastern DRC, results are mixed. The IPs in Kituva resulted in higher food security and reduced poverty, while those in

Rubare led to lower food security and a higher poverty count. There were no significant impacts of IPs on poverty in Bweremana and Rumangabo (Pamuk, Bulte, and Adekunle 2012). The mixed outcomes of IPs in the context of the IAR4D program and SSA-CP are consistent with results obtained by other authors in other contexts (Kilelu, Klerkx, and Leeuwix 2013; van Paassen et al. 2014). There is limited information on why some IPs were more successful, except that Pamuk, Bulte, and Adekunle (2014) indicated that the level of initial social capital in a village was positively associated with the success of an IP.

Also, the IPs that were implemented for two years or longer had a higher percentage reduction in poverty than those implemented for only one year (Pamuk, Bulte, and Adekunle 2014). The more established IPs on average experienced a 20 percent reduction in poverty compared to an increase of 5 percent for immature IPs. This is consistent with our findings on the performance of CARGs: regression results show that the age of a CARG helps explain its performance. Combining both studies, there seems to be negative startup effects that may capture the investment component of building a functional platform and its short-term (opportunity) costs, and that medium-term benefits seem to materialize after several years of establishment.

Other lessons can also be derived from other IPs in other contexts. For example, Plataformas in Ecuador, which center on alliances among small-scale farmers and a range of service providers to link smallholders to high-value agricultural markets, have been quite successful (Cavatassi et al. 2011). Overall, there has been more success in IPs when the private sector is involved. When economic incentive and market orientation are strong, participation and contribution by the private sector, as well as producers and other actors, become stronger. CARGs can help set up and facilitate innovation clusters and IPs focusing on priority value chains. Sanyang et al. (2016) shows that several or different IPs and innovation clusters can be formed within a value chain, such as on technology, marketing, and policy. This argument is also supported by the multiple barriers in a value chain (technological, institutional, lack of resources, administrative, organizational, opposition, among others) and therefore requires multiple innovation champions described by Klerkx et al. (2013). As in the case of oil palm and cocoa in Ghana and rice in Benin, several innovation champions worked toward actions within each of these value chains: technology champion (for technological barriers); power champion (for institutional barriers, ignorance, oppositions, or lack of resources); process champion (for institutional barriers, administrative, bureaucratic); and network champion (for organizational, cooperation problems).

Extension Services in the Food Production, Processing, and Marketing Project

In 2011, in collaboration with the United States Agency for International Development (USAID), the DRC government launched the Food Production, Processing, and Marketing (FPPM) project—which aimed to raise incomes and improve food security in the target areas by improving agricultural productivity, market efficiency, and the capacity of producers to respond to market signals. Provision of agricultural and market extension services are key interventions under the project. In August–October 2013 and February–March 2014, halfway through the project's implementation, a midline survey of 3,110 consisting of a random selection of project beneficiaries in project sites and a random selection of households in comparable nonproject areas was conducted to assess progress with respect to intermediate outcomes (see Ragasa, Nkonya et al. 2016 for more details on the sampling). The study pays close attention to accurate attribution of observed changes to the project and employs a double-difference method that compares the changes in indicators before the project and at the time of the survey (project midline) between the beneficiaries and comparable control groups.

We highlight the main impacts of the project on extension services. We see some positive impact on participation in farmer field schools in Bandundu province only but not in the other two provinces: 59 percent of farmers in FPPM villages had access to FFS in 2013, up from 49 percent of farmers in 2010, compared with 33 percent of farmers in control villages (Bandundu only). We also see some positive impact of the FPPM project on the access to agricultural extension services in the three provinces: a total of 35 percent to 50 percent of FPPM beneficiaries, while only 15 percent to 30 percent of the control group reported access to extension services in the three provinces in 2013. Positive change from 2010 to 2013 is observed among FPPM beneficiaries, while a decrease in extension services access is reported among the control group.

These changes are statistically different between FPPM beneficiaries and the control group in the three provinces. However, we see less impact on market extension: the FPPM intervention did facilitate increased access to market information in Bandundu but had no significant effect in Kinshasa and Bas-Congo. While we see some changes in access to extension services, we do not see positive impacts on access and use of productivity-enhancing inputs and improved technologies being promoted by the project. We see minimal improvement in productivity in Bandundu, but no significant impact on the change in productivity is observed in the other two provinces.

Overall, the survey results suggest weak impact on most of the outcome indicators, and they highlight challenges in implementing small-scale farmers' capacity building within the context of weak institutions and a fragile political context. Given the long-term nature of the FPPM objectives of increasing crop productivity, market efficiency, and farmers' capacity to respond to market signals, we did not expect that significant impacts of the FPPM intervention would be evident in the early stages of the project's implementation. However, weak impacts at midline show challenges in the implementation and some areas that need improvements and focus.

One implication is the need for complementary investments on the supply, accessibility, and affordability of inputs and technologies for the training and extension services to show impacts. For instance, a major focus of the project is on seed multiplication and distribution as reflected in the bulk of activities during the first two to three years of its implementation. However, there seems to be weak impact to date, with only a modest increase in adoption of improved varieties of maize in Bandundu and beans in Bas-Congo and no impact in the other focus province or on the other focus crops. The project has run into problems related to the foundation seed purchase and certified seed production and multiplication components in the earlier years of implementation. To fix this, efforts should focus on better selection and enforcement of contracts among seed multipliers and stricter monitoring of services and seed quality. Traders, extension agents, and farmers interviewed expressed the need for more certified seeds to be distributed.

A second implication is the need to approach the issues and investments within a value-chain perspective and focusing investments on developing different aspects of the value chain. To improve market efficiency, our findings suggest the necessity of intensive investment in developing collective marketing by linking producers and traders. Building partnerships with transporters, exploring different transport schemes, and rehabilitating aggregation centers to reduce transport costs and better support the marketing activities of local producers, traders, and consumers will also improve market efficiency substantially. There is a need to intensify activities geared toward setting up the village savings and loan cooperatives and linking farmers with formal credit institutions. More training activities on business management skills development are also needed.

Integrated Multipronged Communication Approaches

Eastern DRC is one of the project areas for the Consortium for Improving Agriculture-Based Livelihoods in Central Africa (CIALCA), a research-for-development consortium led by the Tropical Soil Biology and Fertility Research Area of the International Center for Tropical Agriculture (TSBF-CIAT), the International Institute of Tropical Agriculture (IITA) and Bioversity International. It aims to improve the livelihoods of rural households in Central Africa through the identification, evaluation, and promotion of technological options to enhance the productivity of banana-, maize-, cassava- and legume-based systems and to create an enabling environment for their adoption. CIALCA promoted complete integrated soil fertility management (ISFM) comprising the use of improved germplasm, mineral fertilizer, appropriate organic resource management, and local adaptation.

Vanlauwe et al. (2012) describe CIALCA as an integrated, multipronged communication approach using a mix of tools to disseminate and promote ISFM at a large scale. The first steps toward ISFM were fertilizer and improved varieties. An essential condition for their adoption is access to farm inputs, markets, and financial resources. CIALCA has worked on dissemination strategies, including ways to facilitate access to the required inputs, simple information fliers spread through extension networks, and knowledge on how to avoid less responsive soils. CIALCA implemented campaigns that addressed farmers' constraints by offering them information, technology demonstrations, product exhibits, financial incentives, and opportunities to develop their skills within their own farms.

As efforts to promote the seed and fertilizer strategy were under way, activities such as FFSs were initiated to guide farming communities toward complete ISFM. The CIALCA Knowledge Resource Centre was established in the African Great Lakes region to identify and leverage new impact pathways for ISFM technologies. By working closely with extension agents and outreach partners, targeted information tools can be developed to support adoption of practices by farmers in specific settings.

According to Vanlauwe et al. (2012), a particular challenge was to develop innovative knowledge products that take into account the low rates of adult literacy and formal education prevalent in the region. Rural radio, one tool that was used, offered a wide reach and was very useful for raising awareness around a particular issue. However, it is less suitable as a training tool, particularly as knowledge complexity increases. With increasing complexity of knowledge, CIALCA focuses on rigorous and in-depth farmer facilitation and training. CIALCA has intervened in markets by working with farmers' organizations to achieve a marketable production scale. Capacity building on collaborative action, marketing and business planning skills, and management of credit and finances have ensured that farmers are now able to bulk their produce, wait for better prices, and earn higher incomes from their produce. In addition to farmers, training also targets the institutions and organizations that support farmers' organizations, such as NGOs and national research staff, to ensure postproject sustainability.

Based on Vanlauwe et al. (2012), farmers in South Kivu were able to raise their sales revenues by 50 percent through strategic storage facilitated by inventory credit schemes (*warrantage*): farmers did not have to sell immediately after harvest but were able to store their produce collectively, awaiting better prices for their products. Through group efforts, farmers were able to acquire credit for their ISFM-based farming activities and, because they had targeted production to key markets, were able for the first time to borrow funds without collateral. In addition, farmers working in groups have been able to initiate mutual savings schemes that supplement other sources of finance, particularly for investment in new technologies.

8.7 Conclusion

As pointed out earlier in the chapter, 10 percent of the country 2013–2020 NAIP under CAADP is allocated to agricultural research and extension. However, per the recent review by the DREA (AU-DREA 2017), the country is not on track to achieving the CAADP/Malabo commitments. This means that the ambitious strategy adopted by the government of the DRC, as part of the NAIP, is still very much relevant. This serves as basis for more concerted efforts to rebuild the agriculture sector and provide the needed extension and other services to the rural population. Rebuilding the agriculture sector is crucial for food security, restoring livelihoods, and economic recovery of the Democratic Republic of the Congo, as well as revitalizing the extension system, which will be extremely important for rebuilding the agriculture sector. Strategies to rebuild the agricultural extension service in the DRC must look at both demand-side and supply-side constraints and should follow a defined sequence.

This chapter highlights four lessons. First, it highlights that public-sector extension remains crucial even in areas with very weak government institutions and fragile states like the DRC. As widely observed in the DRC and other fragile countries, many donor and NGO-led projects tend to bypass and not involve government institutions in their project design, implementation, and capacity-strengthening activities. An important strategy therefore is to include public-sector extension agents and subject matter specialists in capacity-strengthening and learning programs instead of focusing only on

NGOs. It is crucial to work with the government counterparts (agriculture ministry, extension system, research institute, and seed inspection systems) and not bypass them in agricultural and rural projects. Even when there is some distrust and perceived inefficiency in the government institutions, they continue to have the cadre of human resources scattered throughout the country and have long-term structures, as compared to the limited coverage and more ad hoc nature of NGO activities, therefore they should be a crucial part of any capacity-strengthening and extension activities.

Second, in the case of a pluralistic system, focusing on coordination, quality control, and regulation to avoid conflicting extension messages and duplication of efforts will be a priority role for the government and this would be equally or even more important than providing extension services themselves. Although there remains pressure to continue service provision by the public sector, there has to be major decisions on how to invest limited resources more effectively. The public sector could focus more on coordination and regulation of extension services and freeing more resources from the actual service provision, which could be done mainly by, or contracted out to, the private sector and nongovernmental organizations.

Third, given the greater knowledge brokering and facilitation role of extension, linkages and partnerships become extremely crucial for performance. As shown in this chapter, there are many constraints to linkages, and therefore part of the strategy is to find ways to help extension agents and organizations link more to each other and to other relevant actors. This chapter has shown that providing means of transportation and reducing the time and transaction costs for extension agents to go to the field and visit farmers and to interact with other actors are some of the ways to encourage greater interactions and linkages.

Lastly, this chapter complements findings by Davis et al. (2010) and Ragasa et al. (2013) on Ethiopia that number or quantity of agents is not a sufficient measure of performance, but an effective system needs to focus on building quality of agents and enabling environment for them to be motivated to work as mandated. Enabling conditions that are found to be significant are external funding, enforcement of performance targets, systems of rewards and sanctions, and skills development and training. In measuring performance, it is important not to focus only on the number of staff and getting a good ratio of extension agents to farmers, but more important, it is key to look at systems in the organizations to ensure that performance targets are set and being followed, that there are right incentives based on the performance, and funding, skills sets, and mobility are present for the agents to perform their tasks.

In terms of priority action areas, this chapter proposes the following: First, the enabling environment for extension services is a priority. In particular, improving input and output market efficiency and increasing access to credit, inputs, markets, land, and equipment/tools are the most common and most consistently mentioned constraints among farmers, based on the perspectives of extension agents, extension organization heads, and farmers themselves. Increasing the productivity and incomes of the rural community does not require changes in the extension system alone; rather, a holistic approach is essential to address constraints in the input distribution system and technology adoption. The results warrant complementary review and reform in the policies and investments governing these inputs and services.

Second, within the extension system, governance issues need to be addressed, including clarity in policy and strategy, coordination, government commitment, and funding. Lack of clear direction and vision, coupled with measurable targets, are the weakest points of the DRC extension system. In the most immediate term, the priority is to have a unified policy and clearer strategy for extension services in the country. This requires designing and communicating a clear mission and mandate to more than 11,000 and more extension agents scattered throughout the country. With clear mandate, functions, and performance targets of extension services staff, it may trigger the change in perception and trust among farmers that extension agents are no longer "monitors" but knowledge brokers and technical advisers. The national workshop on agricultural extension and productivity held in June 2012 offered a good start, and the government can continue to bring the workshop's recommendations into its agricultural strategy planning and policy formulation.

Moreover, institutional coordination, quality control, and regulation of extension services should be strengthened. Greater coordination, quality control, and regulation are critical to communicate consistent extension messages across many different extension service providers. This involves encouraging development organizations to work with local MINAGRI offices, INERA, SENASEM, and SNV, to avoid inconsistencies and duplication of efforts, as well as to ensure capacity building and continuity of activities. The capacity of SNV to play its role of coordination and technical backstopping should be strengthened. There is a need to mobilize substantive funds and commitment from the government and its partners. As discussed above, the government often defaults on its commitments in the implementation of key policies and programs. No amount and degree of institutional reforms or approaches will work without sustained funding, especially from the government. The

National Agricultural Investment Plan, developed in 2013, details a series of reforms on the extension system for the coming years. This initiative is encouraging and deserves bold and firm commitment from the government and its partners.

Third, management systems, especially incentive systems based on a credible performance M&E system, need to be strengthened at the organizational level. The enforcement of an incentive system (reward or sanction) based on performance is currently very weak with only a few extension organizations having systems of reward for good performance and sanctions for nonperformance. It does not really matter whether extension organizations or agents do good or bad. Moreover, performance indicators and targets are often not set and enforced in most organizations; and for those few that have these targets, they stop at the level of input or output and almost none monitors on the level of outcomes and impacts (yield, income, or nutrition improvements).

Fourth, human as well as physical capacity need to be strengthened. Priority lies on the streamlining of the number of extension services staff by retiring the older agents to hire and retrain younger and more dynamic ones. This could remake the salary structure and free valuable resources that can be shifted to much needed operating and capital components. The plan has already been started and the government can facilitate its implementation for quicker impacts.

Fifth, there is need to address the demand side to strengthen the role of farmers' organizations at the village level, and the role of CARGs at the sector and territory levels, in demand articulation and the capacity for planning and monitoring service provision. The role of CARGs in the policymaking process should be strengthened so that they can advocate effectively for greater investment and commitment to agricultural extension.

Last, in the medium term, complementary investment in reforming the agricultural education and training institutes will be a priority as they are the critical institutions that train and nurture the new wave of agricultural extension agents and officers. This includes a review and reform of extension curriculum. And, given the serious food and nutrition insecurity in the Democratic Republic of the Congo, inclusion of nutrition messages in agricultural extension services is an important strategy. The scarcity of female extension workers requires investing in and supporting girls' education and mobilizing more female agents. Given that extension services are just one of the factors determining productivity growth, complementary investments and policy reforms to facilitate access to markets, and affordability of inputs to complement the knowledge and extension provision, are also important strategies.

References

Anderson, J. R., and G. Feder. 2004. "Agricultural Extension: Good Intentions and Hard Realities." *World Bank Research Observer* 19 (1): 41–60.

AU-DREA (African Union Commission, Department of Rural Economy and Agriculture). 2017. *The 2017 Progress Report to the Assembly—Highlights on Intra-African Trade for Agriculture Commodities and Services: Risks and Opportunities.* Addis Ababa: African Union Commission.

Bernard, T., M.-H. Collion, A. de Janvry, and P. Rondot. 2008. "Do Village Organizations Make a Difference in African Rural Development? A Study for Senegal and Burkina Faso." *World Development* 36 (11): 2188–2204.

Birner, R., K. Davis, J. Pender, E. Nkonya, P. Anandajayasekeram, J. Ekboir, A. Mbabu, D. Spielman, D. Horna, and S. Benin. 2009. "From Best Practice to Best Fit: A Framework for Analyzing Agricultural Advisory Services Worldwide." *Journal of Agricultural Extension and Education* 15 (4): 341–355. DOI: 10.1080/13892240903309595.

CARG (Agricultural and Rural Management Council). n.d. CARG brochure (in English). Kinshasa.

Cavatassi, R., M. Gonzalez-Flores, P. Winters, J. Andrade-Piedra, P. Espinosa, and G. Thiela. 2011. "Linking Smallholders to the New Agricultural Economy: The Case of the Plataformas de Concertacion in Ecuador." *Journal of Development Studies* 47 (10): 1545–1573.

Davis, K., B. Swanson, D. Amudavi, D. A. Mekonnen, A. Flohrs, J. Riese, C. Lamb, and E. Zerfu. 2010. *In-Depth Assessment of the Public Agricultural Extension System of Ethiopia and Recommendations for Improvement.* International Food Policy Research Institute (IFPRI) Discussion Paper 1041. Washington, DC: IFPRI.

DRC, MINAGRI (Ministry of Agriculture). 2005. *Plan de restructuration du Ministère de l'Agriculture, Pêche et Elevage.* Democratic Republic of the Congo.

———. 2013. *Plan National d'Investissement Agricole (PNIA): 2013–2020.* Democratic Republic of the Congo.

FAO (Food and Agriculture Organization of the United Nations). 2010. FAOSTAT database. www.fao.org/faostat.

Faysse, N. 2006. "Troubles on the Way: An Analysis of the Challenges Faced by Multi-Stakeholder Platforms." In *Natural Resources Forum,* 219–229. Wiley Online Library.

IFPRI. 2016. *Global Hunger Index.* Washington, DC: IFPRI.

Karami, E., and K. Rezaei-Moghaddam. 2005. "Modeling Determinants of Agricultural Production Cooperatives' Performance in Iran." *Agricultural Economics* 33 (3): 305–314.

Kilelu, C. W., L. Klerkx, and C. Leeuwis. 2013. "Unravelling the Role of Innovation Platforms in Supporting Co-Evolution of Innovation: Contributions and Tensions in a Smallholder Dairy Development Programme." *Agricultural Systems* 118: 65–77.

Klerkx, L., S. Adjei-Nsiah, R. Adu-Acheampong, A. Saidou, E. Zannou, L. Soumano, O. Sakyi-Dawson, A. van Paassen, and S. Nederlog. 2013. "Looking at Agricultural Innovation Platforms through an Innovation Champion Lens: An Analysis of Three Cases in West Africa." *Outlook on Agriculture* 42 (3): 185–192.

Maystadt, J.-F., G. De Luca, P. G. Sekeris, and J. Ulimwengu. 2014. "Mineral Resources and Conflicts in DRC: A Case of Ecological Fallacy?" *Oxford Economic Papers* 66 (3): 721–749.

Moellenkamp, S., M. Lamers, C. Huesmann, S. Rotter, C. Pahl-Wostl, K. Speil, and W. Pohl. 2010. "Informal Participatory Platforms for Adaptive Management: Insights into Niche-Finding, Collaborative Design and Outcomes from a Participatory Process in the Rhine Basin." *Ecology and Society* 15 (4): 41. www.ecologyandsociety.org/vol15/iss4/art41/.

Musopole, R., F. Nkana, A. BenYishay, M. Mobarak, M. Jones, and F. Kondylis. 2013. *Communicating with Farmers through Social Networks.* Innovations for Poverty Action (IPA) Policy Brief. Lilongwe, Malawi.

Pamuk, H., E. Bulte, and A. Adekunle. 2012. "Decentralized Innovation Systems and Poverty Reduction: Experimental Evidence from Central Africa." Wageningen University, Development Economics Group.

———. 2014. "Do Decentralized Innovation Systems Promote Agricultural Technology Adoption? Experimental Evidence from Africa." *Food Policy*: 227–236.

Pfingu, C. 2011. "Determinants of Child Malnutrition in the Democratic Republic of Congo." PhD Thesis, School of Economics, University of Missouri, Columbia, MO, US.

Ragasa, C., S. C. Babu, and J. Ulimwengu. 2014. "Institutional Reforms and Agricultural Policy Process: Lessons from Democratic Republic of Congo." *Agricultural and Food Economics* 2: 4.

Ragasa, C., T. Badibanga, and J. Ulimwengu. 2016. "Effectiveness and Challenges of Participatory Governance: The Case of Agricultural and Rural Management Councils in the Western Democratic Republic of the Congo." *Food Security* 8: 827–854.

Ragasa C., and J. Golan. 2014. "The Role of Rural Producer Organizations in Agricultural Service Provision in Fragile States." *Agricultural Economics* 45: 1–17.

Ragasa, C., E. M. Nkonya, J. Ulimwengu, and J. Randriamamonjy. 2016. *Challenges in Implementing a Small-Scale Farmers' Capacity-Building Program: The Case of the Food Production, Processing, and Marketing Project in the Democratic Republic of Congo.* IFPRI Discussion Paper 1501. Washington, DC: IFPRI.

Ragasa, C., J. Ulimwengu, J. Randriamamonjy, and T. Badibanga. 2013. *Assessment of the Capacity, Incentives, and Performance of Agricultural Extension Agents in Western Democratic Republic of Congo.* IFPRI Discussion Paper 1283. Washington, DC: IFPRI.

———. 2016. "Factors Affecting Performance of Agricultural Extension: Evidence from Democratic Republic of Congo." *Journal of Agricultural Extension and Education* 22 (2): 113–143.

Sanyang, S., S. J.-B. Taonda, J. Kuiseu, N. T. Coulibaly, and L. Konaté. 2016. "A Paradigm Shift in African Agricultural Research for Development: The Role of Innovation Platforms." *International Journal of Agricultural Sustainability* 14 (2): 187–213. http://dx.doi.org/10.1080/14735903.2015.1070065.

Schut, M., L. Klerkx, M. Sartas, D. Lamers, M. Mc Campbell, I. Ogbonna, P. Kaushik, K. Atta-Krah, and C. Leeuwis. 2016. "Innovation Platforms: Experiences with Their Institutional Embedding in Agricultural Research for Development." *Experimental Agriculture* 52 (4): 537–561. doi:10.1017/S001447971500023X.

Tecsult International. 2009. *Étude du Secteur Agricole Rapport Bilan Diagnostic et Note d'Orientation.* Ministère de l'Agriculture et du Développement rural. Rapport 0517395. Democratic Republic of the Congo.

Tollens, E. 2004. *Current Situation of Food Security in the D.R. Congo: Diagnostic and Perspectives.* Working Paper 2003/80. Leuven, Belgium: Katholieke Universiteit Leuven.

Ulimwengu, J., C. Robets, and J. Randriamamonjy. 2012. *Resource-Rich Yet Malnourished: Analysis of the Demand for Food Nutrients in the Democratic Republic of Congo.* IFPRI Discussion Paper 1154. Washington, DC: IFPRI.

UNDP (United Nations Development Programme). 2010. *Human Development Report.* Accessed July 15, 2011. http://hdrstats.undp.org.

Van Paassen, A., L. Klerkx, R. Adu-Acheampong, and S. Adjei-Nsiah. 2014. "Agricultural Innovation Platforms in West Africa: How Does Strategic Institutional Entrepreneurship Unfold in Different Value Chain Contexts?" *Outlook on Agriculture* 43 (3): 193–200.

Vanlauwe, B., P. Pypers, E. Birachi, M. Nyagaya, B. van Schagen, J. Huising, E. Ouma, G. Blomme, and P. van Asten. 2012. "Integrated Soil Fertility Management in Central Africa: Experiences of the Consortium for Improving Agriculture-Based Livelihoods in Central Africa (CIALCA)." In *Eco-Efficiency: From Vision to Reality*, edited by C. H. Hershey, 1–17. Cali, Colombia: Centro Internacional de Agricultura Tropical (CIAT).

Varughese, G., and E. Ostrom. 2001. "The Contested Role of Heterogeneity in Collective Action: Some Evidence from Community Forestry in Nepal." *World Development* 29 (5): 747–765.

World Bank. 2006. *Democratic Republic of the Congo: Agricultural Sector Review*. Main Report, April 15. Report No. 30215-ZR, Washington, DC: World Bank.

———. 2007. *Enhancing Agricultural Innovation: How to Go Beyond the Strengthening of Research Systems*. Washington, DC: World Bank.

———. 2012. *Agricultural Innovations Systems: An Investment Sourcebook*. Washington, DC: World Bank.

———. 2018. *Democratic Republic of Congo Systematic Country Diagnostic: Policy Priorities for Poverty Reduction and Shared Prosperity in a Post-Conflict Country and Fragile State*. Africa Region. Report No. 112733-ZR. Washington, DC: World Bank.

World Bank and IFPRI. 2009. *Gender and Governance in Rural Services*. Washington, DC: World Bank.

Chapter 9

CONCLUSIONS AND POLICY IMPLICATIONS

Kristin Davis, Catherine Ragasa, and Suresh C. Babu

The two objectives of this book are to assess extension and advisory services in a cross-country comparative context in the following ways:

1. Provide a description and comparison of the existing extension services in the regions and countries under consideration focused on a set of characteristics; and
2. Provide an assessment of the performance of extension services provision and impact evaluation of extension approaches in selected case study countries.

Objective 1 addresses specific extension characteristics from the best-fit framework that include governance structures, capacity, management, and advisory methods. Part 1 offers a global overview based on primary data as well as regional and country cases based on primary and secondary data. Objective 2 undertakes in-depth country-level case studies based on primary and secondary data (Part 2). The best fit conceptual framework allows some comparability of selected characteristics and results across geographies. The book generates specific policy recommendations that will be of use to government agencies, development partners, academics, and coordinating bodies for agricultural extension services. With the best-fit framework that allows for comparative analysis of selected characteristics (governance structures, capacity, management, and advisory methods), the book also demonstrates the process of extension system assessment through country-level case studies.

This final chapter presents conclusions organized according to the best-fit framework characteristics and provides policy recommendations for improving extension globally. Finally, it sets out some future research areas. To summarize the best-fit approach by which we organize the conclusions, the framework looks at characteristics of extension and advisory services.

The extension characteristics used for analysis include the governance structures, capacities, management, and advisory methods. "Governance structures" refer to the institutional setup of extension services and strongly depend on the existing policy environment. Governance structures depend on the role of the public sector in extension services, level of privatization, degree of decentralization, funding mechanisms, and capacity of farmers to influence extension services. "Capacities" refer to the human and organizational competencies as well as financial and physical infrastructure and assets. "Management" is the way in which extension services are managed within the respective governance structures. Organizational capacity and management have been grouped together in several chapters since they are closely related. "Advisory methods" are approaches used by extension services field staff in their interactions with farmers. Methods can be classified according to various aspects, such as the number of clienteles involved (individuals, groups); the types of decisions on which advice is provided (specific to the production of certain crops or livestock, managerial decisions, group activities); and the nature of media used (radio, television, Internet, video, print, face-to-face interactions). We also examine cross-cutting issues in each of the chapters such as gender, nutrition, and climate change to draw specific conclusions.

9.1 Governance Structures

Conclusions relating to the governance structures are grouped under policies and strategies, funding, and institutional pluralism and linkages.

Policies and Strategies

Explicit policies and strategies for extension systems are important for several reasons. Policies and strategies affect governance structures and thus organization and delivery of extension services, the level of decentralization, and coordination and linkages within the agricultural innovation system. The majority of countries did not have a specific policy for extension and advisory services. Any strategy or directive was usually rooted in their broader agricultural sector development policies. Some countries had a policy that is outdated or not well implemented. In some countries, while policies existed on paper, follow-up with adequate funding was lacking. For example, the provision of extension services in Malawi has been guided by its agricultural extension policy, which was put in place in 2000 and is under revision. The 2000 policy vision is that "all farmers are able to demand and have access to high quality extension services from those best able to deliver them" (Malawi, MoAIWD 2016). Thus a first step in promoting and coordinating a pluralistic extension system is to have an explicit policy on extension service provision.

Existence of specific policies and strategies also affects financing and implementation. Financing of extension services was quite diverse in Latin America. In Africa and Central Asia and the Caucasus, financing was donor-dependent and public sector–led. For example, while Rwanda and Ethiopia did use donor funds, their extension programs were driven by their extension and agriculture strategies rather than by donor priorities. Countries may have subsectoral policies and strategies as well that compete with extension for resources. Funding for extension should be explicitly identified in the national agricultural investment plans; without such clear earmarking, extension funds could be diverted to other purposes that are politically more attractive. For example, the challenge in Malawi has been low investment in extension services, as most of the budget allocated to the agricultural sector has been spent on fertilizer subsidies. In addition, this spending was included in total agricultural investment and shown to meet the African Union's Comprehensive Africa Agriculture Development Programme (CAADP) agreement of 10 percent allocation of annual budget to agriculture. Here, the political economy of agricultural expenditure with respect to extension services development suggests the need for further analysis.

Issues of coordination, decentralization, and implementation of pluralistic extension needs explicit policies and strategies. All the countries studied had some level of pluralistic extension systems with a large number of public, private, and civil society providers. The public sector, usually ministries of agriculture, was mainly responsible for the overall coordination and regulation of extension. Many countries were decentralizing extension services to lower levels of governance. Coordination between extension service providers with other actors in the agricultural innovation system (such as research, education, and vocational training) remains a challenge in most countries, although we found some mechanisms in place to deal with this challenge in Malawi and Brazil. Explicit policy for extension systems would help.

Among the countries studied in this book, Ethiopia provides an example of specific policy- and strategy-driven, government-led extension services promotion that makes a difference in the agriculture sector. The Ethiopian government has been substantially committed to agriculture in general, and to extension services in particular, as seen from the implemented policies. Ethiopia's investment in agriculture has mainly focused on the provision of "advisory and training services" through a public extension services structure that spans from the federal ministry to the regions and down to the lowest administrative units through frontline extension agents. The Ethiopian government has heavily invested in its public extension system, training over

45,000 extension agents and building more than 11,000 farmer training centers in the past 10 years. Results show that in addition to increasing extension services investments, support to Ethiopian farmers in general can be further improved through supporting systems that provide fertilizer, seed, and credit, and through support to private-sector development. These findings reinforce the need for complete liberalization of input supply markets along with extension system development to support smallholder efforts to intensify production activities. The Ethiopia case (Chapter 6) recognizes the call for continued public engagement in input markets and extension services, while carving out new space for private investment in the efficient provision of goods and services for smallholders. Greater flexibility in organizing how inputs and extension services are provided and giving more choice for smallholders can open up new markets and technological opportunities in the agriculture sector. All these are possible when a specific policy for extension exists that is linked to overall agriculture sector policy for better coordination and implementation.

The need for an explicit policy or strategy is exemplified by other countries studied. For example, the case of the Democratic Republic of the Congo (Chapter 8) highlights the fact that public-sector extension remains crucial even in areas with very weak government institutions and in fragile states like the DRC. As widely observed in the DRC and other fragile countries, many donor and NGO-led projects tend to bypass and not involve government institutions in their project design, implementation, and capacity-strengthening activities (Ragasa and Ulimwengu 2017). An important explicit extension strategy therefore is to include public-sector extension agents and subject matter specialists in capacity-strengthening and learning programs instead of focusing only on NGOs. It will be crucial to work with the government counterparts and not bypass them in agricultural and rural projects (DLEC 2019). Even when there is some distrust and perceived inefficiency in the government institutions, they continue to have a cadre of human resources scattered throughout the country and have long-term structures, as compared to the limited coverage and more ad hoc nature of NGO activities. Therefore they are a crucial part of any capacity-strengthening and extension activities.

Finally, restructuring and modernization have been objectives of extension system reforms in some countries; these also require explicit policy or strategy. For example, the government of Cambodia attempted to modernize its extension system through a new policy that encouraged pluralism for better-linked research, extension services, farmers' organizations, agricultural value chains, and improved laboratories. Since 2014, Viet Nam's state extension system has focused on its agriculture sector restructuring plan to develop rural areas, increase food safety, and promote food safety and hygiene. The national extension system in Viet Nam is responsible for enhancing the capacity of extension services staff and collaborators, strengthening the application of information technology in extension services, increasing the connection between farmers and enterprises, and formulating policies on extension services in line with requirements and production practices. However, implementation of these plans is weak in these countries. Absence of explicit policy and strategy, along with limited funding and commitment, may prevent or delay the full implementation of such reform process.

Funding

Adequate funding for extension systems is key for their effective implementation. Much of the decline seen in public extension services in the past three decades has been due to the reduction of public-sector funding. For example, case studies from the Asian countries indicate that public-sector funding has been decreasing for several decades. However, in Latin America each provider has a different source of funding, which potentially creates more stability in the extension system, even though individual programs and territories may experience major changes in extension services coverage over time. Yet the country-level reporting indicates that most of the services have unstable funding over time, resulting in insufficient operational support. In most country cases, extension services rely on external (donor) project funding. In some of these projects, funding lasted less than one year, an insufficient time to provide quality extension services or to demonstrate development impact.

The combination of the public sector paying public extension salaries and the donor-assisted NGO projects paying for operational costs has become a common trend and a funding model in several African countries, including the case studies presented in this book. In sum, funding for extension services shows a mixed picture as pluralism increases in the regions. As more private-sector support is brought in for specific commodities and value chains, the role of the public sector is likely to be reduced. This is particularly a dominant scenario in Latin America and, to a lesser extent, in Central Asia and Africa. Each country has a different combination of funding sources, which potentially creates more stability across country systems although individual programs and territories experience vast shifts in coverage. As argued in several chapters, poor farmers, those in remote areas, and those relying on staple crops are less likely to commercialize and less likely to access private extension services. The public sector has to play a major role in funding, coordinating, and ensuring equitable access to quality extension services. Overall, optimizing funding and harmonizing the provision of extension services among the actors will require continuous innovation in extension services.

While sustained funding for extension services is key, this book also highlights how combinations of different methods of extension can increase financial sustainability of extension services. In Latin America, for example, the study showed the average cost per farmer reached from US$44 to US$2,400 per year. Differences were based on the use of individual versus group methods and operational costs between countries. Additional and new innovative methods such as information and communications technologies and the Internet of Things (van Campenhout et al. 2018) as well as lead farmers as extensionists (for example, in Malawi and Rwanda [see Franzel et al. 2016]) can help enhance financial sustainability.

Institutional Pluralism and Linkages

Part 1 showed that extension and advisory services have become much more pluralistic over the decades. There are more than one million extension agents today; however, they belong to a variety of different types of organizations from the private, nongovernmental, and public sectors, and even include volunteer farmer extension agents. Increasing pluralism in extension has implications for coordination of the services and for better understanding of the comparative advantage of different players. For instance, while the public sector is dealing with public-goods types of services and addressing national food security objectives, the private sector can help by linking commercially oriented farmers to domestic and export markets through value-chain operations.

In such pluralistic systems, focusing on coordination, quality control, and regulation to avoid conflicting extension messages and duplication of efforts will be a priority role for the government. Such a role would be equally or even more important than providing extension services themselves. For example, most countries in Latin America saw that the changes in the institutional context of extension delivery left a gap in the coordination of services, leading to breaks in coverage, duplication of efforts, contradictory messages, and weakened service delivery. In addition, the need to increase linkages between innovation actors, including the private sector, was noted.

Relatedly, there is a need for a greater knowledge brokering and facilitation role of extension, where linkages and partnerships become crucial for performance. As shown throughout this book, there are many constraints to linkages; therefore, organizations must find ways to help extension agents and organizations link more to each other and to other relevant actors. Moreover, while more organizations are providing extension services, the country cases show that pluralism of service delivery is far from complete. Service providers are dominated by international NGOs, often with limited human resources and relying heavily on government extension workers. It is common to find international NGOs addressing government resource limitations by providing mobility and operating funds to government extension agents, while the government provides the human resources. As a result of this limited pluralism, there is little competition among service providers and hardly any expansion of options or choices for farmers.

9.2 Organizational Capacity and Management

Organizational capacity and management issues are closely related, and we summarize the findings together. Organizational capacity and management issues varied among the cases examined in the book. Yet the issues relating to the coordination and harmonization of extension services activities by various entities in a pluralistic setting depend on how the systems share their extension service goals, jointly set priorities for their activities, generate and manage their funding, assess performance of their staff, and maintain a functional monitoring and evaluation system that provides opportunities for learning and improvement.

Within the realm of capacities and management, mind-sets, roles, and expectations matter, and in many cases need to be changed. For instance, in Malawi it is common to hear extension workers express something like, "The project was a success, so it is the farmers' fault that they do not continue to adopt the technologies promoted." This type of mind-set is worrisome in terms of lack of understanding of farmer realities and their role as extension workers. It also may reflect how they are supervised and monitored. It reveals the ad hoc nature of most projects, the culture of "checking boxes" among these projects, and the lack of evaluation and learning. It echoes a supply-driven approach of pushing technologies, rather than putting farmers at the center and working together to support them.

Relatedly, we saw in Chapter 8 how management systems, especially incentive systems in the Democratic Republic of the Congo, need to be strengthened at the organizational level. Furthermore, the lack of extension performance indicators was a critical vacuum in the organization and management of Latin American extension systems. As noted in Chapter 3, most of the case studies presented report human resources as one of the major constraints in increasing the effectiveness of the extension system. Capacity weaknesses include staff skills, the infrastructure and equipment to carry out the job, and foundational and continuing education for extension personnel. Extension services capacity applies both to frontline workers and the organizations for which they work. The capacity at the personnel level includes quantity and quality of those officials and staff providing extension services.

Numbers of public-sector extension agents were seldom sufficient for the job at hand, and there were high vacancy rates and turnover in some countries. The foundational training for extension agents usually focused on technical topics and may miss the functional skills that are also needed for extension. Exceptions were seen in several Central and South American countries that appeared to have well-established foundational and continuing education systems for extension providers. Although a few countries provided incentives for extension agents (for example, Ethiopia, Mozambique), in most cases extension agents were demotivated and lacked basic transportation and equipment. Salaries were normally low, especially in the public sector, and there were few rewards or recognition.

The analysis of the extension system in Ethiopia (Chapter 6) showed three key constraints that play against the greater contributions of extension services to productivity growth and agricultural transformation. First, with limited institutional innovations and poor coordination with research centers—hence the limited injection of new knowledge into the system—development agents are left with little leverage to convince lead and other farmers. Extension agents' operation under poorly resourced work conditions (given the amount and diversity of work they are tasked with) implies that the link between research and extension services remains weak, which leads to diminishing returns to the technical support of development agents for technology adoption (Krishnan and Patnam 2014). Second, the fact that extension agents were overburdened by activities beyond their regular mandates provides little time for them to search for additional knowledge and information. While the current system can be commended for having one of the highest extension-to-farmer ratios, it is overly standardized (one-size-fits-all) and lacks the flexibility to adapt to local conditions. Third, the efficacy of farmer training centers is also constrained because they are generally underresourced and scattered, with little focus and scale. While evidence suggests that the number of farm households reached with extension services has substantially increased, these constraints negate sustaining future gains. It is unlikely, therefore, that the increased farmers' access to the system, as it is now, can be translated into productivity gains.

In cases such as the Democratic Republic of the Congo, the priority should be on streamlining the number of extension staff by retiring those currently in their jobs beyond their retirement age and hiring and retraining younger and more dynamic staff. This would have the added benefit of revamping the salary structure and freeing valuable resources that can be shifted to much-needed operating and capital components in the extension subsector. There has been little updating of extensionist competencies or in-service training provided to extensionists. More efforts are needed to educate extension professionals on issues relating to new technologies, soft skills, markets, nutrition, natural resources, and climate change. With the exception of the promising efforts in Peru, little work has been done to certify the competencies of professional extensionists or community-level farmer leaders. Support is required to develop a strategic regional framework for professionalization, registration, and certification of extension agents.

Beyond the individual level, capacities are needed at the organizational and system level (FAO 2012). For example, as India moves toward pluralistic extension provision, the government needs to develop systemwide capacity for coordination, monitoring, and evaluation of extension provision.

9.3 Advisory Methods

Advisory methods include the different means to serve clients and provide information, empower farmers, and otherwise fulfill the extension services mandate in each country. In general, many countries are moving from top-down to more participatory methods, and to market-oriented approaches rather than a sole focus on production. In Latin America, for instance, the breadth and depth of participatory extension methodologies was increasing in the region (rural promoters, farmer to farmer, farmer field schools, learning groups, and local innovation networks). The global assessments (Chapter 2) showed that more methods are available today, especially digital ones. Private-sector extension and the use of volunteer farmers is advocated and used worldwide.

In the Latin American context we saw little use of modern information and communication technologies by extension (Chapter 3). In Central Asia and the Caucasus, measures are needed to establish better communications, include creating and strengthening agricultural information centers and enabling ICT infrastructure, for directly connecting to the farmers or through their cooperatives (Dosov 2018). Thus understanding the challenges in scaling up successful projects would be important. Although new innovations using information and communications technologies have been adopted in all the countries and the regions studied, largely due to the penetration of mobile phones, there is still a need for maintaining traditional approaches to extension services. The context-specific nature of the knowledge sharing requires local specific solutions that can be delivered only through reaching out to farmers on an individual or group basis. While investments in such traditional methods will continue for some time, new approaches such as the Internet of Things, will be more effective in replacing some of the traditional approaches to extension services.

9.4 Cross-Cutting Issues

Increasingly, a number of cross-cutting issues are brought in as part of extension system delivery at the national level. These issues include climate change and climate-smart technologies, nutrition in agriculture, and gender and youth in agriculture. There has been debate on how extension and advisory services organizations can adapt and respond to emerging development challenges (Davis and Sulaiman 2014). Expanding or changing the current extension service functions to cover emerging thematic issues depends on the context, which will require a commensurate level of investment in capacity development. For instance, Brazil's extension services, despite having a policy to reach disadvantaged groups, did not always perform accordingly, which leads to the recommendation that extension training in the country should focus in part on issues relating to gender, youth, and ethnicity.

In a broader sense, the new extension service worker is supposed to be a facilitator and problem solver in the rural community, able to help the farming communities in various challenges that they face, not only in their production enterprises but also in other social and cultural dimensions of rural life.

9.5 Policy Recommendations, Specific Actions, Lessons from Assessment Methods and Framework, and Future Research Needs

Based on these conclusions, we provide here a summary of policy recommendations to decisionmakers and implementers of extension services to strengthen their extension systems. The conclusions provide policy implications across the best-fit characteristics—implementing policies and programs to improve governance, reforming curricula to strengthen capacity, providing incentives to extension staff at various levels to improve management, and using appropriate advisory methods that enhance financial sustainability and achieve greater scale. Specific policy insights include the following:

1. Explicit policy or strategy for extension services is a key first step for better governance, funding, coordination, decentralization, and for overall effective design and implementation of extension services. Capacity for developing such country-specific extension strategies is weak or does not exist and thus needs strengthening in the national systems.
2. Public funding for extension services has been declining, current funding remains unstable, and most of the poorer countries continue to depend on donor funding to run their extension systems. Although private-sector organizations, NGOs, and farmer-based organizations play an increasingly important role in extension services, the public role cannot be overestimated for reaching remote areas and marginal groups. A combination of funding sources and mixing innovative modern approaches with traditional mechanisms can increase the sustainability of extension services.
3. Given that extension and advisory services have become more pluralistic in developing countries, the need for coordination, quality control, and avoiding duplication of efforts becomes a key policy and programmatic challenge for the public sector. In addition, identifying the gaps left by the multiplicity of actors as they focus on specific target groups or value chains, and defining and coordinating the roles and responsibilities of these actors is a major challenge for policymakers. Investment in such coordination capacity needs immediate attention in the public extension system.
4. The organizational and institutional capacity to effectively reach the farmers in a country context is important to improve the efficiency and effectiveness of the extension services provided by the pluralistic actors. The first set of policy and programmatic concerns relate to the supply-driven nature of the extension services in several countries studied. To make them more demand-driven, there is need for identifying information needs of the producers in an organizational context, setting priorities for the extension services, and sharing the goals and approaches among the extension services providers. This implies building capacities of producers and especially producer organizations to identify and prioritize their needs. These activities will further help in the management of the limited resources for extension that is currently split among the pluralistic providers.

5. Assessing the performance of extension in meeting their clientele's needs requires sound monitoring and evaluation systems, which remain weak in many countries. The learning and improvement and the trade-off between the quantity and quality of extension professionals can be addressed only through feedback provided by a well-capacitated monitoring and evaluation system.

6. Organizational and institutional innovations are needed on a regular basis for improving the capacity of the extension professionals, improving their focus on the farmers' problems, reducing their overburden with multiple objectives, and increasing the reach of extension in different agroecological zones. Continuous improvement in the technical, managerial, and leadership capacities of extension professionals that goes beyond foundational training should be a key part of the national extension strategy.

7. Innovative policy and programmatic approaches are needed to reduce the top-down approaches to information sharing. Increasing farmer participation in decision-making on what they would like the extension services to provide will help move from a technology transfer orientation to market-driven approaches to the extension services. Use of the private sector, traders, volunteer farmers, lead farmers, and youth as entrepreneurs to increase the reach of advisory services will require adequate attention at the policy level.

8. Moving toward digital technology and its use for reaching producers on topics such as weather, technology, markets, prices, and other real-time information for solving farmers' problems also requires adequate policy and institutional arrangements at various levels. Digital technologies and use of the Internet of Things can save resources at the extension system and at the farmers' level. Their use can be intensified both as extension delivery tools and as mechanisms for demand articulation, monitoring, and greater transparency and accountability.

9. Finally, the future of extension systems crucially depends on how they are built to meet the emerging challenges such as climate change, precision agriculture, nutrition and health goals, youth and gender, and other challenges related to the transformation of food systems and to resilience building. In this context the extension worker is seen more as a problem solver and a facilitator of services in the rural areas. Developing a policy environment to strengthen the capacity of the extension system

to meet these emerging needs remains the most important development concern.

Some of the specific examples of actions are identified below for each of the characteristics studied using the best-fit framework.

GOVERNANCE AND FINANCING

1. Make and implement policies that promote pluralism and inclusion, allow the flexibility to deal with emerging issues, use new outreach tools, and promote sustainable models.
2. Experiment with more innovative financing models. Table 9.1 shows examples of innovative financing mechanisms for extension.

CAPACITY

3. Conduct needs assessments and tracer studies to find out what skills are needed by extension professionals from the public and private sectors.
4. Partner with educational institutions and use practical means (webinars, blended learning, short courses, on-the-job training, peripatetic trainers) to increase the skills of existing extension staff.

MANAGEMENT

5. Ease the transportation time (for example, providing more motorcycles or vehicles) and accessibility to technologies and tools (for example, radio, video, and smartphones) to make extension delivery more efficient.
6. Avoid giving nonextension assignments to staff (for example, collecting taxes or enforcing loan repayments).
7. Increase incentives for extension agents, including salaries, career paths, continuing education, and rewards and recognition, including incentives for extension providers to operate in remote areas.
8. Develop a reliable system of monitoring and evaluation to ensure optimal allocation of resources toward extension programs and tracking of the benefits of extension on the farmers.
9. Partner and experiment with development, collection, and analysis of key indicators for extension and advisory services to provide information to decisionmakers about extension (see Chapter 3 and below).

TABLE 9.1 Models and examples of innovative financing for extension

Model	Examples
Farmer organization–owned services with public subsidies and farmer payments	Danish Agricultural Advisory Service (DAAS) and National Small-holder Farmers' Association of Malawi (NASFAM). Advisers are employed by Farmer Organizations and financed partly by public subsidies, partly by farmers' own contributions.
Decentralized services with public financing of demand-driven services	Farmers articulate their demands and define who provides services, for example, Tanzania's District Development Funds; Coordinadora Nacional de las Fundaciones Produce (COFUPRO) in Mexico; Fadama II in Nigeria
Public sector–driven privatization of services through competitive grants and contracts	Chilean Institute for Agricultural Development (INDAP); Innovation and Competitiveness Programme for Peruvian Agriculture (INCA-GRO); Brazil's Decentralized System of Rural Extension
Producer cooperative–based embedded services fully financed by processing and marketing revenue	Nariño Dairy Products Cooperative (Colácteos) in Colombia or Colombian Coffee Growers Federation

Source: Blum and Chipeta (2016).

ADVISORY METHODS

10. Use appropriate digital tools that enhance efficiency while still allowing inclusion of disadvantaged groups

11. Use lead farmers (with appropriate support), information and communications technologies, and group approaches to enhance efficiency and relevance of extension services delivery.

Below are recommendations for assessing extension systems and their reforms—including potential indicators—within the best-fit framework.

Extension System Assessment Methodology

Continuous reforms are needed to optimize the contribution of an extension system to agricultural development and transformation processes. In designing reform options, it is essential to understand the status of the extension services and assess or evaluate their impacts. Evaluation of an extension service allows for structured learning from experience and can help in reforming extension to increase overall impact (Christoplos, Sandison, and Chipeta 2012). Examining extension reform efforts are also imperative to provide feedback to the program implementers and the decision-making process, to determine if new or revised approaches are worth scaling up. Such assessments can also provide valuable lessons to other countries undergoing, or about to undergo, similar reforms (GFRAS 2012). Extension evaluations further provide insights into what extent and how the goals and objectives of the advisory services are achieved from the perspective of various stakeholders, throughout the commodity value chains and the innovation systems in which they are situated.

From the assessment methodological perspective, the process encountered by the research studies presented in this book was challenging due to the diversity and the pluralistic nature of the extension services provided; however, the experiences do offer some useful lessons for future assessments. First, multidisciplinary teams and strong engagement of local partners are critical to the credibility of the assessment process. Second, participatory assessment processes and inputs of key stakeholders in defining and selecting performance indicators are important. Third, triangulation of data and results and combination of different data sources and methods are crucial for credible assessment processes. For example, self-reporting by service providers of their effectiveness may need to be accompanied by other sources of data and information. High satisfaction and ratings by farming households on services received may need verification with actual changes in behavior or adoption of promoted technologies. Fourth, the assessment process shows the difficulties in evaluating outcomes or impacts of extension systems and highlights the need for complementarity between qualitative and quantitative methods and indicators. Such assessments require discussion with different technical departments or sectors within or outside of agriculture to understand the contributions of different services and programs in achieving development outcomes and to understand where major gaps are.

Operationalizing the Framework in National Extension Systems

Chapter 3 introduced a common set of indicators proposed by a number of institutions interested in measuring extension systems. Following the development of a proposed list of indicators, what should happen next is further discussion with governments and other stakeholders interested in collecting such data to see what is needed, what is demanded, and what is feasible. Then a pilot program should take place to see how best to operationalize this. The pilot would adapt and validate the common framework and indicators for extension metrics, as well as the metrics themselves, by testing their use with several governments or other providers to measure their public extension programs. The pilot would seek to understand the following three questions:

1. What is the ideal number of common extension indicators with buy-in of governments and other users, and are these data currently available or only aspirational?

2. How can users most effectively gather, analyze, and share data on extension indicators, and what capacity is needed?
3. What are the best ways to integrate the data across different countries or providers to enable cross-learning at a global level?

The pilot program would use a consultative process built on previous work to refine the proposed common framework and metrics and engage with users. The program would adapt the framework and metrics to the needs of the users. At the end of a first phase, users would have a list of metrics (potentially both aspirational and based on currently available data) that they need to collect data on and use to make data-driven decisions. In a second phase, once the common framework and metrics have been identified, users could begin collecting data on the identified metrics. Through this cycle the pilot program should identify capacity gaps in data collection and analysis, and create a strategy to fill such gaps. Finally, once the data have been collected, they would be shared and integrated across countries to enable cross-learning by developing guidelines for data integration.

One possible model for collection of agricultural indicators is ASTI (formerly Agricultural Science and Technology Indicators). ASTI collects information regularly from a wide range of agricultural research organizations using a coherent methodology. The information is published via ASTI country reports and in the form of regional and global assessments. The availability of the ASTI data has enabled important studies on the role of agricultural research. Extension data are difficult to obtain in part due to the lack of a similar Frascati manual that contains agreed-upon definitions and conventions on what constitutes research and how to measure it (OECD 2015). But once the definitions and measurement issues are addressed, the ASTI model can be a model to follow for data collection and sharing.

Future Research Needs

We finish with a set of research questions for extension and advisory services that remain to be addressed. Research is needed to continuously update what works and why in the context of innovations that are introduced in extension service delivery. In particular, the sustainability and cost-effectiveness of extension programs and approaches will become increasingly important. Extension and advisory services will be facing increasing challenges and will require transformation as the food and agriculture sectors are rapidly evolving.

Technology change in the agriculture sector is gaining increasing momentum, and the adoption of the advanced technology in the developed world is much faster due to private-sector investments in spreading technology. For example, gene editing and drone technology are changing the ways research is done and how such technologies will reach smallholder farmers. Technologies that are introduced are often to reduce costs and external input use. At the same time, there are also movements in developing countries that emphasize a "going back to basics" approach to agriculture that reduces the cost and input use. How does the extension system adapt and adjust to the two types of polarized agriculture?

Research systems are changing to private-sector approaches when food systems move toward high-value agriculture. Research is needed to address the question of how traditional extension systems should evolve to meet this trend. Training extensionists in new innovations is key for the extension services to be effective in reaching the target audience. At the same time, extension staff need functional skills to enable them to effectively communicate. How can large numbers of existing staff be equipped with the requisite knowledge and skills?

Information should be accessible in the context of water crisis problems and climate change issues emerging in developing countries. Groundwater depletion is an ongoing concern for which we need local and community-based solutions and new models for technical assistance. How will these models be developed and deployed as part of the extension services? How do the extension services go beyond the traditional subject matter to incorporate planning and execution of technical services and processes to address such natural resource–based and climate-related problems?

Markets are more integrated now and agriculture marketing through electronic means is gaining momentum. How can these elements be captured in extension systems' service delivery by the public sector, private sector, and civil society organizations?

Engaging with research and other actors such as farmer-based organizations in developing new partnerships may deliver better extension services than traditional service providers and partnerships. Further, new generations of farmers—mostly youth in rural areas—are moving toward becoming entrepreneurs while offering extension services. Research is needed on how to engage these youth as both providers and clientele of extension services. Data are needed as to the efficiency and effectiveness of new models of service provision, such as those services provided by young "agripreneurs." How can extension systems help and use the rural youth to benefit smallholder farmers through agribusiness entrepreneurship?

Reliance on public extension services is diminishing while private equipment dealers and input dealers are taking over in some areas. Extension faces new competition in the information marketplace as well (King 2018). Will this be the end of the public system altogether, or should the public extension delivery system reinvent itself? If so, how? Relatedly, how can the combination of information and communications technologies and the Internet of Things change extension actors and delivery methods? What is the ideal ratio of agents to farmers in an Internet age?

Big data–driven extension services are also emerging and linking the findings to risk reduction and resilience-building measures. Yet it is not clear how developing country extension services will incorporate big data in their service delivery. Furthermore, venture capitalists are increasingly engaging in agricultural and biological research that relate to genomics, big data, satellites, and precision farming. However, it is not clear how the products coming from these investments will reach farmers and their communities. What type of extension system will emerge in the future to meet the requirements of these institutional innovations? These selected issues have been touched on but are not fully addressed in the chapters of this book and may require additional research in the future in the context of developing and implementing effective extension and rural services.

Finally, we call for more ambitious objectives in the assessment and evaluation of extension and advisory services—that is, to *explain* variation of performance and impact by assessing to what extent the characteristics of the extension services and their position in the agricultural innovation system fit with the overall prevailing environment. The best-fit framework was originally developed with the aim of supporting such assessments. Much more research is needed along these lines. We call for more and better evaluations of extension and advisory services that clearly link outcomes and impact to the services themselves. These kinds of findings will go a long way in garnering support to the extension services that are critical to more than one billion small-scale and marginalized farmers worldwide.

References

Blum, M., and S. Chipeta. 2016. *Innovative Financing Mechanisms for Demand-Driven Agricultural Advisory Services*. Note 21. GFRAS Good Practice Notes for Extension and Advisory Services. Lausanne, Switzerland: Global Forum for Rural Advisory Services (GFRAS).

Christoplos, I., P. Sandison, and S. Chipeta. 2012. *Guide to Evaluating Agricultural Extension.* Lindau, Switzerland: Global Forum for Rural Advisory Services (GFRAS).

Davis, K., and R. V. Sulaiman. 2014. "The 'New Extensionist': Roles and Capacities to Strengthen Extension and Advisory Services." *Journal of Agricultural Education and Extension* 21 (3): 6–18. doi: 10.5191/jiaee.2014.21301.

DLEC (Developing Local Extension Capacity) Project. 2019. *Strengthening Private Sector Extension and Advisory Services Portfolio Review.* Washington, DC: USAID.

Dosov, B. 2018. "Central Asia and the Caucasus." Unpublished report, IFPRI, Washington, DC.

FAO (Food and Agriculture Organization of the United Nations). 2012. *FAO Corporate Strategy on Capacity Development.* Rome: FAO.

GFRAS. 2012. "GFRAS Evaluation Initiative Background Document." Unpublished document. Lindau, Switzerland: GFRAS.

King, D. 2018. "Hey, Siri, What Is the Future of Extension?" *Journal of Extension* 56 (5).

Krishnan, P., and M. Patnam. 2014. "Neighbors and Extension Agents in Ethiopia: Who Matters More for Technology Adoption?" *American Journal of Agricultural Economics* 96 (1): 308–327.

Malawi, MoAIWD (Ministry of Agriculture, Irrigation and Water Development). 2016. *National Agricultural Policy.* Lilongwe, Malawi.

OECD (Organization of Economic Cooperation and Development). 2015. *Frascati Manual 2015: Guidelines for Collecting and Reporting Data on Research and Experimental Development, the Measurement of Scientific, Technological and Innovation Activities.* Paris: OECD Publishing. https://doi.org/10.1787/9789264239012-en.

Ragasa, C., and J. Ulimwengu. 2017. "Challenges in Rebuilding Agricultural Extension System in the Democratic Republic of Congo." In *Building Agricultural Extension Capacity in Post-Conflict Settings,* edited by P. McNamara and A. Moore, 35–61. Wallingford, UK: CABI.

CONTRIBUTORS

Kristin Davis is a senior research fellow in the Development Strategy and Governance Division of the International Food Policy Research Institute (IFPRI), Pretoria, South Africa.

Suresh C. Babu is a senior research fellow and the head of Capacity Strengthening at IFPRI, Washington, DC, and extraordinary professor of Agricultural Economics at the University of Pretoria, South Africa.

Catherine Ragasa is a senior research fellow in the Development Strategy and Governance Division of IFPRI, Washington, DC.

Gashaw Tadesse Abate is a research fellow in the Markets, Trade, and Institutions Division of IFPRI, Addis Ababa, Ethiopia.

Gary Alex was program manager, Farmer to Farmer, at the United States Agency for International Development, Washington, DC.

Thomas Woldu Assefa is a doctoral student at the University of Georgia, Athens, US.

Bernard Bashaasha is a professor of Agricultural Economics and principal of the College of Agricultural and Environmental Sciences, Makerere University, Kampala, Uganda.

Sonia Maria Pessoa Pereira Bergamasco is a senior professor at Campinas State University (UNICAMP), Brazil.

Guush Berhane is a research fellow in the Development Strategy and Governance Division and the Ethiopia Strategy Support Program of IFPRI, Addis Ababa, Ethiopia.

Ricardo Serra Borsatto is an associate professor at São Carlos Federal University (UFSCar), Brazil.

Edward Kato is a senior research analyst in the Environment and Production Technology Division of IFPRI, Washington, DC.

Nana Afranaa Kwapong is a lecturer in the Department of Agricultural Extension, University of Ghana, Accra.

Margaret Mangheni is an associate professor of Agricultural Extension Education at Makerere University, Kampala, Uganda.

Catherine Mthinda is a senior lecturer in Agricultural Extension and Gender at the Lilongwe University of Agriculture and Natural Resources, Malawi.

Ephraim Nkonya was a senior research fellow in the Environment and Production Technology Division of IFPRI, Washington, DC, at the time he contributed to this work. He is currently a nonresident fellow in the Environment and Production Technology Division of IFPRI, Kansas City, MO, US.

Patience Rwamigisa is a commissioner for Agricultural Extension Services in the Ministry of Agriculture, Animal Industry and Fisheries, Entebbe, Uganda.

Carolina Rios Thomson is a doctoral student at UNICAMP, Brazil.

John Ulimwengu is a senior research fellow in the Africa Division of IFPRI, Washington, DC.

INDEX

Page numbers for entries occurring in figures are followed by an *f*; those for entries in notes, by an *n*; those for entries in a text box, by a *b*; and those for entries in tables, by a *t*.

Made in the USA
Las Vegas, NV
03 November 2026

57422208R00226